Microbial Biodegradation of Xenobiotic Compounds

Editor

Young-Cheol Chang
Department of Applied Sciences
College of Environmental Technology
Muroran Institute of Technology
Muroran, Hokkaido
Japan

CRC Press
Taylor & Francis Group
Boca Raton London New York

CRC Press is an imprint of the
Taylor & Francis Group, an **informa** business

A SCIENCE PUBLISHERS BOOK

Cover illustrations reproduced by kind courtesy of Dr. Young-Cheol Chang (editor)

CRC Press
Taylor & Francis Group
6000 Broken Sound Parkway NW, Suite 300
Boca Raton, FL 33487-2742

First issued in paperback 2021

© 2019 by Taylor & Francis Group, LLC
CRC Press is an imprint of Taylor & Francis Group, an Informa business

No claim to original U.S. Government works

Version Date: 20181105

ISBN-13: 978-0-367-78036-4 (pbk)
ISBN-13: 978-1-138-74392-2 (hbk)

Library of Congress Cataloging-in-Publication Data
Names: Chang, Young-Cheol, 1968- editor.
Title: Microbial biodegradation of xenobiotic compounds / editor Young-Cheol Chang (Department of Applied Sciences, College of Environmental Technology, Muroran Institute of Technology, Muroran, Hokkaido, Japan).
Description: Boca Raton, FL : CRC Press, [2018]	"A science publishers book."	Includes bibliographical references and index.
Identifiers: LCCN 2018044523	ISBN 9781138743922 (hardback)
Subjects: LCSH: Xenobiotics--Biodegradation.	Chemicals.	Biodegradation.
Classification: LCC QP529 .M53 2018	DDC 615.9--dc23
LC record available at https://lccn.loc.gov/2018044523

Visit the Taylor & Francis Web site at
http://www.taylorandfrancis.com

and the CRC Press Web site at
http://www.crcpress.com

Preface

This book contains information about the application of recent microbial technologies for the removal of xenobiotic compounds. Xenobiotic compounds are chemical substances, which are foreign to a living organism and cause high potential risk to human and animal health. This book focuses on biodegradation and bioremediation of xenobiotic compounds (aromatic compounds, heavy metals, dyes, and other industrial pollutants) using plants and microbes (bacteria, fungi, and microalgae). In some chapters, identification and characterizations of novel genes and enzymes that degrade the xenobiotic compounds are discussed. In addition, molecular models of the interaction between hydrophobic organic contaminants and soil organic matter have been addressed. It also describes comprehensive wastewater management technologies via contaminant removal and alternative energy generation using bacteria through fermentation and microbial fuel cell technologies. This book can be readily comprehended by students and scientists and makes for a more informed selection of microbial biodegradation of xenobiotic compounds for environmental cleanup.

Contents

1

Degradation of Aromatic Compounds and their Conversion into Useful Energy by Bacteria

M. Venkateswar Reddy,[1,]* *Rui Onodera*[2] and
Young-Cheol Chang[2]

INTRODUCTION

Aromatic compounds can be defined as organic molecules containing one or more aromatic rings, specifically benzene rings. Different aromatic compounds co-exist as complex mixtures in petroleum refinery and distillation sites (Seo et al. 2009). There are three major categories of toxic aromatic compounds in nature, that is, polycyclic aromatic hydrocarbons (PAHs), heterocyclic compounds, and substituted aromatics (Fig. 1). PAHs are a group of chemicals that contain two or more fused aromatic rings in linear, angular, or cluster arrangements (Cheung and Kinkle 2001). Physical and chemical properties of PAHs vary with the number of rings and hence their molecular weight. Chemical reactivity, aqueous solubility, and volatility of PAHs decrease with increasing molecular weight. Thus, PAHs differ in their transport, distribution, and fate in the environment and their effects on biological systems (Seo et al. 2009). The US EPA has identified 16 PAHs as priority pollutants. Some of

---

[1] Institut für Molekulare Mikrobiologie und Biotechnologie, Westfälische, Wilhelms Universität Münster, Corrensstr. 3, 48149 Münster, Germany.
[2] Department of Applied Sciences, College of Environmental Technology, Muroran Institute of Technology, 27-1 Mizumoto, Muroran, Hokkaido 050-8585, Japan.
* Corresponding author: mvr_234@yahoo.co.in, reddy1234@uni-muenster.de

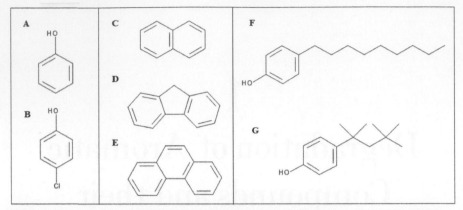

Fig. 1: Structure of various toxic aromatic compounds. (A) phenol, (B) 4-chloro phenol, (C) naphthalene, (D) fluorene, (E) phenanthrene, (F) nonylphenol, and (G) 4-*tert*-octylphenol.

these PAHs are considered to be possible or probable human carcinogens, and hence their distributions in the environment and possible exposure to humans have been of concern (Menzie et al. 1992). High molecular-weight PAHs are paid particular attention as they are recalcitrant (Cheung and Kinkle 2001). In general, PAHs are relatively stable and recalcitrant in soils and less easy to degrade than many other organic compounds (Seo et al. 2009).

They are difficult to remove from contaminated soil using the treatments that have been used successfully to clean soils contaminated with more degradable or volatile organic compounds (Pitter and Chudoba 1990). Three major sources of PAHs are petrogenic, pyrogenic, and biogenic. Petrogenic PAHs are from petroleum and petroleum-derived products, and are often marked as in abundance of alkyl substituted PAHs such as alkyl naphthalenes, alkyl phenanthrenes, and alkyl dibenzothiophenes (Seo et al. 2009). Pyrogenic PAHs are produced from combustion processes and are comprised of predominantly unsubstituted PAHs. Biogenic aromatic compounds including aromatic amino acids, lignin compounds, and their derivatives are of biotransformation origin. PAHs may accumulate in high concentrations in terrestrial environments near coal gasification sites and tar oil distillation plants (Capotori et al. 2004). Major sources of PAHs are incomplete combustion of organic materials, gas production, wood treatment facilities, and waste incineration (Kim et al. 2003). PAHs are formed naturally during thermal geologic reactions associated with fossil-fuel and mineral production, and during burning of vegetation in forest and bush fires (Juhasz and Naidu 2000).

Anthropogenic sources, particularly fuel combustion, automobiles, spillage of petroleum products, and waste incinerators are significant sources of PAHs into the environment (Seo et al. 2009). Tobacco cigarette smoking is a significant source of PAH exposure to smokers and secondary smokers. PAHs generated during anthropogenic combustion activities are primarily transported via atmospheric deposition (Freeman et al. 1990). Petroleum refining and transport activities are major contributors to localized loadings of PAHs into the environment. Such loadings may occur through discharge of industrial effluents and through accidental release of raw and refined products (Seo et al. 2009).

Heterocyclic compounds including dibenzothiophene and carbazole are components of creosote, crude oils, and shale oils and often co-exist in the environment with PAHs and other aromatic compounds (Nestler 1974). Dibenzothiophene is a sulfur heterocyclic compound and is quite persistent in the environment. Little information about dibenzothiophene toxicity is available in the literature. Carbazole, a nitrogen heterocycle, is carcinogenic and toxic (Tsuda et al. 1982). Dibenzofuran and its substituted analogues are found in several woody plants as stress chemicals, so called phytoalexins (Gottstein and Gross 1992).

However, most of the environmental concerns with dibenzofuran are related to its halogenated analogues, especially its chloro/bromo derivatives. Persistent organic pollutants (POPs) are among the most concerned environmental pollutants because they persist in the environment, bioaccumulate through the food web, and pose a risk of causing adverse effects to the environment and human health (Seo et al. 2009). POPs are also referred to as persistent, bioaccumulative, and toxic chemicals (PBTs). POPs include aldrin, brominated flame retardants, chlordane, DDT, dieldrin, endrin, mirex, organometallic compounds such as tributyltin, PAHs, heptachlor, hexachlorobenzene, polychlorinated biphenyls (PCBs), polychlorinated dibenzo-p-dioxins (PCDDs), polychlorinated dibenzofurans (PCDFs), and toxaphene. PCDD/Fs are formed unintentionally from human activities. One of the main sources of PCDD/Fs is municipal waste incinerators (Olie et al. 1977). PCDD/Fs formed are absorbed on the fly ash, and then enter into the environment (Seo et al. 2009).

Microbial degradation of aromatic compounds, which represent about 20% of the earth biomass, has been extensively studied owing to its importance in the biogeochemical carbon cycle (Diaz et al. 2013). Since many aromatic compounds are major environmental pollutants, their detection and removal from contaminated sites are of great biotechnological interest. Moreover, the use of aromatic compounds, for example, lignin-derived compounds, as feedstock for the bioproduction of several substances in the pharmaceutical, industrial, agricultural, food, and health sectors stresses the study of aromatic bioconversion processes (Fuchs et al. 2011; Ramos et al. 2011; Vilchez-Vargas et al. 2010). Two major biochemical strategies are used by bacteria to activate and cleave the aromatic ring depending primarily on the availability of oxygen (Diaz et al. 2013). Whereas in the absence of oxygen, reductive reactions take place, in the aerobic catabolism, oxygen is not only the final electron acceptor but also a co-substrate for some key catabolic processes (Diaz et al. 2013; Fuchs et al. 2011).

The modern 'omic' tools have enabled to investigate the metabolism of aromatic compounds from a systems biology perspective (Ramos et al. 2011; Vilchez-Vargas et al. 2010). Thus, the catabolic operons are tightly connected with the global metabolism of the recipient cell and they are subject to varied, host-dependent influences. Since many aromatic compounds are not only nutrients but also important chemical stressors for the bacteria, they constitute a nice model system to study different aspects about the evolution/adaptation mechanisms in life systems (Jimenez et al. 2010). The mechanisms developed by microbial cells to assimilate aromatic compounds were fixed and optimized by natural selection, giving raise to the current enzymes, their organization into functionally separable modules, and to the general trend of a catabolic funnel-like topology (Diaz et al. 2013). Thus, a wide diversity of aromatics are activated via different peripheral pathways to a few key central intermediates that suffer dearomatization and further conversion to intermediary metabolites, such as acetyl-CoA, succinyl-CoA or pyruvate, via some central pathways that are conserved in evolution and function (Fuchs et al. 2011). Following the first metabolic reconstruction of aromatic acids metabolism in *Pseudomonas putida* KT2440 (Jimenez et al. 2002), other aerobic aromatic-degrading bacteria have been evaluated at genome-scale, for example, *Cupriavidus necator* JMP134 and *Burkholderia* sp. (Perez-Pantoja et al. 2012), and *Corynebacterium glutamicum* (Shen et al. 2012).

Replacement of traditional plastics by bioplastics is gaining interest as a way to increase the overall sustainability of plastics production. One of the most promising groups of bioplastics is the polyhydroxyalkanoate (PHA) family (Khanna and Srivastava 2005). The most studied part of this family concerns the short-chain length PHA (PHASCL), namely poly3-hydroxybutyrate (PHB) and poly(3-hydroxybutyrate-co-3-hydroxyvalerate (PHBV) (Keshavarz and Roy 2010). Several improvements in the production of these polymers which have recently been achieved are solid-state fermentations, triggering of nutrition medium, continuous cultures, 3-hydroxyvalerate (3-HV) co-polymer enhancement, and utilisation of inexpensive 3-HV precursors (Berezina et al. 2015). However, the popularisation and widespread use of these polymers has been limited by their production cost, which remains rather high, with raw materials being responsible for most of the cost (Berezina et al. 2015). One promising route for sustainable development is the combination of bioremediation with the production of valuable products. The production of PHA from aromatic compounds is an example of this approach. Some pioneering work in this field has already been reported (Nair et al. 2009; Nikodinovic et al. 2008; Ward et al. 2005; Kenny et al. 2008).

Aromatic Compounds Degradation

Various treatment methods were used for the removal of toxic compounds. Excavation, pumping, stabilization, solvent extraction, thermal desorption, and adsorption are popular physical treatment methods. Incineration, chemical oxidation, ozonation, and photocatalytic degradation are mostly used chemical methods. Clear description about physical and chemical treatment methods were outlined in Table 1. Many bacteria were explored for degradation of toxic compounds by biological methods (Table 2). Biological treatment methods have been shown to be the most economical, practical, and versatile approach as it leads to complete mineralization of toxic compounds (Nair et al. 2009). Hence, in the following sections it was discussed in detail.

Mono-aromatic compounds degradation

Substantial amount of phenol and phenolic compounds are discharged through effluents from a variety of industries including leather processing, textiles, pharmaceutical, and oil plants (Haddadi and Shavandi 2013). Phenol pollution is of great alarm since this chemical is toxic, mutagenic, and carcinogenic. Due to its

Table 1: Short description of types of physical and chemical methods used for aromatics compounds degradation.

Physical methods	
Excavation, pumping, or dredging	A contaminated waste such as soil, water, sediment is removed from contaminated site using excavator, pumping system, or dredger. These methods are usually used in combination with another cleanup method.
Solidification and stabilization	This method involves mixing a binding reagent into the contaminated waste to immobilize contaminants within the contaminated waste. Binding reagents commonly used include cement and limestone.
Solvent extraction	Organic contaminants are removed from contaminated soil/water using an individual solvent or mixture solvents. Various type of organic solvent such as organic material, surfactant-aided water, and cyclodextrins are used as flushing agent.
Thermal desorption	It is a physical separation process applying heat to increase the volatility of contaminants such that they can be separated and removed from the waste contaminant such as soil, sludge, and sediments.
Adsorption	Aromatics compounds in the contaminated air or water are accumulated on the surface of adsorbent such as activated carbon or peat.
Chemical methods	
Incineration	Organic compounds in a contaminated soil/sludge are destroyed with high temperatures ranging from 900 to 1200°C.
Chemical oxidation	The oxidation reaction converts aromatics compounds to other compounds. Different types of oxidants are investigated such as Fenton's reagent, potassium permanganate ($KMnO_4$), and hydrogen peroxide (H_2O_2).
Ozonation	In this method, degradation of aromatics compounds occurs through direct oxidation by O_3 or indirect oxidation by ·OH from O_3.
Photocatalytic degradation	The photocatalytic degradation method uses photo catalysts such as titanium dioxide (TiO_2) to promote oxidizing reactions which destroy organic contaminants in the presence of light radiation.

several toxic effects, removal of phenol from industrial wastewaters before their release is considered to be obligatory (Lu et al. 2012). Phenol and its derivatives are toxic and classified as hazardous materials (Nair et al. 2009). Wastewaters originated from oil refineries, pulp and paper manufacturing plants, steel, and pharmaceutical industries contain phenols and phenolic compounds. As a water soluble compound, phenol is generally found to contaminate streams, rivers, and lakes that are situated near the industrial areas (Nair et al. 2009). Wastewaters including phenols and phenolic compounds must be treated to prevent serious ecological risks. Biological treatment has been shown to be economical, practical, and most versatile approach as it leads to complete mineralization of phenol (Nair et al. 2009).

Several authors worked for phenol degradation using various bacteria. Haddadi and Shavandi (2013) reported 100% degradation of phenol using the bacteria *Halomonas* sp. within seven days at 1100 mg/l concentration. Reddy et al. (2015) reported about the degradation of phenol (96 ± 5%) and 4-chlorophenol (15 ± 3%) by using *Cupriavidus* sp. CY-1. Maximum phenol removal efficiency was determined as 68% in the batch studies. Reddy et al. (2015a) reported about the phenol (93 ± 5%) degradation using *Bacillus* sp. CYR1. Kuang et al. (2013) used *B.*

Table 2: Degradation of various aromatics compounds using bacteria.

Bacteria used	Toxic compound	Concentration (mg/l)	Degradation (%)	Time (days)	References
Halomonas sp.	Phenol	1100	100	7	Haddadi and Shavandi (2013)
Cupriavidus sp. CY-1	Phenol	100	96	6	Reddy et al. (2015)
Bacillus sp. CYR1	Phenol	100	93	6	Reddy et al. (2015a)
B. fusiformis	Phenol	200	62	0.29	Kuang et al. (2013)
Bacillus sp.	Phenol	200	43	-	Fayidh et al. (2015)
B. cereus	Phenol	200	94	-	Fayidh et al. (2015)
Modicisalibacter tunisiensis	Phenol	280	100	5	Bonfa et al. (2013)
Pseudomonas aeruginosa	Naphthalene	10	78	-	Qi et al. (2008)
Bacillus sp. CYR1	Naphthalene	100	26	6	Reddy et al. (2015a)
Cupriavidus sp. CY-1	Naphthalene	100	100	6	Reddy et al. (2015a)
Pseudomonas aeruginosa	Phenanthrene	15	86	-	Qi et al. (2008)
Pseudomonas USTB-RU	Phenanthrene	100	86	8	Masakorala et al. (2013)
Pseudoxanthomonas DMVP2	Phenanthrene	300	100	5	Patel et al. (2012)
Bacillus sp. CYR1	4-nonylphenol 4-t-butylphenol 4-s-butylphenol	100 100 100	51 47 41	6	Reddy et al. (2015a)
Cupriavidus sp. CY-1	4-s-butylphenol	100	29	6	Reddy et al. (2015)
Sphingobium fuliginis	4-butylphenol 4-t-butylphenol 4-s-butylphenol 4-nonylphenol 4-t-octylphenol	75 75 75 110 103	100 100 100 50 50	3	Ogata et al. (2013)
Stenotrophomonas IT-1	4-t-octylphenol	1000	100	5	Toyama et al. (2011)
Mycobacterium neoaurum	4-s-butylphenol	250	42	7	Hahn et al. (2013)
Sphingobium sp. WZ2	4-nonylphenol	100	89	13	Wang et al. (2014)
Rhizobium sp. WZ1	4-nonylphenol	100	80	17	Wang et al. (2014)
Acinetobacter OP5	4-t-octylphenol	30	91	8	Tuan et al. (2013)

fusiformis as biocatalyst and achieved 62% of phenol removal within 7 h at 200 mg/l concentration. Fayidh et al. (2015) reported 43% and 94% removal of phenol at 200 mg/l concentration using *Bacillus* sp. and *B. cereus* respectively. Bacteria *Rhodococcus ruber* was able to effectively degrade phenol within 1.5 days (He et al. 2014). Maza-Márquez and his co-workers (2013) reported 93% degradation of phenol within six days using the bacteria *Raoultella terrigena*. Bonfa et al. (2013) reported 100% degradation of phenol within five days at 280 mg/l using *Modicisalibacter tunisiensis*.

Poly-aromatic compounds degradation

Polycyclic aromatic hydrocarbons (PAHs) are classified as priority pollutants by the United States Environmental Protection Agency and are ubiquitous in the environment. They pose a significant threat to human health due to their mutagenic and carcinogenic properties (Xia et al. 2015). Wastewaters with these organic compounds can be treated mainly by physicochemical processes, but these methods are costly and in-appropriate for large wastewaters volumes (Lobo et al. 2013). Biological degradation has been utilized as alternative, since it has low associated costs and is more effective in degradation of organic compounds (Park et al. 2013).

Many bacteria have the capacity for degradation of PAH compounds. Bacterial strains such as *B. pumilus*, and *B. subtilis* were able to grow and utilize PAH compounds as a carbon source (Toledo et al. 2006). Bioremediation of crude oil using anaerobic mixed culture was done by Devi et al. (2011), and they reported the presence of bacteria belongs to *Bacillus* group in the mixed culture. Bioremediation of crude oil was also reported by Sathishkumar et al. (2008) using *Bacillus* sp. IOS1-7. Qi et al. (2008) reported 78% and 86% degradation of naphthalene and phenanthrene using the bacteria *Pseudomonas aeruginosa* at 10 and 15 mg/l, respectively. *Pseudomonas* USTB-RU showed 86% degradation of phenanthrene within eight days at 100 mg/l (Masakorala et al. 2013). Patel et al. (2012) reported 100% degradation of phenanthrene within 5 days using *Pseudoxanthomonas* DMVP2 at 300 mg/l. Reddy et al. (2015a) reported about the degradation of naphthalene using *Bacillus* sp. CYR1. Jones et al. (2005) reported nag genes of *Ralstonia* sp. strain U2 code for the enzymes for the catabolism of naphthalene, which converts naphthalene to fumarate and pyruvate via salicylate and gentisate and they are organized in a continuous sequence of adjacent genes. The common degradation pathways of naphthalene and phenanthrene were mentioned in Fig. 2.

Alkylphenols degradation

Alkylphenols are toxic compounds and are considered as endocrine disrupters that can cause various harmful effects, including reproductive effects by imitating the typical female sex hormone, estrogen, in aquatic life and in humans (Ogata et al. 2013; Toyama et al. 2011). Therefore, the industrial effluents containing alkylphenols should be properly treated to remove them and many bacteria can degrade these compounds effectively. *Bacillus* sp. CYR1 degraded the different alkylphenols supplied in the synthetic medium on the 6th day of operation (Reddy et al. 2015a).

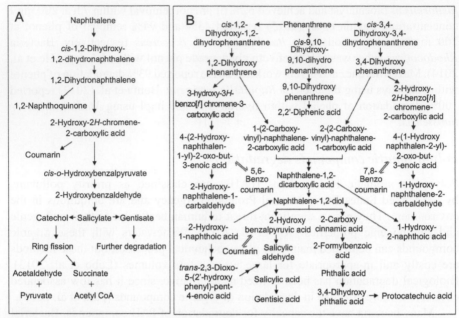

Fig. 2: Metabolic pathways involved in degradation of (A) naphthalene, and (B) phenanthrene (Figure was generated with the information from Seo et al. 2009).

Bacteria showed the degradation of 4-NP (51 ± 4%), 4-t-BP (47 ± 2%), 4-s-BP (41 ± 3%), and 4-t-OP (8 ± 2%). Bacteria were unable to degrade some of the alkylphenols like 4-BP. Interestingly, the Tween-80 addition showed improvement in degradation of alkylphenols like 4-BP (92 ± 6%), 4-s-BP (66 ± 5%), and 4-NP (54 ± 4%) by *Bacillus* sp. CYR1. Ogata et al. (2013) reported 100% degradation of 4-BP, 4-s-BP, and 4-t-BP using the bacteria *Sphingobium fuliginis* within three days at 75 mg/l concentration. They reported the 50% degradation of 4-NP and 4-t-OP at the concentration of 110 and 103 mg/l using the same bacteria. Toyama et al. (2011) reported 100% degradation of 4-t-OP using the bacteria *Stenotrophomonas* strain IT-1 within five days at 1000 mg/l concentration. Reddy et al. (2015) reported about the degradation of different alkylphenols by using *Cupriavidus* sp. CY-1. CY-1 showed 29 ± 4% degradation of 4-s-BP. Hahn et al. (2013) reported 42% degradation of 4-s-BP within seven days at 250 mg/l using *Mycobacterium neoaurum*. Wang et al. (2014) reported 89% and 80% removal of 4-NP at 100 mg/l using *Sphingobium* sp. WZ2, and *Rhizobium* sp. WZ1, respectively, with the incubation time of 13 days and 17 days, respectively. Tuan et al. (2013) reported 91% degradation of 4-t-OP at 30 mg/l using *Acinetobacter* strain OP5 within eight days.

Metabolic Pathways and Enzymes

Diaz et al. (2013) reported that in the classical aerobic catabolism, the hydroxylation and oxygenolytic cleavage of the aromatic ring is carried out by hydroxylating oxygenases and ring-cleavage dioxygenases, respectively. Most classical aerobic pathways converge to catecholic substrates which undergo either ortho or meta cleavage by intradiol or extradiol (types I and II) dioxygenases, respectively (Diaz et al. 2013). However, several bacterial degradation pathways generate non-catecholic intermediates, for example, gentisate, homogentisate, mono hydroxylated aromatic acids, heteroaromatic flavonols, that are subject to ring cleavage by devoted type III extradiol dioxygenases (Fetzner 2012). Ring cleavage of N-heteroaromatic compounds can be carried out by additional types of dioxygenases, for example, the 3-hydroxy-4-oxoquinaldine 2,4-dioxygenase (HOD) and 3-hydroxy-4-oxoquinoline 2,4-dioxygenase (QDO), are homologous metal-independent and cofactor-independent dioxygenases that possess a classical hydrolase fold core domain evolved to host and control dioxygen chemistry (Steiner et al. 2010; Diaz et al. 2013).

Despite many classical central pathways having been known for long, the gene encoding of some of these pathways are still uncharacterized. Some examples of recently uncovered genes clusters encoding the central pathways for meta cleavage of gallate and 2,3-dihydroxybenzoate in *Pseudomonas* strains have been reported (Nogales et al. 2011; Marin et al. 2012). Aromatic peripheral pathways appeared later in the evolution, are usually more tightly regulated, and show broader substrate specificity than central pathways (Diaz et al. 2013). The complex polymer lignin was considered an almost exclusive substrate of fungal laccases and peroxidases. Recent works have shown that lignin can be the substrate of bacterial dedicated peripheral pathways releasing low molecular weight phenolic products that finally lead to protocatechuate or some derivatives of the latter (Bugg et al. 2011). In the biphenyl-degrading *Rhodococcus jostii* RHA1 strain, an extracellular peroxidase has been shown to be active for lignin degradation, and it has homologues in a wide range of actinomycetes and proteobacteria (Diaz et al. 2013; Bugg et al. 2011; Ahmad et al. 2011). The catabolism of aromatic compounds also plays an essential role in the degradation of some terpenoids (such as steroids and resin acids) that generate an aromatic intermediate during their degradation. Cholesterol degradation has been shown to be critical to the pathogenesis of *Mycobacterium tuberculosis*, and the close similarity between some of the enzymes involved in the catabolism of cholesterol and aromatic-degrading enzymes may facilitate the design of new drugs to deal with tuberculosis (Diaz et al. 2013; Garcia et al. 2012).

A second aerobic strategy for cleaving the aromatic ring relies on the use of oxygenases, but solely to form a nonaromatic epoxide. Since these aerobic pathways share features of anaerobic catabolism, they are called aerobic hybrid pathways (Diaz et al. 2013). Thus, in these hybrid pathways, as in the anaerobic catabolism, all metabolites are activated to CoA thioesters through the action of an initial CoA ligase, the ring cleavage is carried out hydrolytically rather than oxygenolytically, and further metabolism of the non-aromatic CoA thioesters involves β-oxidation-like reactions yielding ketoadipyl-CoA, a common intermediate with the conventional β-ketoadipate pathway (Diaz et al. 2013; Fuchs et al. 2011) (Fig. 3).

The aerobic hybrid pathways are widely distributed among bacteria for the degradation of benzoate (box pathway) and phenylacetate (paa pathway) (Fuchs et al. 2011; Teufel et al. 2010; Law et al. 2011). In some bacteria, the classical benzoate degradation pathway coexists with the box pathway, and it was suggested that the latter could be advantageous under less favourable energetic conditions, that

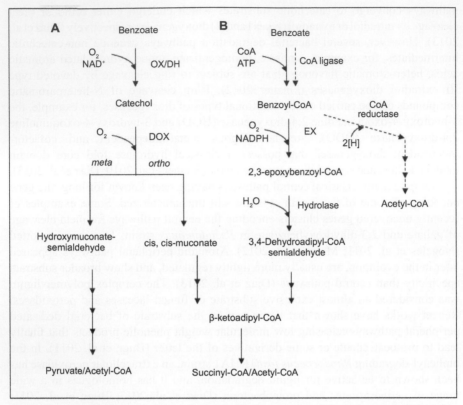

Fig. 3: Scheme of the biochemical strategies involved in benzoate degradation in bacteria by (A) aerobic, and (B) anaerobic pathway (Figure was generated with the information from Diaz et al. 2013).

is, low oxygen and benzoate concentrations (Fuchs et al. 2011). Key enzymes of aerobic hybrid pathways are class I di-iron proteins that catalyze the ring epoxidation (dearomatization) of the first intermediate of the catabolic pathway, that is, benzoyl-CoA and phenyl acetyl-CoA from benzoate and phenylacetate, respectively (Grishin et al. 2011; Rather et al. 2011). The multicomponent phenyl acetyl-CoA epoxidase (PaaABCDE) not only transforms phenyl acetyl-CoA into 1,2-epoxyphenylacetyl-CoA but also mediates an unprecedented NADPH dependent deoxygenation of the epoxide regenerating phenyl acetyl-CoA when oxygen becomes limiting (Diaz et al. 2013). Presumably, this bifunctionality plays an important biological role to avoid toxic intracellular epoxide levels if the subsequent catabolic steps are impeded (Teufel et al. 2012). On the contrary, the two-component benzoyl-CoA epoxidase (BoxAB) converts benzoyl-CoA into 2,3-epoxybenzoyl-CoA, a reaction significantly enhanced by the interaction with the BoxC ring-cleavage dihydrolase (Fig. 3) (Rather et al. 2011).

As part of a general detoxification strategy, aerobic hybrid pathways have developed salvage mechanisms to avoid the accumulation of CoA-thioesters and depletion of the intracellular CoA-pool, for example, the PaaI and PaaY thioesterases of the phenylacetate degradation pathway (Teufel et al. 2012). The aromatic acid H^+ symporters (AAHS) of the major facilitator superfamily are the main family of transporters involved in the uptake of aromatic acids, and they become essential for growth on some aromatic compounds that are easily oxidized and difficult to uptake by passive diffusion, for example, the GalT permease for gallic acid uptake and chemotaxis in *P. putida* (Nogales et al. 2011). Genes encoding multicomponent ABC transporters are also commonly found within or close to catabolic clusters, and in some cases they were shown to be involved in the uptake of aromatic acids (Hara et al. 2010; MacLean et al. 2011). In the proteobacterium *Rhodopseudomonas palustris*, periplasmic solute-binding proteins belonging to four different ABC-type transporters of aromatic compounds have been biochemically characterized (Pietri et al. 2012). Other family of periplasmic solute-binding proteins (the Bug family) has been identified in β-proteobacteria, where they can be involved in aromatic compound detection by tripartite transporters, for example, phthalates transporters in *Comamonas* strains (Fukuhara et al. 2010). Transporters of aromatic compounds in Gram-negative bacteria that are adapted to nutrient-poor conditions are often accompanied by nearby specific outer membrane substrate-specific channels (porins) that accelerate transport (Nogales et al. 2011). The recent structural and functional characterization of some porins of the OccK subfamily in *Pseudomonas aeruginosa*, such as the OccK1 porin that shows high structural similarity to the putative BenF benzoate channel from *Pseudomonas fluorescens* (Parthasarathy et al. 2010), reveal that they mediate an efficient uptake of a substantial number of monocyclic carboxylic compounds and certain medium-chain fatty acids (adipate, octanoate, etc.) (Diaz et al. 2013; Eren et al. 2012).

Energy Generation via Aromatic Compounds Degradation

Increasing concern for the environment has recently highlighted two major problems to be resolved, namely pollution and scarcity of resources. Pollution is a serious issue as a cause of many health-related problems (Hall et al. 2010) and can occur in numerous forms, including water (Vorosmarty et al. 2010), soil (Samecka-Cymerman et al. 2009), and air or atmospheric pollution (Elperin et al. 2011). Of all the various pollutants, aromatic substances are of particular concern (Kondratyeve et al. 2012) because their chemical structure makes them particularly hazardous and extremely resistant to different types of bioremediation and other treatments (Zhao et al. 2011). On the other side, a conventionally large fraction of synthetic plastic waste is buried in the soil due to its non-biodegradable property. Bioplastics in the form of polyhydroxyalkanoates (PHA) are an alternative to synthetic plastics and use of these bioplastics could lower the contribution of synthetic plastics to municipal landfills. Poly-3-hydroxybutyrate (PHB) is a type of PHA produced by many species of bacteria that serves as a storage form of carbon and energy (Reddy and Mohan 2012; Keshavarz and Roy 2010; Venkata Mohan et al. 2010). Development of the method to produce PHB from toxic aromatic compounds is more attractive to solve the problems of toxic compounds degradation and plastic disposal (Fig. 4).

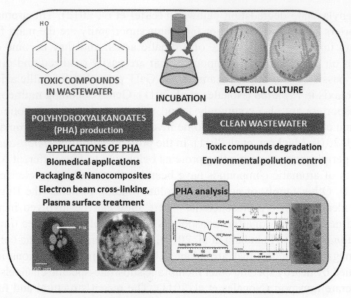

Fig. 4: Schematic representation of toxic compounds conversion in to PHA.

Although the isolation of PHB producing bacteria from environments contaminated with different pollutants has been studied, only a few reports describe the microbial production of PHB using pollutants (Table 3).

PHA production from monoaromatic hydrocarbons

Ward et al. (2005) reported that, styrene, the starting material for polystyrene synthesis, is a major toxic environmental pollutant. Worldwide, millions of kilograms of styrene are released each year as industrial effluents into the environment. In the United States alone, over 25 million kilograms of styrene waste is annually released, primarily into the air and through underground injection (U.S. Inventory of Toxic Compounds 2001). Styrene is associated with respiratory tract irritation, central nervous system depression, muscle weakness, and narcosis in humans and other mammals (U.S. Inventory of Toxic Compounds 2001). The conversion of styrene to PHA by *Pseudomonas putida* CA-3 provides a new and unique link between an aromatic environmental pollutant and aliphatic PHA accumulation. Investigations into the biodegradation of styrene have resulted in the elucidation of biochemical pathways and molecular control of styrene degradation. *P. putida* CA-3 is capable of the complete mineralization of styrene (O'Connor et al. 1995). It does so by epoxidation of styrene and isomerization of the epoxide to phenylacetaldehyde, which is further oxidized to phenylacetic acid (O'Connor et al. 1995). Phenyl acetic acid is converted to phenyl acetyl coenzyme A (CoA), which is further oxidized to acetyl-CoA (O'Leary et al. 2002).

Mono-aromatic hydrocarbons such as styrene and phenyl acetic acid have been evaluated as feedstock for the production of PHA by *Pseudomonas putida* CA-3 (Ward et al. 2005). Nikodinovic et al. (2008) reported degradation and conversion

Table 3: Aromatics compounds degradation and PHB production using bacteria.

S. No.	Bacteria	Aromatic compound type	Concentration	PHB production	Time	References
1	*Alcaligenes* sp. d2	Phenol	150 mg/l	22% DCW	24 h	Nair et al. 2009
2	*Cupriavidus necator*	Benzoic acid	2.5 g/l	35% DCW	48 h	Berezina et al. 2015
3	*Cupriavidus* sp. CY-1	4-t-butylphenol Phenol Naphthalene	100 mg/l 100 mg/l 100 mg/l	23% DCW 48% DCW 42% DCW	72 h	Reddy et al. 2015
4	*Bacillus* sp. CYR1	4-nonylphenol Phenol Naphthalene	100 mg/l 100 mg/l 100 mg/l	29% DCW 51% DCW 42% DCW	72 h	Reddy et al. 2015a
5	*Aromatoleum aromaticum*	Ethyl benzene Toluene	0.32 mM 0.74 mM	10.3% DCW 5.2% DCW	-	Trautwein et al. 2008
6	*Pseudomonas putida*	Styrene Phenylacetic acid	2000 mg/l 2000 mg/l	25% DCW 30% DCW	48 h	Ward et al. 2005
7	*Pseudomonas putida* S12	Styrene Phenylacetic acid	5 mM 5 mM	14% DCW 12% DCW	24 h	Tobin and O'Connor 2005
8	*Pseudomonas putida* CA-1	Styrene Phenylacetic acid	5 mM 5 mm	8% DCW 4% DCW	24 h	Tobin and O'Connor 2005
9	*Pseudomonas fluorescens* B3	Phenylacetic acid	5 mM	12% DCW	24 h	Tobin and O'Connor 2005
10	*Comamonas testosteroni*	Naphthalene	5 mM	85% DCW	48 h	Thakor et al. 2003
11	*Burkholderia cepacia*	3% xylose + Furfural	5000 mg/l	35% DCW	48 h	Pan et al. 2012
12	*Burkholderia cepacia*	3% xylose + Levulinic acid	1000 mg/l	54% DCW	48 h	Pan et al. 2011

of BTEX (benzene, toluene, ethylbenzene, and xylene) compounds in to PHA using single and defined mixed cultures. Berezina and Paternostre (2010) established that only mono-substituted aromatic compounds were actually transformed into PHB by *Cupriavidus necator*. Berezina et al. (2015) conducted batch experiments to improve the bioremediation of aromatic compounds by *C. necator*, different initial concentrations, ranging from 1 to 10 g/l of benzoic acid were used. The limitation of initial concentration appeared at a rather low level, indeed any concentration above 2.5 g/l was found to be toxic to the microorganism. However, at 2.5 g/l of benzoic acid the microorganism performed well, and no specific lag phase or saturation were observed. The final cells' dry mass (CDM) reached 1.4 g/l compared to 0.55 g/l obtained with the experiment with the initial 1 g/l benzoic acid concentration. Also, with the initial 2.5 g/l concentration of benzoic acid, the PHA content in the cells reached

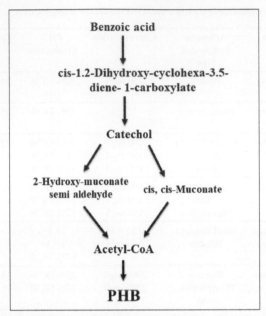

Fig. 5: Transformation of toxic compound benzoic acid in to PHB by *C. necator* (Figure was generated with the information from Berezina et al. 2015).

35% at 48 hours which is higher than the 21% reached with the initial concentration of benzoic acid of 1 g/l (Fig. 5). PHB production could be considered as one of the solutions for the bioremediation of wastewaters containing phenolic compounds (Nair et al. 2009). Nair et al. (2009) isolated *Alcaligenes* sp. d2 from soil which was reported as a potent phenol degrading organism. The organism could carry out 100% degradation of phenol in 32 h and could progressively result in early accumulation of PHB intracellularly from 8 h onwards. The various conditions optimized for the maximum accumulation of intracellular PHB were pH 7.0, incubation time 24 h, phenol concentration 15 mg/100 ml, and ammonium sulfate concentration 25 mg/100 ml.

Narancic et al. (2012) reported that medium-chain-length PHA was accumulated by *Pseudomonas* sp. TN301 when monoaromatic hydrocarbons and mixtures thereof were supplied to the culture as a vapor from a central column as previously described (Ward et al. 2005). While biomass levels were generally similar, mcl-PHA was accumulated to different levels ranging from 3% to 25% CDW. The best accumulation of biopolymer from a single monoaromatic substrate was achieved when o-xylene (19.2%) and 3-methylbenzene (19.2%) were used as carbon sources, and it was 6·4-fold higher in comparison with PHA accumulation from styrene (3%), which also supported the lowest levels of growth (0.2 g/l). *Pseudomonas* sp. TN301 accumulated the highest level of mcl-PHA (25% CDW) when supplied with a BTEX mixture. Narancic et al. (2012) reported that the best biomass yields were achieved when a mixture of xylenes (mixture of o-, m- and p-xylene) was used as carbon source, and it was 1.2 to 1.9 folds higher in comparison with all other single substrates. The PHA accumulation productivity from the mixture of xylenes and mixture of benzenes (mixture of benzene, ethylbenzene, 3-methylbenzene, and butylbenzene) was similar, but 2.4-fold lower than the PHA productivity from BTEX mixture. The predominant monomer of the mcl-PHA from all monoaromatic substrates tested was the C10 monomer ((R)-3-hydroxydecanoic acid) with the molar percentage ranging from 63% to 75%. The distribution of other monomers varied from 0% to 3% for C6, 9% to 19% for C8 and 6% to 28% for C12. Recent studies reported that PHB was accumulated by *Bacillus* sp. CYR1 and *Cupriavidus* CY-1 when various aromatic compounds were supplied to the culture. They reported that best accumulation of PHB was achieved when phenol was used in MSM (Reddy et al. 2015, 2015a). Both bacteria were able to produce PHB using some of the alkylphenols as carbon sources. Although all toxic substrates supported biomass growth to the comparable levels, but some substrates were poorer substrates for the PHB accumulation. This may be due to the different bioavailability of these substrates to the medium and different efficiency in utilization of these substrates by the *Bacillus* sp. CYR1 (Reddy et al. 2015a). Hoffmann et al. (2000) reported many bacteria can consume a large number of aromatic compounds, but they cannot always accumulate PHB.

PHA production from PAH compounds

Narancic et al. (2012) reported that the seven isolates were screened for PHA accumulation in shake-flask experiments when naphthalene and glucose (10 mmol/l) were supplied as carbon sources. For the purpose of PHA screening, naphthalene crystals (12 mg) were supplied directly to medium. Owing to naphthalene hydrophobicity and reported toxicity (Pumphrey and Madsen 2007), they used the amount of naphthalene that corresponds to 2 mmol/l which completely dissolved in the growth medium. When naphthalene was used as a carbon source, the best biomass yields were achieved by *Pseudomonas* sp. TN301, which was between 1.35-fold to 2-fold higher in comparison with all other strains. PHA was accumulated as 39% of the cell dry weight (CDW). GC-MS analysis of the accumulated PHA revealed that the PHA accumulated by the isolate TN301 contained monomers (R)-3-hydroxyhexanoate (C6), (R)-3-hydroxyoctanoate (C8), (R)-3-hydroxydecanoate (C10), and (R)-3-hydroxydodecanoate (C12) in the following percentage ratio C6/C8/C10/C12 of 4:27:60:9. While the C10 monomer was predominant in mcl-PHA accumulated from both glucose and naphthalene, there was 1.2-fold more C10 monomer present in PHA accumulated from glucose, compared with PHA accumulated by naphthalene grown cells. *Pseudomonas* sp. TN301 was also tested for the ability to grow in the presence of heavy metals such as Cd, Hg, Ni, Cu, and Fe as they often co-occur at the petrochemically contaminated sites (Pepi et al. 2009).

As PAH are highly hydrophobic and insoluble in water, it was of great importance to find a delivery method for PAH to the growth media. Experiments conducted by Narancic et al. (2012) using naphthalene as a model PAH compound as it was used as the selective pressure in the enrichment isolation and the initial screen for PHA accumulation. They analysed growth and mcl-PHA accumulation of *Pseudomonas* sp. TN301 when different solvents (methanol, benzene, acetone, and 1-pentanone) were used to dissolve and supply naphthalene to the medium. These solvents have previously been proposed to increase the solubility of aromatic hydrocarbons in organic solvent and water mixtures (Dickhut et al. 1989). They also included surfactant (Tween 80), as the application of surfactants has been suggested as a possible way to increase bioavailability of PAH (Hickey et al. 2007). Appropriate controls (a) naphthalene without solvents or surfactants and (b) solvents or surfactants were used. The growth and accumulated PHA were compared with naphthalene added directly to media without any solvents or surfactant. Interestingly, while the similar levels of growth were achieved when 2 mmol/l naphthalene was supplied to the medium during the initial PHA screen in comparison with when the total amount of naphthalene supplied was 15 mmol/l, no mcl-PHA was detected in the culture when 15 mmol/l naphthalene was used. The best growth was observed when Tween 80 was added to the media together with naphthalene. Achieved cell dry weight after 60 h when synthetic surfactant Tween was present in the medium was from 1.1 fold to 2.8 folds higher in comparison with all other methods of supply of the naphthalene to the medium. *Pseudomonas* sp. TN301 could grow in the mineral medium when Tween 80 was supplied as a sole source of carbon; however, the biomass yields were 5.8 folds lower in comparison with when naphthalene was added. No mcl-PHA was detected in the TN301 culture grown on Tween 80; thus, the improvement of 2.6 folds in biomass and mcl-PHA yield of 0.1 g/l was achieved by adding synthetic surfactant to the culture. Dissolving naphthalene in methanol, acetone, and 1-pentanone resulted in apparent improved biomass yields of 2.27 folds, 1.38 folds, and 1.22-folds, respectively, in comparison when naphthalene crystals were supplied to the medium directly.

However, *Pseudomonas* sp. TN301 could grow on methanol as a sole source of carbon and energy, with the same amount of methanol supporting 2.5 folds lower growth in comparison when naphthalene was present. Thus, the actual improvement in naphthalene consumption when methanol was used as solvent was 1.38 folds. *Pseudomonas* sp. TN301 could also accumulate low levels of mcl-PHA when methanol was supplied as a sole source of carbon and energy to the medium. Poor growth and no mcl-PHA accumulation were supported by acetone and 1-pentanone. Interestingly, when total amount of benzene (200 μl) was supplied sequentially directly to the medium, 2.5 folds increase in the mcl-PHA accumulation was observed in the comparison to when benzene was supplied as a vapor, while the levels of accumulated biomass were similar. When naphthalene was dissolved in benzene there was a decrease in both CDW and PHA accumulation in comparison with when benzene alone was supplied as a vapor, while similar growth levels were observed in comparison with growth when naphthalene crystals were supplied directly to the medium. Thus, supplementation of the medium by synthetic surfactant Tween 80 at inoculation time was chosen for further experiments. Based on the improved growth and PHA accumulation through the addition of Tween 80 with naphthalene, other PAH compounds with Tween 80 were evaluated for PHA production. *Pseudomonas* sp. TN301 grew in liquid medium on all substrates tested, the best growth was achieved when naphthalene was carbon source and it was from 2.4–1.5 folds higher in comparison with all other substrates. Under conditions tested, PHA was not detected in the cultures when 1-ethylnaphthalene, 2-methylnaphthalene, dimethylnaphthalene, and the mixture of naphthalene and phenanthrene were used as substrates. In the case of 1-ethylnaphthalene, 2-methylnaphthalene, and dimethylnaphthalene growth yields were between 0.2 and 0.3 g/l, which is comparable with growth obtained on monoaromatic hydrocarbons.

mcl-PHA was accumulated when naphthalene, phenanthrene, and chrysene were used as substrates to 22.9%, 3.5%, and 5% of total CDW, respectively. Although the biomass yield from naphthalene was 1.5–2 folds lower in comparison with phenanthrene and chrysene, mcl-PHA productivity from these substrates was 6.5 folds and 4.6 folds lower compared with cells grown on naphthalene. The C10 was predominant monomer in mcl-PHA accumulated from all three PAH, with the C6 monomer detected in mcl-PHA from naphthalene only. When the mixtures of PAH were used as carbon sources, biomass yields were 1.8 to 2.1 folds lower than that achieved with naphthalene supplied as a sole carbon and energy source. PHA accumulation was not observed when naphthalene/phenanthrene mixture was supplied. A 4.8 to 20.8 folds lower mcl-PHA productivity was observed for other substrate combinations compared with when naphthalene was alone. Monomer composition of the PHA obtained from the mixtures of PAH was similar and comparable with the monomer composition obtained when single monoaromatic or polyaromatic substrates were used as a carbon source.

Recent studies reported that PHB was accumulated by *Bacillus* sp. CYR1 and *Cupriavidus* CY-1 when various PAH were supplied to the culture. They reported that best accumulation of PHB was achieved when naphthalene was used in MSM (Reddy et al. 2015, 2015a). Due to its biodegradable nature, PHA has a broad range of applications, including medical applications such as wound management, drug delivery, and tissue engineering (Zinn et al. 2001). Furthermore, PHA is composed of chiral hydroxy acids that have potential as synthons for anti-human immunodeficiency virus drugs, anticancer drugs, antibiotics, and vitamins (de Roo et al. 2002).

PHA Production Proteins Response Under Aromatic Compounds

Trautwein et al. (2008) reported that many bacteria accumulate PHA as an insoluble energy and carbon storage compound when an excess of a carbon source is accompanied by a limitation in another nutrient. In *Ralstonia eutropha*, the initial formation of acetoacetyl-CoA from two acetyl-CoA units is catalyzed by acetyl-CoA acetyltransferase (PhbA). Subsequent reduction catalyzed by acetoacetyl-CoA reductase (PhbB) yields D-3-hydroxyacyl-CoA, which is polymerized to PHB by PHA synthase (PhaC) (Trautwein et al. 2008). The product accumulates as granules which are coated with phospholipids and proteins, mostly phasins. PHA forming cells synthesize phasins only in amounts that can be bound to the granules (Wieczorek et al. 1995). Therefore, the occurrence of phasins can be used as a marker for PHA synthesis (York et al. 2001).

Trautwein et al. (2008) reported that in the proteomic analysis, the levels of PhbB and PhaC were increased three times during growth with 0.74 mM toluene. In addition, three phasin-like proteins (EbA1323, EbA6852, and EbA5033) were up-regulated in the same cultures (up to 27.4 folds) and also during growth with 0.32 mM ethylbenzene (up to 5.7 folds). The sequence similarities of these predicted phasins to their orthologs PhaP1 to PhaP4 in *R. eutropha* (Potter et al. 2004) are 42 to 66%. In succinate-utilizing, solvent-shocked cultures of strain EbN1, where the aromatic compounds were not catabolized, either PHB-related proteins were not detected or their abundances were not increased. The proteomic findings were corroborated by chemical determination of PHA levels. Only during growth with 0.21 and 0.32 mM ethylbenzene and with 0.74 mM toluene did PHB accumulate at levels that accounted for up to 2.0, 10.3, and 5.2% of the cell dry weight, respectively. PHAs other than PHB were not observed. However, the absence of PHB in ethylbenzene- and toluene-shocked cells may have been due to an insufficient incubation time (maximum of 10 h after alkylbenzene addition), since phasin-like proteins were first detected after 20 h in cultures utilizing 0.07 mM toluene (Trautwein et al. 2008).

Under all applied growth conditions, the potential electron supply provided by the organic substrate exceeded the electron accepting capacity of nitrate. PHB accumulation was observed only in alkylbenzene utilizing cultures displaying reduced growth, nitrate consumption, and nitrite turnover. Importantly, in these cultures the electron accepting nitrate and nitrite were not limiting. Thus, PHB formation presumably does not result from imbalanced nutrient supply but rather results from an impaired coupling of alkylbenzene catabolism and denitrification. Thus, one may speculate that in alkylbenzene utilizing cells of strain EbN1 alkylbenzene derived acetyl-CoA is rerouted from oxidation via the tricarboxylic acid cycle to PHB synthesis accompanied by a decrease in the NAD(P)H pool and recycling of free CoA. In such a scenario, PHB could be considered a sink for reducing equivalents, ensuring continuous alkylbenzene degradation (Trautwein et al. 2008). PHB could also function as a kind of hydrophobic trap or sink for aromatic compounds, since PHB granules are known to accumulate hydrophobic compounds, such as the fluorescent dye Nile red (Spiekermann et al. 1999). Considering the absence of PHB and related proteins in succinate-utilizing, solvent-shocked cells of strain EbN1, this would be restricted to long-term solvent exposure and/or utilization.

Conclusions

PHA is gaining importance as a promising biodegradable plastic, and other side pollution caused by aromatic compounds creating many problems for the environment and human health. To solve these two problems, researchers working for the identification of new bacteria which can degrade the toxic compounds and produce bioplastics simultaneously. Even though prominent research employed throughout the world, many questions still exist on how to shape the bacterial cultures and, if successful, whether the system is stable to withstand operating problems that may occur during industrial processing. The other areas that need to study are the bioplastics extraction technologies without using expensive and hazardous solvents. Another driver may be the use of this technology in large-scale fermenters, but further scale-up work is necessary before practical application will become important. As an emerging field of study, many research questions exists, such as at what toxic compound concentration does the bacteria become inhibited, what is the optimum temperature and pH to maximize the production rates, and how to best extract and separate the PHA from the media.

References

Ahmad, M., J.N. Roberts, E.M. Hardiman, R. Singh, L.D. Eltis and T.D.H. Bugg. 2011. Identification of DypB from *Rhodococcus jostii* RHA1 as a lignin peroxidase. Biochemistry 50: 5096–5107.

Berezina, N. and L. Paternostre. 2010. Transformation of aromatic compounds by *C. necator*. Chem. Eng. Trans. 20: 259–263.

Berezina, N., B. Yada and R. Lefebvre. 2015. From organic pollutants to bioplastics: insights into the bioremediation of aromatic compounds by *Cupriavidus necator*. N. Biotechnol. 32: 47–53.

Bonfa, M.R.L., M.J. Grossman, F. Piubeli, E. Mellado and L.R. Durrant. 2013. Phenol degradation by halophilic bacteria isolated from hypersaline environments. Biodegradation 24: 699–709.

Bugg, T.D.H., M. Ahmad, E.M. Hardiman and R. Rahmanpour. 2011. Pathways for degradation of lignin in bacteria and fungi. Nat. Prod. Rep. 28: 1883–1896.

Capotori, G., P. Digianvincenzo, P. Cesti, A. Bernardi and G. Guglielmetti. 2004. Pyrene and benzo[a]pyrene metabolism by an *Aspergillus terreus* strain isolated from a polycyclic aromatic hydrocarbons polluted soil. Biodegradation 15: 79–85.

Cheung, P.Y. and B.K. Kinkle. 2001. Mycobacterium diversity and pyrene mineralization in petroleum contaminated soils. Appl. Environ. Microbiol. 67: 2222–2229.

de Roo, G., M.B. Kellerhals, Q. Ren, B. Witholt and B. Kessler. 2002. Production of chiral *R*-3-hydroxyalkanoic acids and *R*-3-hydroxyalkanoic acid methylesters via hydrolytic degradation of polyhydroxyalkanoate synthesized by Pseudomonads. Biotechnol. Bioeng. 77: 717–722.

Devi, M.P., M.V. Reddy, A. Juwarkar, P.N. Sarma and S.V. Mohan. 2011. Effect of co-culture and nutrients supplementation on bioremediation of crude petroleum sludge. Clean—Soil, Air, Water 39: 900–907.

Diaz, E., J.I. Jimenez and J. Nogales. 2013. Aerobic degradation of aromatic compounds. Curr. Opin. Biotechnol. 24: 431–442.

Dickhut, R.M., A.W. Andren and D.E. Armstrong. 1989. Naphthalene solubility in selected organic solvent/water mixtures. J. Chem. Eng. Data. 34: 438–443.

Elperin, T., A. Fominykh, B. Krasovitov and A. Vikhansky. 2011. Effect of rain scavenging on altitudinal distribution of soluble gaseous pollutants in the atmosphere. Atmos. Environ. 45: 2427–2433.

Eren, E., J. Vijayaraghavan, J. Liu, B.R. Cheneke, D.S. Touw, B.W. Lepore et al. 2012. Substrate specificity within a family of outer membrane carboxylate channels. PLoS Biol. 10: e1001242.

Fayidh, M.A., S. Kallary, P.A.S. Babu, M. Sivarajan and M.A. Sukumar. 2015. A rapid and miniaturized method for the selection of microbial phenol degraders using colourimetric microtitration. Curr. Microbiol. 6: 898–906.

Fetzner, S. 2012. Ring-cleaving dioxygenases with a cupin fold. Appl. Environ. Microbiol. 78: 2505–2514.

Freeman, D.J. and D.W.S. Cattell. 1990. Wood burning as a source of atmospheric polycyclic aromatic hydrocarbons. Environ. Sci. Technol. 24: 1581–1585.

Fuchs, G., M. Boll and J. Heider. 2011. Microbial degradation of aromatic compounds—from one strategy to four. Nat. Rev. Microbiol. 9: 803–816.

Fukuhara, Y., K. Inakazu, N. Kodama, N. Kamimura, D. Kasai, Y. Katayama et al. 2010. Characterization of the isophthalate degradation genes of *Comamonas* sp. strain E6. Appl. Environ. Microbiol. 76: 519–527.

Garcia, J.L., I. Uhia and B. Galan. 2012. Catabolism and biotechnological applications of cholesterol degrading bacteria. Microb. Biotechnol. 5: 679–699.

Gottstein, D. and D. Gross. 1992. Phytoalexins from woody plants. Trees 6: 55–68.

Grishin, A.M., E. Ajamian, L. Tao, L. Zhang, R. Menard and M. Cygler. 2011. Structural and functional studies of the *Escherichia coli* phenylacetyl-CoA monooxygenase complex. J. Biol. Chem. 286: 10735–10743.

Haddadi, M. and M. Shavandi. 2013. Biodegradation of phenol in hypersaline conditions by *Halomonas* sp. strain PH2-2 isolated from saline soil. Int. Biodet. Biodeg. 85: 29–34.

Hahn, V., K. Sünwoldt, A. Mikolasch and F. Schauer. 2013. Biotransformation of 4-*sec*-butylphenol by Gram-positive bacteria of the genera *Mycobacterium* and *Nocardia* including modifications on the alkyl chain and the hydroxyl group. Appl. Microbiol. Biotechnol. 97: 8329–8339.

Hall, J., V. Brajer and F. Lurmann. 2010. Air pollution, health and economic benefits—lessons from 20 years of analysis. Ecol. Econ. 69: 2590–2597.

Hara, H., G.R. Stewart and W.W. Mohn. 2010. Involvement of a novel ABC transporter and monoalkyl phthalate ester hydrolase in phthalate ester catabolism by *Rhodococcus jostii* RHA1. Appl. Environ. Microbiol. 76: 1516–1523.

He, Z., C. Niu and Z. Lu. 2014. Individual or synchronous biodegradation of di-n-butyl phthalate and phenol by *Rhodococcus ruber* strain DP-2. J. Haz. Mat. 273: 104–109.

Hickey, A.M., L. Gordon, A.D. Dobson, C.T. Kelly and E.M. Doyle. 2007. Effect of surfactants on fluoranthene degradation by *Pseudomonas alcaligenes* PA-10. Appl. Microbiol. Biotechnol. 74: 851–856.

Hoffmann, N., A. Steinbuchel and B.H.A. Rehm. 2000. Homologous functional expression of cryptic phaG from *Pseudomonas oleovorans* establishes the transacylase mediated polyhydroxyalkanoate biosynthetic pathway. Appl. Microbiol. Biotechnol. 54: 665–670.

Jimenez, J.I., J. Nogales, J.L. García and E. Díaz. 2010. A genomic view of the catabolism of aromatic compounds in Pseudomonas. pp. 1297–1325. *In*: Timmis, K.N. [ed.]. Handbook of Hydrocarbon and Lipid Microbiology. Springer. Berlin, Heidelberg.

Jones, R.M., B. Britt-Compton and P.A. Williams. 2005. The naphthalene catabolic genes (nag) genes of *Ralstonia* sp. strain U2 are an operon that is regulated by nagR, a LysR-type transcriptional regulator. J. Bacteriol. 185: 5847–5853.

Juhasz, A.L. and R. Naidu. 2000. Bioremediation of high molecular weight polycyclic aromatic hydrocarbons: a review of the microbial degradation of benzo[a]pyrene. Int. Biodeter. Biodegr. 45: 57–88.

Kenny, S., J.N. Runic, W. Kaminsky, T. Woods, R.P. Babu, C.M. Keely et al. 2008. Up-cycling of PET (polyethylene terephthalate) to the biodegradable plastic PHA (polyhydroxyalkanoate). Environ. Sci. Technol. 42: 7696–701.

Keshavarz, T. and I. Roy. 2010. Polyhydroxyalkanoates: Bioplastics with a green agenda. Curr. Opin. Microbiol. 13: 321–326.

Khanna, S. and A. Srivastava. 2005. Recent advances in microbial polyhydroxyalkanoate. Process Biochem. 40: 607–619.

Kim, T.J., E.Y. Lee, Y.J. Kim, K.S. Cho and H.W. Ryu. 2003. Degradation of polyaromatic hydrocarbons by *Burkholderia cepacia* 2A-12. World J. Microbiol. Biotechnol. 19: 411–417.

Kondratyeve, L., N. Fisher and V. Bardyuk. 2012. Bioindication of transboundary pollution of the Amur river with aromatic hydrocarbons after the technologenic accident in China. Contemp. Probl. Ecol. 2: 245–252.

Kuang, Y., Y. Zhou, Z. Chen, M. Megharaj and R. Naidu. 2013. Impact of Fe and Ni/Fe nanoparticles on biodegradation of phenol by the strain *Bacillus fusiformis* (BFN) at various pH values. Bioresour. Technol. 136: 588–594.

Law, A. and M.J. Boulanger. 2011. Defining a structural and kinetic rationale for paralogous copies of phenylacetate-CoA ligases from the cystic fibrosis pathogen *Burkholderia cenocepacia* J2315. J. Biol. Chem. 286: 15577–15585.

Lobo, C.C., N.C. Bertola and E.M. Contreras. 2013. Stoichiometry and kinetic of the aerobic oxidation of phenolic compounds by activated sludge. Bioresour. Technol. 136: 58–65.

Lu, D., Y. Zhang, S. Niu, L. Wang, S. Lin, C. Wang et al. 2012. Study of phenol biodegradation using *Bacillus amyloliquefaciens* strain WJDB-1 immobilized in alginate-chitosan-alginate (ACA) microcapsules by electrochemical method. Biodegra. 23: 209–219.

MacLean, A.M., W. Haerty, G.B. Golding and T.M. Finan. 2011. The LysR-type PcaQ protein regulates expression of a protocatechuate inducible ABC-type transport system in Sinorhizobium meliloti. Microbiology 157: 2522–2533.

Marin, M., I. Plumeier and D.H. Pieper. 2012. Degradation of 2,3-dihydroxybenzoate by a novel meta-cleavage pathway. J. Bacteriol. 194: 3851–3860.

Masakorala, K., J. Yao, M. Cai, R. Chandankere, H. Yuan and H. Chen. 2013. Isolation and characterization of a novel phenanthrene (PHE) degrading strain *Psuedomonas* sp. USTB-RU from petroleum contaminated soil. J. Haz. Mat. 263: 493–500.

Maza-Márquez, P., M.V. Martínez-Toledo, J. González-López, B. Rodelas, B. Juárez-Jiménez and M. Fenice. 2013. Biodegradation of olive washing wastewater pollutants by highly efficient phenol-degrading strains selected from adapted bacterial community. Int. Biodet. Biodeg. 82: 192–198.

Menzie, C.A., B.B. Potochi and J. Santodonato. 1992. Exposure to carcinogenic PAHs in the environment. Environ. Sci. Technol. 26: 1278–1284.

Mohan, S.V., M.V. Reddy, G.V. Subhash and P.N. Sarma. 2010. Fermentative effluents from hydrogen producing bioreactor as substrate for poly (β-OH) butyrate production with simultaneous treatment: An integrated approach. Bioresour. Technol. 101: 9382–0386.

Nair, I.C., S. Pradeep, M.S. Ajayan, K. Jayachandran and Shankar Shashidhar. 2009. Accumulation of intracellular polyhydroxybutyrate in *Alcaligenes* sp. d_2 under phenol stress. Appl. Biochem. Biotechnol. 159: 545–552.

Narancic, T., S.T. Kenny, L. Djokic, B. Vasiljevic, K.E. O'Connor and J.N. Runic. 2012. Medium-chain-length polyhydroxyalkanoate production by newly isolated *Pseudomonas* sp. TN301 from a wide range of polyaromatic and monoaromatic hydrocarbons. J. Appl. Microbiol. 113: 508–520.

Nestler, M.F.H. 1974. Characterization of wood-preserving coal-tar creosote by gas-liquid chromatography. Anal. Chem. 46: 46–53.

Nikodinovic, J., S.T. Kenny, R.P. Babu, T. Woods, W.J. Blau and K.E. O'Connor. 2008. The conversion of BTEX compounds by single and defined mixed cultures to medium-chain-length polyhydroxyalkanoate. Appl. Microbiol. Biotechnol. 80: 665–673.

Nogales, J., A. Canales, J. Jimenez-Barbero, B. Serra, J.M. Pingarron, J.L. Garcia et al. 2011. Unravelling the gallic acid degradation pathway in bacteria: the gal cluster from *Pseudomonas putida*. Mol. Microbiol. 79: 359–374.

O'Connor, K., C.M. Buckley, S. Hartmans and A.D.W. Dobson. 1995. Possible regulatory role for non-aromatic carbon sources in styrene degradation by *Pseudomonas putida* CA-3. Appl. Environ. Microbiol. 61: 544–548.

Ogata, Y., T. Toyama, N. Yu, X. Wang, K. Sei and M. Ike. 2013. Occurrence of 4-*tert*-butylphenol (4-*t*-BP) biodegradation in an aquatic sample caused by the presence of Spirodela polyrrhiza and isolation of a 4-*t*-BP-utilizing bacterium. Biodegradation 24: 191–202.

O'Leary, N., W.A. Duetz, A.D.W. Dobson and K.E. O'Connor. 2002. Induction and repression of the sty operon in *Pseudomonas putida* CA-3 during growth on phenylacetic acid under organic and inorganic nutrient limiting continuous culture conditions. FEMS Microbiol. Lett. 208: 263–268.

Olie, K., P.L. Vermeulen and O. Hutzinger. 1997. Chlorodibenzo-p-dioxins and chlorodibenzofurans are trace components of fly ash and flue gas of some municipal incinerators in the Netherlands. Chemosphere 6: 455–459.

Pan, W., J.A. Perrotta, A.J. Stipanovic, C.T. Nomura and J.P. Nakas. 2012. Production of polyhydroxyalkanoates by *Burkholderia cepacia* ATCC 17759 using a detoxified sugar maple hemicellulosic hydrolysate. J. Ind. Microbiol. Biotechnol. 39: 459–469.

Park, M.R., D.J. Kim, J.W. Choi and D.S. Lim. 2013. Influence of immobilization of bacterial cells and TiO_2 on phenol degradation. Water Air Soil Pollut. 224: 1–9.

Parthasarathy, S., F. Lu, X. Zhao, Z. Li, J. Gilmore, K. Bain et al. 2010. Structure of a putative BenF-like porin from *Pseudomonas fluorescens* Pf-5 at 2.6 A° resolution. Proteins 78: 3056–3062.

Patel, V., S. Cheturvedula and D. Madamwar. 2012. Phenanthrene degradation by *Pseudoxanthomonas* sp. DMVP2 isolated from hydrocarbon contaminated sediment of Amlakhadi canal, Gujarat. India. J. Haz. Mat. 201: 43–51.

Pepi, M., A. Lobianco, M. Renzi, G. Perra, E. Bernardini, M. Marvasi et al. 2009. Two naphthalene degrading bacteria belonging to the genera Paenibacillus and Pseudomonas isolated from a highly polluted lagoon perform different sensitivities to the organic and heavy metal contaminants. Extremophiles 13: 839–848.

Perez-Pantoja, D., R. Donoso, L. Agullo, M. Cordova, M. Seeger, D.H. Pieper et al. 2012. Genomic analysis of the potential for aromatic compounds biodegradation in *Burkholderiales*. Environ. Microbiol. 14: 1091–1117.

Pietri, R., S. Zerbs, D.M. Corgliano, M. Allaire, F.R. Collart and L.M. Miller. 2012. Biophysical and structural characterization of a sequence diverse set of solute-binding proteins for aromatic compounds. J. Biol. Chem. 287: 23748–23756.

Pitter, P. and J. Chudoba. 1990. Biodegradation of Organic Substances in the Aquatic Environment. CRC Press: Boca Raton, Florida.

Potter, M., H. Muller, F. Reinecke, R. Wieczorek, F. Fricke, B. Bowien et al. 2004. The complex structure of polyhydroxybutyrate (PHB) granules: four orthologous and paralogous phasins occur in *Ralstonia eutropha*. Microbiology 150: 2301–2311.

Pumphrey, G.M. and E.L. Madsen. 2007. Naphthalene metabolism and growth inhibition by naphthalene in *Polaromonas naphthalenivorans* strain CJ2. Microbiology 153: 3730–3738.

Qi, J., B. Wang, J. Li, H. Ning, Y. Wang, W. Kong et al. 2008. Genetic determinants involved in the biodegradation of naphthalene and phenanthrene in *Pseudomonas aeruginosa* PAO1. Environ. Sci. Pollut. Res. 22: 6743–6749.

Ramos, J.L., S. Marqués, P. van Dillewijn, M. Espinosa-Urgel, A. Segura, E. Duque et al. 2011. Laboratory research aimed at closing the gaps in microbial bioremediation. Trends Biotechnol. 29: 641–647.

Rather, L.J., T. Weinert, U. Demmer, E. Bill, W. Ismail, G. Fuchs et al. 2011. Structure and mechanism of the diiron benzoylcoenzyme A epoxidase BoxB. J. Biol. Chem. 286: 29241–29248.

Reddy, M.V. and S.V. Mohan. 2012. Effect of substrate load and nutrients concentration on the polyhydroxyalkanoates (PHA) production using mixed consortia through wastewater treatment. Bioresour. Technol. 114: 573–582.

Reddy, M.V., Y. Mawatari, Y. Yajima, C. Seki, T. Hoshino and Y.C. Chang. 2015a. Poly-3-hydroxybutyrate (PHB) production from alkylphenols, mono and poly-aromatic hydrocarbons using *Bacillus* sp. CYR1: A new strategy for wealth from waste. Bioresour. Technol. 192: 711–717.

Reddy, M.V., Y. Mawatari, Y. Yajima, C. Seki, T. Hoshino and Y.C. Chang. 2015b. Degradation and conversion of toxic compounds into useful bioplastics by *Cupriavidus* sp. CY-1: relative expression of the PhaC gene under phenol and nitrogen stress. Green Chem. 17: 4560–4569.

Samecka-Cymerman, A., A. Stankiewicz, K. Kolon and A. Kempers. 2009. Self-organizing feature map (neural networks) as a tool to select the best indicator of road traffic pollution (soil, leaves or bark of *Robina pseudoacacia* L.). Environ. Pollut. 157: 2061–2065.

Sathishkumar, M., A.R. Binupriya, S.H. Baik and S.E. Yun. 2008. Biodegradation of crude oil by individual bacterial strains and a mixed bacterial consortium isolated from hydrocarbon contaminated areas. Clean—Soil, Air, Water 36: 92–96.

Seo, J.S., Y.S. Keum and Q.X. Li. 2009. Bacterial degradation of aromatic compounds. Int. J. Environ. Res. Public Health 6: 278–309.

Shen, X.H., N.Y. Zhou and S.J. Liu. 2012. Degradation and assimilation of aromatic compounds by *Corynebacterium glutamicum*: another potential for applications for this bacterium? Appl. Microbiol. Biotechnol. 95: 77–89.

Spiekermann, P., B.H.A. Rehm, R. Kalscheuer, D. Baumeister and A. Steinbuchel. 1999. A sensitive, viable-colony staining method using Nile red for direct screening of bacteria that accumulate polyhydroxyalkanoic acids and other lipid storage compounds. Arch. Microbiol. 171: 73–80.

Steiner, R.A., H.J. Janssen, P. Roversi, A.J. Oakley and S. Fetzner. 2010. Structural basis for cofactor-independent dioxygenation of N heteroaromatic compounds at the a/b-hydrolase fold. Proc. Natl. Acad. Sci. USA 107: 657–662.

Teufel, R., V. Mascaraque, W. Ismail, M. Voss, J. Perera, W. Eisenreich et al. 2010. Bacterial phenylalanine and phenylacetate catabolic pathway revealed. Proc. Natl. Acad. Sci. USA 107: 14390–14395.

Teufel, R., T. Friedrich and G. Fuchs. 2012. An oxygenase that forms and deoxygenates toxic epoxide. Nature 483: 359–362.

Thakor, N.S., M.A. Patel, U.B. Trivedi and K.C. Patel. 2003. Production of poly(β-hydroxybutyrate) by Comamonas testosteroni during growth on naphthalene. World J. Microbiol. Biotechnol. 19: 185–189.

Tobin, K. and K.M O'Connor. 2005. Polyhydroxyalkanoate accumulating diversity of Pseudomonas species utilising aromatic hydrocarbons. FEMS. Microbiol. Lett. 253: 111–118.

Toledo, F.L., C. Calvo, B. Rodelas and J. Gonzalez-Lopez. 2006. Selection and identification of bacteria isolated from waste crude oil with polycyclic aromatic hydrocarbons removal capacities. Syst. Appl. Microbiol. 29: 244–252.

Toyama, T., M. Murashita, K. Kobayashi, S. Kikuchi, K. Sei, Y. Tanaka et al. 2011. Acceleration of Nonylphenol and 4-*tert*-Octylphenol degradation in sediment by *Phragmites australis* and associated Rhizosphere bacteria. Environ. Sci. Technol. 45: 6524–6530.

Trautwein, K., K. Simon, W. Lars, H. Thomas, K. Kenny, S. Alexander et al. 2008. Solvent Stress Response of the Denitrifying Bacterium "*Aromatoleum aromaticum*" Strain EbN1. Appl. Environ. Microbiol. 74: 2267–2274.

Tsuda, H., A. Hagiwara, M. Shibata and N. Ito. 1982. Carcinogenic effect of carbazole in the liver of a (C57BL/6NxC3H/HeN) F1 mice. J. Nat. Cancer Ins. 69: 1389–1393.

Tuan, N.N., Y.W. Lin and S.L. Huang. 2013. Catabolism of 4-alkylphenols by *Acinetobacter* sp. OP5: Genetic organization of the *oph* gene cluster and characterization of alkylcatechol 2, 3-dioxygenase. Bioresour. Technol. 131: 420–428.

U.S. Inventory of Toxic Compounds. 2001. TRI92. Toxics release inventory public data. Office of Pollution Prevention and Toxics, U.S. EPA, Washington, D.C. http://www.epa.gov/tri/.

Vilchez-Vargas, R., H. Junca and D.H. Pieper. 2010. Metabolic networks, microbial ecology and 'omics' technologies: towards understanding *in situ* biodegradation processes. Environ. Microbiol. 12: 3089–3104.

Vorosmarty, C., P. McIntyre, M. Gessner, D. Dudgeon, A. Prusevich, P. Green et al. 2010. Global threats to human water security and river biodiversity. Nature 467: 555–561.

Wang, Z., Y. Yang, W. Sun and S. Xie. 2014. Biodegradation of nonylphenol by two alphaproteobacterial strains in liquid culture and sediment microcosm. Int. Biodet. Biodeg. 92: 1–5.

Ward, P.G., G. de Roo and K.E. O'Connor. 2005. Accumulation of polyhydroxyalkanoate from styrene and phenylacetic acid by *Pseudomonas putida* CA-3. Appl. Environ. Microbiol. 71: 2046–2052.

Wieczorek, R., A. Pries, A. Steinbuchel and F. Mayer. 1995. Analysis of a 24-kilodalton protein associated with the polyhydroxyalkanoic acid granules in *Alcaligenes eutrophus*. J. Bacteriol. 177: 2425–2435.

Xia, X., N. Xia, Y. Lai, J. Dong, P. Zhao, B. Zhu et al. 2015. Response of PAH-degrading genes to PAH bioavailability in the overlying water, suspended sediment, and deposited sediment of the Yangtze River. Chemosph. 128: 236–244.

York, G.M., B.H. Junker, J. Stubbe and A.J. Sinskey. 2001. Accumulation of the PhaP phasin of *Ralstonia eutropha* is dependent on production of polyhydroxybutyrate in cells. J. Bacteriol. 183: 4217–4226.

Zhao, G., L. Jiang, Y. He, J. Li, H. Dong, X. Wang et al. 2011. Sulfonated grapheme for persistent aromatic pollutant management. Adv. Mater. 23: 3959–3963.

Zinn, M., B. Witholt and T. Egli. 2001. Occurrence, synthesis and medical application of bacterial polyhydroxyalkanoate. Adv. Drug Delivery Rev. 53: 5–21.

2

Bioaugmentation and Biostimulation Remediation Technologies for Heavy Metal Lead Contaminant

K. Jayaprakash,[1] M. Govarthanan,[2,3,]* R. Mythili,[3]
T. Selvankumar[3] and Young-Cheol Chang[2]

INTRODUCTION

Bioremediation is defined as the increase of the rate of the natural metabolic process using microorganisms to alter and break down organic molecules into other substances. According to the United States EPA, bioremediation is defined as treatment that uses naturally occurring organisms to break down hazardous substances into less toxic or non-toxic substances.

Bioremediation is an ecologically advanced technique that employs natural biological activities employing microorganisms, fungi, green plants or their enzymes to return the natural environment altered by contaminants to its original condition. With rapid industrialization all over the world, the pollution rate also increases.

[1] Department of Biotechnology, Shanmuga Industries Arts and Science College – Tiruvannamalai-606601, India.
[2] Department of Applied Sciences, College of Environmental Technology, Muroran Institute of Technology, 27-1, Mizumoto, Muroran, Hokkaido 050-8585, Japan.
[3] PG & Research Department of Biotechnology, Mahendra Arts and Science College (Autonomous), Kalippatti, Namakkal-637501, Tamil Nadu, India.
* Corresponding author: gova.muthu@gmail.com

One of the modes through which all types of pollutants enter the biosphere is that of industrial effluents. To completely eliminate the toxic compounds occurring in sludges, and ground water is effectively remediating with low level of residual contamination compared to the source pollution. Compared with other technologies, such as thermal desorption and incineration (which require excavation and heating), thermally enhanced recovery (which requires heating), chemical treatment (which may require relatively expensive chemical reagents), and *in situ* soil flushing (which may require further management of the flushing water), bioremediation may enjoy a cost advantage. Not all contaminants, however, are easily treated by bioremediation using microorganisms. Bioremediation technologies are phytoremediation, bioleaching, landfarming, bioreactor, bioaugmentation, rhizofiltration, and biostimulation (Baker and Brooks 1989). Bioremediation technologies can be generally classified as *in situ* or *ex situ*. *In situ* bioremediation involves the placement of conversion directly into a contaminated area, whereas *ex situ* bioremediation transfers the contaminated source to a selected site for treatment. Biostimulation is one of the method, by adding the amendments to accelerate the growth of indigenous microbial population in the contaminated site. Recent bioremediation process has also proven successful via the addition of indigenous microbe strains to the medium to boost the resident microbe population's ability to break down contaminants.

General Heavy Metal Pollution

Heavy metals pollution has become one of the most serious ecological problems today. Many heavy metals with toxic properties have been brought into the environment through human activity. With the fast development of industries such as metal plating facilities, mining operations, fertilizer industries, tanneries, batteries, sheet manufacture, and pesticides, etc. The toxicity level of the heavy metals are varied and its danger to human health. Heavy metals are directly or indirectly release into the environment progressively; especially in developing countries. Release of heavy metals without proper treatment pass a significant threat to public health because of its persistence and accumulation in the food chain. Unlike organic contaminants, heavy metals are not biodegradable and tend to accumulate in living organisms and abundant heavy metal ions are recognized to be poisonous or carcinogenic (Medici et al. 2010; Yang et al. 2009). Each heavy metal has unique toxicity or function. For example, zinc and copper can enhance microbial development at low concentrations, but represses growth at high concentrations (Ge et al. 2009). Heavy metals such as copper, iron, chromium, and nickel are fundamental minerals since they play an important role in biological systems, where cadmium and lead are non-major metals, they are toxic, even in trace amounts (Ibrahim et al. 2013). Some conventional methods to remediate sites contaminated with heavy metals are excavation and solidification/stabilization which are suitable in controlling contamination but not permanently remove heavy metals (Bahi et al. 2012). However, these technologies are expensive and can lead to incomplete decomposition of contaminants. Therefore, unconventional methods like using microorganisms which help in reducing the toxicity of harmful effluents have been explored. Microbial communities respond to heavy metals depending upon the concentration and availability of heavy metals and are also a complex process which is controlled by factors such as, the type of metal, the nature of the medium, and microbial species (Goblenz et al. 1994). Frequent attempts have been made to design genetically modified microorganisms for environmental release as agents for the bioremediation of heavy metal pollutants. However, these microorganisms do not behave in a predictable fashion under conditions that are quite different from the controlled ones of the laboratory.

Bioremediation of Heavy Metals

Heavy metals are chemical elements with a specific gravity that is at least five times the specific gravity of water. The specific gravity of water is 1 at 4°C (39°F). Some well-known toxic metallic elements are arsenic, cadmium, iron, lead, and mercury. Based on the toxicological point of view, heavy metals can be divided into two types. The first type is an essential heavy metal, where its presence in a certain amount is needed by living organisms, but in excessive quantities can cause toxic effects. Examples of the first kind is Pb, Zn, Cu, Fe, Co, Mn, etc., while the second type includes the heavy metals that are not essential and toxic, whose presence in the body have no known benefits or may even be toxic, such as Hg, Cd, Pb, Cr, and others. Heavy metals can affect human health effects depending on which part of the heavy metal is found in the body. Various organisms have the ability to bind metals with a very high capacity, namely marine algae, fungi, and molds that have been reported to be able to accumulate various metals.

Lead: Environmental Forms and Sources of Pollution

Lead (Pb) exists in many forms in the natural sources throughout the world and is now one of the most widely and evenly distributed trace metals. Pb^{2+} was found to be acutely toxic to human beings when present in high amounts. Since Pb^{2+} is not biodegradable, once soil has become contaminated, it remains a long-term source of Pb^{2+} exposure. Metal pollution has a harmful effect on biological systems and does not undergo biodegradation (Pehlivan et al. 2009).

Soil can be contaminated with Pb from several other sources such as industrial sites, from leaded fuels, old lead plumbing pipes, or even old orchard sites in production where lead arsenate is used. Lead accumulates in the upper 8 inches of the soil and is highly immobile. Contamination is long term. Without remedial action, high soil lead levels will never return to normal (Traunfeld and Clement 2001). In the environment, lead is known to be toxic to plants, animals, and microorganisms. Effects are generally limited to especially contaminated areas. Pb contamination in the environment exists as an insoluble form, and the toxic metals pose serious human health problem, namely, brain damage and retardation.

Sources of Pb Pollution

The primary industrial sources of Pb contamination include:

➢ Metal smelting and processing,

➢ Secondary metals production,

➢ Lead battery manufacturing,

➢ Pigment and chemical manufacturing,

➢ Lead contaminated wastes, and

➢ Lead contaminated mines.

Toxicity of Pb

Lead (Pb) has been commonly used since ancient time and plays important role in the industrial economy. Lead (Pb) is known as toxic element in environment. Besides, its spreading in the environment is also connected to both agricultural and urban activities, such as land application of sewage sludge, smelting operations, and use of leaded petrol (Marquita 2004; International Lead association 2009). Therefore, to control environmental pollution by Pb, it is necessary to restrict maximum content of Pb in the waste water that discharged into the environment. The concentration of Pb higher than the standards would be harmful to living organisms, especially indirect impact on the human health, it can damage the brain which reduce the intelligence of children (Ekere et al. 2014). Lead causes interference in the nervous system, reproductive system, and urinary tract (Stancheva et al. 2013). Besides this, lead is also a neurotoxic metal, affecting visual/motor performance, memory, attention, and verbal comprehension. Subtle changes in neuropsychological functions have in fact been seen in inorganic lead workers with blood lead levels as low as 40 µg/100 ml. Moreover, chronic workplace exposure increases the likelihood of high blood pressure, can damage the nervous system and kidneys, and sometimes leads to anemia and infertility (Marquita 2004; Gidlow 2004). As the body treats Pb much as it does with calcium, lead accumulates in the skeleton and can remain in the bodies for decades. While blood level can show recent exposure, bone lead level reflects exposure over a lifetime (Collins et al. 2005). Actually about 90% of a person's lead intake is eventually stored in the skeleton and lead levels in modern human skeletons and teeth are hundreds of times greater than those found in pre-industrial-age skeletons (Marquita 2004). Lead is only weakly mutagenic, but *in vitro* it inhibits DNA repair and acts synergistically with other mutagens. Nevertheless, there are at present insufficient data for suggesting that lead compounds are carcinogenic in humans (Gidlow 2004). Lead enters the waters through efflorescence in the air with the help of rain water (Widiyanti et al. 2005). Alternative treatments should be done to avoid such health problems, especially treatment for waste problem.

Researchers have demonstrated the successful use of biosurfactants for facilitating the degradation of organic pollutants in soil and water. The assessment of efficiency of biosurfactants (rhamnolipid) producing microorganisms (Pseudomonas sp.) isolated from a heavy metal contaminated site has been reported (Jayabarath et al. 2009). The release of heavy metals into the environment, mainly as a consequence of anthropogenic activities, constitutes a worldwide environmental pollution problem. Bioremediation of heavy metals is considered to be an economically viable alternative to conventional methods of heavy metal clearance. Soil bioremediation is a complex and costly process that aims to restore contaminated sites to environmentally sustainable conditions using microorganisms.

Bioremediation techniques

Bioremediation strategies employed for *in situ* bioremediation; biostimulation and bioaugmentation.

Bioaugmentation

Bioaugmentation involves the addition of genetically engineered microorganisms or microorganisms with enhanced degradation capabilities to the contaminated site. Bioremediation can also be accelerated through the injection of native or non-native microbes (bioaugmentation) into a contaminated area. Bioaugmentation has been proven successful in cleaning up sites contaminated with aromatic compounds but still faces many environmental problems. One of the most difficult issues is the survival of strains introduced to the soil. It has been observed that the number of exogenous microorganisms has decreased shortly after soil inoculation. Many studies have shown that both abiotic and biotic factors influence the effectiveness of bioaugmentation (Cho et al. 2000; Bento et al. 2005; Wolski et al. 2006).

Bioaugmentation should be applied in soils (1) with a low or non-detectable number of contaminant-degrading microbes, (2) containing compounds requiring multiprocess remediation, including processes detrimental or toxic to microbes, and (3) for small-scale sites for which the cost of non-biological methods exceeds

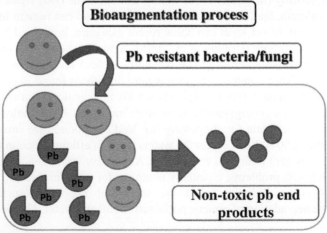

Fig. 1: Schematic representation of the bioaugmentation process.

the cost of bioaugmentation. Moreover, the introduction of microorganisms into soil is particularly recommended for areas polluted with compounds requiring long acclimation or an adaptation period of time. This review addresses the bioaugmentation of soils polluted with aromatic compounds; however, it should be noted that this strategy may be also effective for cleaning up diverse biotops contaminated with heavy metals (Jezequel and Lebeau 2008; Lebeau et al. 2008; Beolchini et al. 2009) and radionuclides (Kumar et al. 2007; Gavrilescu et al. 2009).

Advance Molecular ecological techniques will be useful for the analysis of the diversity of pollutant degrading microorganisms, and for the development of strategies to improve bioremediation (Watanabe 2001; Macnaughton et al. 1999).

The use of genetic engineering to produce microorganisms capable of converting the heavy metal or to enhance such processes in native organisms with such capabilities has become a popular way of increasing the efficiency of bioremediation in laboratory studies. Techniques used can include engineering with single genes, pathway construction, and alteration of the sequences of existing coding and regulatory genes. These applications could further be extended to greenhouse gas control, carbon sequestration, or conversion of wastes to value added eco-friendly products. Regardless, there remains the need for a regulatory safety, or costs benefit-driving force to make these potentials a reality (Sayler and Ripp 2000). Due to an eco-friendly approach and lesser health hazards as compared to physico-chemical based strategies to combat heavy metal pollution; genetic engineering microbes based remediation offered a more promising field. Good microbiological and ecological knowledge, biochemical mechanisms, and field engineering designs would be an essential element for successful *in situ* bioremediation in contaminated sites using engineered bacteria. Various bio safety and environmental concerns like genetic pollution, caused by using genetic engineering microbes should be well accounted before releasing into environment.

Factors influencing bioaugmentation

The most important abiotic factors such as temperature, moisture, pH, and organic matter content are discussed; however, aeration, nutrient content, and soil type also determine the efficiency of bioaugmentation. There are many examples that prove that pH, moisture, aeration, nutrient content, and soil type play important roles in bioaugmentation techniques. For example, Hong et al. (2007), studying the effect of temperature and pH on fenitrothion (nitrophenolic pesticide) degradation by inoculated *Burkholderia* sp. FDS-1, found that optimal parameters for bacteria activity were 30°C and slightly alkaline pH, whereas 10 and 50°C and a highly acidic condition were unsuitable for pesticide detoxification. The effect of water content on the survival of *Achromobacter piechaundii* TBPZ and degradation of tribromophenol (TBP) in soil samples were reported by Ronen et al. (2000). Their results indicated that, at 25% and 50% water content, TBP degradation was rapid whereas in soil with only 10% moisture the degradation proceeded to a small extent. Water potential has been reported to have significant influence on survival and degradative activity of *Pseudomonas stutzeri* P16 lux AB4 in sterile and non-sterile soil amended with phenanthrene (Mashreghi and Prosser 2006). The discussed results indicated that bioaugmentation is not always the best method for cleaning up of contaminated soils and it is difficult to predict the final results of this process. One of the problems connected with soil inoculation is how to deliver the suitable microorganisms to the desired sites. It is easy to disperse inoculants into surface soil but it is difficult or even impossible to do it in subsurface environments. Soils have potential for microbial transport but cell adhesion to organic matter strongly limits their distribution. To avoid these constraints, surfactants, foams, and strains resistant to adhesion may be applied (Wang and Mulligan 2004; Franzetti et al. 2009). Recently, bioaugmentations with encapsulated or immobilized cells for various purposes have been tested (Cassidy et al. 1996; Gardin and Pauss 2001; Gentili et al. 2006).

Microorganisms for bioaugmentation

There are several approaches that allow selection of microorganisms useful for bioaugmentation. Bacteria for this purpose may be isolated from given contaminated soils and after culturing under laboratory conditions, pre-adapted pure bacterial strains return to the same soil. Most experiments dealing with bioaugmentation were carried out using gram-negative bacteria belonging to genus Pseudomonas (Heinaru et al. 2005), Flavobacterium (Crawford and Mohn 1985), Sphingobium (Dams et al. 2007), Alcaligenes (Haluska et al. 1995), and Achromobacter (Ronen et al. 2000). Increasing attention also needs to be directed to gram-positive bacteria belonging to the genera Rhodococcus (Briglia et al. 1990), Mycobacterium (Jacques et al. 2008), and Bacillus (Silva et al. 2009). In turn, fungi potentially useful for bioaugmentation are represented by species from genus Absidia (Garon et al. 2004), Achremonium (Silva et al. 2009), Aspergillus (Dos Santos et al. 2008), Verticillium (Silva et al. 2009), Penicillium (Mancera-Lopez et al. 2008), and Mucor (Szewczyk and Długonski 2009). There are no microorganisms or their groups universally applicable to bioaugmentation. Many microorganisms are metabolically versatile and are capable of degrading a wide spectrum of substrates.

Bioaugmentation with GMOs and genes

Indigenous microorganisms during long-term exposure to xenobiotics evolve to create a capacity to degrade these compounds. The evolution, involving mutations and horizontal gene transfer (HGT), takes place constantly but is relatively slow in nature. Due to this there is a need to improve microbial degradative activity using molecular biology, which offers numerous technologies for engineering or enhancing remediation genes (Halden et al. 1999; De Lipthay et al. 2001; Rodrigues et al. 2001). Recently, special attention has been focused on enhancing the biodegradative potential of microorganisms by transfer of packaged catabolic genes from one or more donor strains to indigenous microflora existing in contaminated areas. Many catabolic pathways are located on plasmids such as TOL/pWW0, TOL/pWW53, TOL/pDK1, BPH/pWW110, NAH/NAH7, and PHE/pVI150, transpozons or other mobile and/or integrative elements (Top et al. 2002; Jussila et al. 2007). Plasmid-encoded pathways are beneficial since they present genetically flexible systems and can be transferred between bacteria species or even genera (Sayler et al. 1990; Reineke 1998; Sayler and Ripp 2000). As most of the degradative plasmids are self-transmissible, conjugation has the highest significance in widespread catabolic genes with respect to bioaugmentation. However, other mechanisms of HGT such as transformation and transduction play important roles in the development and adaptation of microorganisms.

Biostimulation

Biostimulation in which the biodegradation is accelerated by the addition of amendments to contaminated water or soil to encourage the growth and activity of bacteria already existing in the contaminant environment. Amendments include air (oxygen), added by bioventing; oxygen-releasing compounds, which keep the contaminated media aerobic; and reducing agents, such as carbon-rich vegetable oil and molasses, which promote growth of anaerobic microbial populations in wastewater treatment facilities. To date several studies have been focused on the degradative capacities of bacterial population in polluted environments (Cavalca et al. 2000; Juck et al. 2000; Bundy et al. 2002). The important objective was the determination of physiology and function of such diverse catabolic populations in the bioremediation process. However, all the environmental bacteria cannot be cultured yet by conventional laboratory techniques (Torsvik et al. 2002). Therefore, we need to encourage the growth and activity of bacteria existing in the native environments. Microbial communities can adapt to contaminants after prolonged exposure by changing their composition in the native ecosystem. Hence, assessment of the structure of microbial communities is an important step to determine possible indicators of heavy metals. In this aspect, some studies investigated the changes in the indigenous bacterial community structure for addressing the impact of contamination on the microbiology of ecosystems (Macnaughton et al. 1999; Ogino et al. 2001).

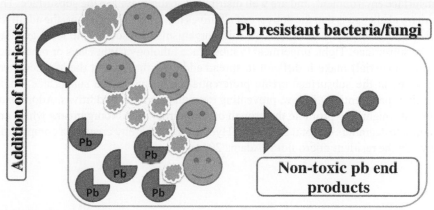

Fig. 2: Biostimulation process.

Frequently bioremediation has been studied in polluted marine environment and biostimulation studies have indicated that can efficiently promote biodegradation (Venosa et al. 1996). However, current knowledge of the impact of this process on the ecosystem is limited. Therefore, a detailed study of the contaminated site in relation to the pollutant, environmental conditions, and the microbial community is still necessary for *in situ* biostimulation to be considered reliable for successful removal of pollutants (Iwamoto et al. 2001). To date, this method has not been very successful when done at the site of the contamination because it is difficult to control site conditions for the optimal growth of the microorganisms added. Scientists have yet to completely understand all the mechanisms involved in bioremediation, and organisms introduced into a foreign environment may have a hard time surviving (Dejonghe et al. 2001).

Factors influencing biostimulation

Heavy metal biodegradation can be limited by many factors, including nutrients, pH, temperature, moisture, oxygen, soil properties, and contaminant presence (Atagana 2008; Al-Sulaimani 2011). This can be done by the addition of various forms of limiting nutrients and electron acceptors, such as phosphorus, nitrogen, oxygen, or carbon (e.g., in the form of molasses), which are otherwise available in quantities low enough to constrain microbial activity (Elektorowicz 1994; Piehler et al. 1999; Rhykerd et al. 1999). Previously, Perfumo et al. (2007) as the addition of nutrients, oxygen, or other electron donors and acceptors to the coordinated site in order to increase the population or activity of naturally occurring microorganisms available for bioremediation. They opined that biostimulation can be considered as an appropriate remediation technique for heavy metal removal in soil and requires the evaluation of both the intrinsic degradation capacities of the autochthonous microflora and the environmental parameters involved in the kinetics of the *in situ* process.

The primary advantage of biostimulation is that bioremediation will be undertaken by already present native microorganisms that are well-suited to the subsurface environment, and are well distributed spatially within the subsurface. The primary challenge is that the delivery of additives in a manner that allows the additives to be readily available to subsurface microorganisms is based on the local geology of the subsurface. Tight, impermeable subsurface lithology (tight clays or other fine-grained material) make it difficult to spread additives throughout the affected area. Fractures in the subsurface create preferential pathways in the subsurface which additives preferentially follow, preventing even distribution of additives. Addition of nutrients might also promote the growth of heterotrophic microorganisms which are not innate degraders of Total Petroleum Hydrocarbon, thereby creating a competition between the resident micro flora (Adams 2014).

Conclusions

Increasing awareness and concern of environmental issues has forced humanity to think above conventional methods of waste treatment. Bioremediation, a need of the present and immediate future, is a powerful tool available to clean up contaminated sites. The success of bioremediation strategies depends on the amenability of the pollutant to get biologically transformed; the accessibility or bioavailability of the contaminant to microorganisms; and the opportunity for optimization of biological activity. It is important to ensure that the contaminated material is suitably detoxified at the end of the treatment. Recent innovative breakthroughs in molecular and '-omics' technologies such as molecular profiling, ultrafast pyro-sequencing, microarrays, mass spectrometry, meta transcriptomics and metaproteomics, and transcriptome and proteome analyses of an entire community along with bioinformatics tools have the potential to gain insights of indigenous microbial communities and their mechanism in bioremediation of environmental pollutants. In future, genetically modified organisms can be developed with chemotaxis power that helps them to approach and degrade toxic compounds in the environment. Therefore, in the future a combination of techniques/microbes can be used for bioremediation purpose. Bioremediation depends for its success on selling the results which not only provide benefit but also remediate the wastes.

References

Adams, G.O., P. Tawari-Fufeyin and E. Igelenyah. 2014. Bioremediation of spent oil contaminated soils using poultry litter. Res. J. Engineer. Appl. Sci. 3: 124–130.

Al-Sulaimani, H., S. Joshi, Y. Al-Wahaibi, S. Al-Bahry, A. Elshafie and A. Al-Bemani. 2011. Microbial biotechnology for enhancing oil recovery: current developments and future prospects. Biotechnol. Bioinf. Bioeng. 1: 147–158.

Atagana, H.I. 2008. Compost bioremediation of hydrocarbon-contaminated soil inoculated with organic manure. Afr. J. Biotechnol. 7: 10, 1516–1525.

Bahi, J., O. Radziah, A. Samsuri, H. Aminudin and S. Fardin. 2012. Bioleaching of heavy metals from mine tailings. Bioremediat J. 16: 57–65.

Baker, A.J.M. and R.R. Brooks. 1989. Terrestrial higher plants which hyperaccumulate metallic elements—A review of their distribution, ecology and phytochemistry. Biorecovery 1: 81–126.

Bento, F.M., F.A.O. Camargo, B.C. Okeke and W.T. Frankenberger. 2005. Comparative bioremediation of soils contaminated with diesel oil by natural attenuation, biostimulation and bioaugmentation. Bioresour. Technol. 96: 1049–55.

Beolchini, F., A. Dell'Anno, L. De Propris, S. Ubaldini, F. Cerrone and R. Danovaro. 2009. Auto- and heterotrophic acidophilic bacteria enhance the bioremediation efficiency of sediments contaminated by heavy metals. Chemosphere 74: 1321–6.

Briglia, M., E.L. Nurmiaho-Lassila, G. Vallini and M.S. Salkinoja-Salonen. 1990. The survival of the pentachlorophenol-degrading *Rhodococcus chlorophenolicus* PCP-1 and *Flavobacterium* sp. in natural soil. Biodegradation 1: 273–281.

Bundy, J.G., G.I. Paton and C.D. Campbell. 2002. Microbial communities in different soil types do not converge after diesel contamination. J. Appl. Microbiol. 92: 276–288.

Cassidy, M.B., H. Lee and J.T. Trevors. 1996. Environmental applications of immobilized microbial cells: a review. J. Ind. Microbiol. Biotechnol. 16: 79–101.

Cavalca, L., P.D. Gennaro, M. Colombo, V. Andreoni, S. Bernasconi, I. Ronco et al. 2000. Distribution of catabolic pathways in some hydrocarbon-degrading bacteria from subsurface polluted soil. Res. Microbiol. 151: 877–887.

Cho, Y.-G., S.-K. Rhee and S.-T. Lee. 2000. Effect of soil moisture on bioremediation of chlorophenol-contaminated soil. Biotechnol. Lett. 22: 915–922.

Coblenz, A., K. Wolf, A. Goblenz, K. Wolf and P. Bauda. 1994. The role of glutathione biosynthesis in heavy metal resistance in the fission yeast *Schizosaccharomyces pombe*. FEMS Microbiol. Rev. 14: 303–308.

Collins, C., M. Endrizzi and G. Shaw. 2005. Final Report-Review of the Environmental Behaviour and Toxicology of Organic Lead and Proposals for its Remediation at the Trento Nord site, Italy, Imperial College London.

Crawford, R.L. and W.W. Mohn. 1985. Microbiological removal of pentachlorophenol from soil using a *Flavobacterium*. Enzyme Microb. Technol. 7: 617–620.

Dams, R.I., G. Paton and K. Killham. 2007. Bioaugmentation of pentachlorophenol in soil and hydroponic systems. Int. Biodeterior. Biodegrad. 60: 171–177.

De Lipthay, J.R., T. Barkay and S.J. Sorensen. 2001. Enhanced degradation of phenoxyacetic acid in soil by horizontal transfer of the tfdA gene encoding a 2,4-dichlorophenoxyacetic acid dioxygenase. FEMS Microbiol. Ecol. 35: 75–84.

Dejonghe, W., N. Boon, D. Seghers, E.M. Top and W. Verstraete. 2001. Bioaugmentation of soils by increasing microbial richness: missing links. Environ. Microbiol. 3: 649–657.

Dos Santos, E.O., C.F.C. Rosa, C.T. Passos, A.V.L. Sanzo, J.F.M. Burkert, S.J. Kalil et al. 2008. Pre-screening of filamentous fungi isolated from a contaminated site in Southern Brazil for bioaugmentation purposes. Afr. J. Biotechnol. 7: 1311–1317.

Ekere, N.R., J.N. Ihedioha, T.I. Oparanozie, F.I. Ogbuehi-Chima and J. Ayogu. 2014. Assessment of some heavy metals in facial cosmetic products. J. Chem. Pharm. Res. 6: 561–564.

Elektorowicz, M. 1994. Bioremediation of petroleum-contaminated clayey soil with pretreatment. Environ. Technol. 15: 373–380.

Forsyth, J.V., Y.M. Tsao and R.D. Bleam. 1995. Bioremediation: when is augmentation needed? Battelle Press, Columbus, OH (United States).

Franzetti, A., P. Caredda, C. Ruggeri, L. La Colla, E. Tamburini, M. Papacchini et al. 2009. Potential applications of surface active compounds by *Gordonia* sp. strain BS29 in soil remediation technologies. Chemosphere 75: 801–807.

Gardin, H. and A. Pauss. 2001. Kappa-carrageenan/gelatin gel beads for the co-immobilization of aerobic and anaerobic microbial communities degrading 2,4,6-trichlorophenol under air-limited conditions. Appl. Microbiol. Biotechnol. 56: 517–523.

Garon, D., L. Sage, D. Wouessidjewe and F. Seigle-Murandi. 2004. Enhanced degradation of fluorene in soil slurry by *Absidia cylindrospora* and maltosyl-cyclodextrin. Chemosphere 56: 159–166.

Gavrilescu, M., L.V. Pavel and I. Cretescu. 2009. Characterization and remediation of soils contaminated with uranium. J. Hazard. Mater. 163: 475–510.

Ge, H.W., M.F. Lian, F.Z. Wen, Y.F. Yun, F.Y. Jian and T. Ming. 2009. Isolation and characterization of the heavy metal resistant bacteria CCNWRS33-2 isolated from root nodule of *Lespedeza cuneata* in gold mine tailings in China. J. Hazard. Mater. 162: 50–56.

Gentili, A.R., M.A. Cubitto, M. Ferrero and M.S. Rodriquez. 2006. Bioremediation of crude oil polluted seawater by a hydrocarbon-degrading bacterial strain immobilized on chitin and chitosan flakes. Int. Biodeterior. Biodegrad. 57: 222–228.

Gidlow, D.A. 2004. Lead toxicity. Occup. Med. 54: 76–81.

Halden, R.U., S.M. Tepp, B.G. Halden and D.F. Dwyer. 1999. Degradation of 3-Phenoxybenzoic Acid in Soil by *Pseudomonas pseudoalcaligenes* POB310(pPOB) and two modified *Pseudomonas* strains. Appl Environ Microbiol. 65: 3354–3359.

Haluska, L., G. Barancikova, S. Balaz, K. Dercova, B. Vrana, M. Paz-Weisshaar et al. 1995. Degradation of PCB in different soils by inoculated *Alcaligenes xylosoxidans*. Sci. Total Environ. 175: 275–285.

Heinaru, E., M. Merimaa, S. Viggor, M. Lehiste, I. Leito, J. Truu et al. 2005. Biodegradation efficiency of functionally important populations selected for bioaugmentation in phenol- and oil-polluted area. FEMS Microbiol. Ecol. 51: 363–373.

Hong, Q., Z. Zhang, Y. Hong and S. Li. 2007. A microcosm study on bioremediation of fenitrothion-contaminated soil using *Burkholderia* sp. FDS-1. Int. Biodeterior. Biodegrad. 59: 55–61.

Ibrahim, S.A., M.M.N. Authman, H.S. Gaber and M.A. El-Kasheif. 2013. Bioaccumulation of heavy metals and their histopathological impact on muscles of *Clarias gariepinus* from El-Rahawy drain, Egypt. Int. J. Environ. Sci. Eng. 4: 51–73.

International Lead association. 2009. http://www.ila-lead.org/.

Iwamoto, T. and M. Nasu. 2001. Current bioremediation practice and perspective. J. Biosci. Bioeng. 92: 1–8.

Jacques, R.J., B.C. Okeke, F.M. Bento, A.S. Teixeira, M.C. Peralba and F.A. Comargo. 2008. Microbial consortium bioaugmentation of a polycyclic aromatic hydrocarbons contaminated soil. Bioresour. Technol. 99: 2637–2643.

Jayabarath, J., S. Shyam, R. Arulmurugan and R. Giridhar. 2009. Bioremediation of heavy metals using biosurfactants. Int. J. Biotechnol. Appl. 1: 50–54.

Jezequel, K. and T. Lebeau. 2008. Soil bioaugmentation by free and immobilized bacteria to reduce potentially phytoavailable cadmium. Bioresour. Technol. 99: 690–698.

Juck, D., T. Charles, L.G. Whyte and C.W. Greer. 2000. Polyphasic microbial community analysis of petroleum hydrocarbon-contaminated soils from two northern Canadian communities. FEMS Microbiol. Ecol. 33: 241–249.

Jussila, M.M., J. Zhao, L. Suominen and K. Lindstrom. 2007. TOL plasmid transfer during bacterial conjugation *in vitro* and rhizoremediation of oil compounds *in vivo*. Environ. Pollut. 146: 510–524.

Kumar, R., S. Singh and O.V. Singh. 2007. Bioremediation of radionuclides: emerging technologies. OMICS: J. Integr. Biol. 11: 295–304.

Lebeau, T., A. Braud and K. Jezequel. 2008. Performance of bioaugmentation-assisted phytoextraction applied to metal contaminated soils: A review. Environ. Pollut. 153: 497–522.

Macnaughton, S., J.R. Stephen, A.D. Venosa, G.A. Davis, Y.J. Chang and D.C. White. 1999. Microbial population changes during bioremediation of an experimental oil spill. Appl. Environ. Microbiol. 65: 3566–3574.

Mancera-Lopez, M.E., F. Esparza-Garcia, B. Chavez-Gomez, R. Rodriguez-Vazquez, G. Saucedo-Castaneda and J. Barrera-Cortes. 2008. Bioremediation of an aged hydrocarbon-contaminated soil by a combined system of biostimulation–bioaugmentation with filamentous fungi. Int. Biodeterior. Biodegrad. 61: 151–60.

Marquita, K.H. 2004. Understanding Environmental Pollution. Cambridge University Press, Nature.

Mashreghi, M. and J.I. Prosser. 2006. Survival and activity of luxmarked phenanthrene-degrading *Pseudomonas stutzeri* P16 under different conditions. Iran J. Sci. Technol. Trans. A. 30: 71–80.

Medici, L., J. Bellanova, C. Belviso, F. Cavalcante, A. Lettino, P.R. Pietro et al. 2011. Trace metals speciation in sediments of the Basento River (Italy). Appl. Clay Sci. 53: 414–442.

Ogino, A., H. Koshikawa, T. Nakahara and H. Uchiyama. 2001. Succession of microbial communities during a biostimulation process as evaluated by DGGE and clone library analysis. J. Appl. Microbiol. 91: 625–635.

Pehlivan, E., A.M.O. Zkan, S. Dinc and S. Parlayici. 2009. Adsorption of Cu^{2+} and Pb^{2+} ion on dolomite powder. J. Hazard. Mater. 167: 1044–1049.

Perfumo, A., I.M. Banat, R. Marchant and L. Vezzulli. 2007. Thermally enhanced approaches for bioremediation of hydrocarbon-contaminated soils. Chemosphere 66: 179–184.

Piehler, M.F., J.G. Swistak, J.L. Pinckney and H.W. Paerl. 1999. Stimulation of diesel fuel biodegradation by indigenous nitrogen fixing bacterial consortia. Microb. Eco. 38: 69–78.

Reineke, W. 1998. Development of hybrid strains for the mineralization of chloroaromatics by patchwork assemble. Annu. Rev. Microbiol. 52: 287–331.

Rhykerd, R.L., B. Crews, K.J. McInnes and R.W. Weaver. 1999. Impact of bulking agents, forced aeration and tillage on remediation of oil-contaminated soil. Bioresour. Technol. 67: 279–285.

Rodrigues, J.L.M., O.V. Maltseva, T.V. Tsoi, R.R. Helton, J.F. Quensen III, M. Fukuda et al. 2001. Development of a Rhodococcus recombinant strain for degradation of products from anaerobic dechlorination of PCBs. Environ. Sci. Technol. 35: 663–668.

Ronen, Z., L. Vasiluk, A. Abeliovih and A. Nejidat. 2000. Activity and survival of tribromophenol-degrading bacteria in a contaminated desert soil. Soil Biol. Biochem. 32: 1643–1650.

Sayler, G.S., S.W. Hooper, A.C. Layton and J.M. King. 1990. Catabolic plasmids of environmental and ecological significance. Microbiol. Ecol. 19: 1–20.

Sayler, G.S. and S. Ripp. 2000. Field applications of genetically engineered microorganisms for bioremediation processes. Curr. Opin. Biotechnol. 11: 286–289.

Silva, I.S., C. Santos Eda, C.R. Menezes, A.F. Faria, E. Franciscon, M. Grossman et al. 2009. Bioremediation of a polyaromatic hydrocarbon contaminated soil by native soil microbiota and bioaugmentation with isolated microbial consortia. Bioresour. Technol. 100: 4669–4675.

Stancheva, M., L. Makedonski and E. Petrova. 2013. Determination of heavy metals (Pb, Cd, As and Hg) in Black Sea grey mullet (*Mugil cephalus*). Bulg. J. Agric. Sci. 1: 30–34.

Szewczyk, R. and J. Długonski. 2009. Pentachlorophenol and spent engine oil degradation by *Mucor ramosissimus*. Int. Biodeterior. Biodegrad. 63: 123–129.

Top, E.M., D. Springael and N. Boon. 2002. Catabolic mobile elements and their potential use in bioaugmentation of polluted soil and waters. FEMS Microbiol. Ecol. 42: 199–208.

Torsvik, V., L. Ovreas and T.F. Thingstad. 2002. Prokaryotic diversity—magnitude, dynamics, and controlling factors. Science 296: 1064–1066.

Traunfeld, J.H. and D.L. Clement. 2001. Lead in Garden Soils. Home and Garden. Maryland Cooperative Extention, University of Maryland. http://www.hgic.umd.edu/ media/documents/ hg18.pdf.

Wang, S. and C.N. Mulligan. 2004. An evaluation of surfactant foam technology in remediation of contaminated soil. Chemosphere 57: 1079–1089.

Watanabe, K. 2001. Microorganisms relevant to bioremediation. Curr. Opin. Biotechnol. 12: 237–241.

Widiyanti, C.A., N.S. Sunarto and dan Handajani. 2005. Kandungan logam berat timbal (Pb) serta struktur mikroanatomi *Ctenidia* dan kelenjar pencernakan (Hepar) *Anodonta woodiana* Lea., di sungai serang hilir waduk kedung ombo. BioSMART 7: 136–142.

Wolski, E.A., S.E. Murialdo and J.F. Gonzalez. 2006. Effect of pH and inoculum size on pentachlorophenol degradation by *Pseudomonas* sp. Water SA 32: 93–98.

Yang, Z., Y. Wang, Z. Shen, J. Niu and Z. Tang. 2009. Distribution and speciation of heavy metals in sediments from the mainstream, tributaries, and lakes of the Yangtze River catchment of Wuhan, China. J. Hazard. Mater. 166: 1186–1194.

3

Water Bioremediation using Aquatic Plants

Maisa'a Wasif Shammout

AN OVERVIEW OF BIOREMEDIATION

"Remediate" means to solve a problem, and "bio-remediate" means to use biological organisms to solve an environmental problem such as contaminated soil or groundwater. Bioremediation is defined as the process whereby organic wastes are biologically degraded under controlled conditions to an innocuous state, or to levels below concentration limits established by regulatory authorities (Mueller et al. 1996). Bioremediation in science means the use of biological agents, such as bacteria, fungi, or green plants, to remove or neutralize contaminants, as in polluted soil or water. Bacteria and fungi generally work by breaking down contaminants such as petroleum into less harmful substances. Plants can be used to aerate polluted soil and stimulate microbial action. They can also absorb contaminants such as salts and metals into their tissues, which are then harvested and disposed of. The use of green plants to decontaminate polluted soil or water is called phytoremediation (The American Heritage Science Dictionary 2002).

Bioremediation is also considered as a waste management technique that involves the use of organisms to neutralize pollutants from a contaminated site. According to the United States EPA, bioremediation is a treatment that uses naturally occurring organisms to break down hazardous substances into less toxic or non-toxic substances. Bioremediation provides a technique for cleaning up pollution by enhancing the same biodegradation processes that occur in nature (King et al. 1997). Depending on the site and its contaminants, bioremediation may be safer than alternative solutions (EI 2009).

Water, Energy and Environment Center, The University of Jordan, Amman 11942 Jordan.
E-mails: maisa_shammout@hotmail.com, m.shammout@ju.edu.jo

Bioremediation applications are categorized as either *in situ* or *ex situ*. *In situ* applications treat contaminated soil or water in the location in which they are found. *Ex situ* bioremediation requires excavation or pumping of contaminated soil or groundwater, respectively, before treatment is initiated. *In situ* techniques are generally less expensive, generate less dust and debris, and release less contaminant than *ex situ* techniques because no excavation processes are required. On the other hand, *ex situ* techniques are generally easier to control, faster, and able to treat a wider range of contaminants and soil types than *in situ* techniques.

Based on the above, with the chapter goal of presenting the potential for aquatic plants in water bioremediation; the specific objectives are as follows:

a. Present an overview and types of bioremediation,

b. Present the importance of aquatic plants in bioremediation,

c. Present a case study from Jordan that includes the Jordan's features, location of aquatic plants (duckweed) among Jordan water bodies with environmental requirements, and duckweed efficiency in water bioremediation.

This chapter is divided into five parts; Part 1 is an overview of bioremediation, Part 2 describes the bioremediation types, Part 3 describes the bioremediation using aquatic plants, and Part 4 describes a case study from Jordan regarding water bioremediation, and Part 5, the discussion and conclusions.

Bioremediation Types

There are far more than nine types of bioremediation (Vidali 2001), but the following are the most common ways in which it is used:

1. Phytoremediation—use of plants to remove contaminants. The plants are able to draw the contaminants into their structures and hold on to them, effectively removing them from soil or water (Garbisu 2002). The primary advantages of using plants in bioremediation are as follows: it is more environmentally friendly; and more aesthetically pleasing than conventional methods. Phytoremediation is a green technology. The success of green technology in phytoremediation, in general, is dependent upon several factors. First, plants must produce sufficient biomass while accumulating high concentrations of metal. Second, the metal-accumulating plants need to be responsive to agricultural practices that allow repeated planting and harvesting of the metal rich tissues (Paz-Alberto and Sigua 2013). Chaudhary and Sharma (2014) indicated that village ponds are rich source of nutrients like nitrate and phosphate which can be recovered by phytoremediation. It is an affordable technology utilizing plants as environmental cleansers in wastewater management.

   Phytoremediation is well suited for use at very large field sites where other methods of remediation are not cost effective or practicable; at sites with a low concentration of contaminants where only polish treatment is required over long periods of time; and in conjunction with other technologies where vegetation is used as a final cap and closure of the site. There are some limitations to the technology that it is necessary to consider carefully before it is selected for site remediation: long duration of time for remediation, potential contamination of the vegetation and food chain, and difficulty establishing and maintaining vegetation at some sites with high toxic levels.

The term "phytoremediation" is relatively new, coined in 1991. Its potential for encouraging the biodegradation of organic contaminants requires further research, although it may be a promising area for the future. There are five types of phytoremediation techniques, classified based on the contaminant fate: phytoextraction, phytotransformation, phytostabilization, phytodegradation, and rhizofiltration, even a combination of these can be found in nature (Vidali 2001).

- Phytoextraction or phytoaccumulation is the process used by the plants to accumulate contaminants into the roots and aboveground shoots or leaves. This technique saves tremendous remediation cost by accumulating low levels of contaminants from a widespread area. Unlike the degradation mechanisms, this process produces a mass of plants and contaminants (usually metals) that can be transported for disposal or recycling (Salt et al. 1995; Clemens et al. 2002).

- Phytotransformation or phytodegradation refers to the uptake of organic contaminants from soil, sediments, or water and, subsequently, their transformation to more stable, less toxic, or less mobile form. Metal chromium can be reduced from hexavalent to trivalent chromium, which is a less mobile and noncarcinogenic form.

- Phytostabilization is a technique in which plants reduce the mobility and migration of contaminated soil. Leachable constituents are adsorbed and bound into the plant structure so that they form a stable mass of plant from which the contaminants will not reenter the environment.

- Phytodegradation or rhizodegradation is the breakdown of contaminants through the activity existing in the rhizosphere. This activity is due to the presence of proteins and enzymes produced by the plants or by soil organisms such as bacteria, yeast, and fungi.

- Rhizodegradation is a symbiotic relationship that has evolved between plants and microbes. Plants provide nutrients for the microbes to thrive, while microbes provide a healthier soil environment.

- Rhizofiltration is a water remediation technique that involves the uptake of contaminants by plant roots. Rhizofiltration is used to reduce contamination in natural wetlands and estuary areas.

2. Bioventing—blowing air through soil to increase oxygen rates in the waste. This is an effective way to neutralize certain oxygen sensitive metals or chemicals.

3. Bioleaching—removing metals from soil using living organisms. Certain types of organisms are draw to heavy metals and other contaminants and absorb them. One new approach was discovered when fish bones were found to attract and hold heavy metals such as lead and cadmium.

4. Landfarming—turning contaminated soil for aeration and sifting to remove contaminants, or deliberately depleting a soil of nitrogen to remove nitrogen based organisms.

5. Bioreactor—the use of specially designed containers to hold the waste while bioremediation occurs.

6. Composting—containing waste so a natural decay and remediation process occurs.

7. Bioaugmentation—adding microbes and organisms to strengthen the same in waste to allow them to take over and decontaminate the area.

8. Rhizofiltration—the use of plants to remove metals in water.

9. Biostimulation—the use of microbes designed to remove contamination applied in a medium to the waste.

The major disadvantage of bioremediation methods that it is limited to those compounds that are biodegradable, where, not all compounds are susceptible to rapid and complete degradation. In addition to that, it is difficult to extrapolate from bench and pilot-scale studies to full-scale field operations. Whereas, the major advantage of the bioremediation methods is that it allows for contamination to be treated, neutralized, or removed and then produces a waste product itself that is more easily disposed of. In some cases, there is no need for disposal at all. In the case of the plants used in phytoremediation and rhizofiltration, the plant is able to do something called bioaccumulation. This means is holds onto the contaminant. As the plant is still growing, there is no need to remove and destroy it. In many ways it is similar to having a rechargeable battery. In the case of contaminated waste, it is the plant that keeps growing to allow for more storage of waste. This is a uniquely cost effective solution for contaminated waste (Conserve Energy Future 2018). Theoretically, bioremediation is useful for the complete destruction of a wide variety of contaminants. This eliminates the chance of future liability associated with treatment and disposal of contaminated material. Instead of transferring contaminants from one environmental medium to another, for example, from land to water or air, the complete destruction of target pollutants is possible. This also eliminates the need to transport quantities of waste off site and the potential threats to human health and the environment that can arise during transportation (Vidali 2001).

Bioremediation is an option that offers the possibility to destroy or render harmless various contaminants using natural biological activity. As such, it uses relatively low-technology techniques, which generally have a high public acceptance and can often be carried out on site. It will not always be suitable, however, as the range of contaminants on which it is effective is limited, the time scales involved are relatively long, and the residual contaminant levels achievable may not always be appropriate. Although the methodologies employed are not technically complex, considerable experience and expertise may be required to design and implement a successful bioremediation program, due to the need to thoroughly assess a site for suitability and to optimize conditions to achieve a satisfactory result (Flathman et al. 1993; Vidali 2001).

Bioremediation has been used at a number of sites worldwide, including Europe, with varying degrees of success. Techniques are improving as greater knowledge and experience are gained, and there is no doubt that bioremediation has great potential for dealing with certain types of site contamination (Vidali 2001; Hinchee et al. 1995).

Bioremediation using Aquatic Plants: Duckweeds

Duckweeds are found worldwide in ponds, ditches, lakes, and canals without any agronomic care, and they primarily reproduce by vegetative growth. They are the smallest and fastest-growing plants in the world, frequently doubling their biomass in two days or less (Culley et al. 1981) under optimum conditions of nutrient availability, sunlight, and temperature (Landolt 1986; Leng et al. 1995). It proved that duckweed survived in outdoor wastewater treatment tanks at below freezing temperatures and resumed growth when the temperature rose above freezing (Classen et al. 2000). Duckweed can tolerate a wide range of pH which varies from 3.0 to 10.0 (Hillman 1976; McLay 1976).

Duckweeds exist as colonies of two or more, called fronds. The fronds are green, leaf like structures, a few millimeters in size, and each new or daughter frond develops from mother frond. Newly formed fronds remain attached to the mother frond during the initial growth phase and detach from each other upon reaching a suitable size with a trend to forming new colonies (Gaigher and Short 1986). The daughter frond repeats the history of its mother frond. This result in exponential growth, where duckweeds' average growth rates tend to decrease from: nutrient scarcity; extreme pH levels; temperature, and crowding by overgrowth of the colony (Skillikorn et al. 1993).

Duckweeds are classified as higher plants or macrophytes, and consist of four genera: *Spirodela*, *Lemna*, *Wolffiella*, and *Wolffia*. *Lemna* is one of the well-known duckweed, while *Wolffia* is the smallest known flowering plant (Dothan 1986; Skillicorn et al. 1993). *Wolffiella* is known only in America and Africa, but *Spirodela*, *Lemna*, and *Wolffia* are found world-wide. *Spirodela* has its greatest diversity in South America, while *Lemna* is most diverse in North America and Southeast Asia. *Wolffia* occurs mainly in America, Southeast Asia, and Australia. *Wolffiella* is mainly subtropical. Duckweeds have also several species. Each species of duckweed has a characteristic distribution. The genus *Lemna* as an example has three species based on their environmental adaptation; where, *Lemna gibba* is found in regions with a Mediterranean climate, *Lemna minor* is found probably in native to cooler regions of North America, Europe, and western Asia (Landolt 1986).

Landolt (1986) conducted extensive investigations of duckweed ecology and found that these plants grow under a wide variety of conditions. The presence of duckweed is an indicator of a rich water nutrient supply, though they are generally not found in oligotrophic water. As nutrient concentration is increased, root elongation decreases. The fronds of duckweed reduced their growth and caused abnormal root elongation, indicative of nutrient deficiency (Culley and EPPA 1973).

Duckweed is potentially an important plant due to its high productivity, ease of cultivation, ability to accumulate large amount of inorganic elements, insecticides,

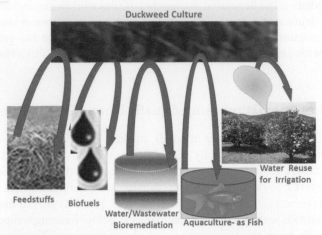

Fig. 1: Applications of duckweed plants.

radioisotopes, and can utilize organic compounds extracted from the water where they grow. These major advantages support duckweed in several applications and uses (Mitsch et al. 1977; Hillman and Culley 1978; Rejmankova 1979; Culley et al. 1981; Dewedar and Bahgat 1995; Alaerts et al. 1996; Cunningham and Ow 1996; Axtell et al. 2003; Horvat et al. 2007; Ansal et al. 2010; Donganlar et al. 2012; Christian et al. 2012; Elma et al. 2012; Demin et al. 2013; Heuzé and Tran 2015; Ansa et al. 2015). Figure 1 shows the duckweed plants applications. The main applications of duckweed are in the following:

1. Duckweed has the ability to purify domestic wastewater at a low cost, and the plant is used in water treatment as a natural bioremediation agent. Duckweed improves water quality by removing nutrients from water (Sutton and Ornes 1975; Oron 1988; Tripathi and Misra 1990; Oron 1994; Ahmet and Erdal 2009). It is also suitable for the treatment of organic wastes, and in the removal of a significant quantity of nitrogen and phosphorous. Duckweed also exhibits efficiency in the removal of biological oxygen demand (BOD_5). Furthermore, a dense cover of duckweed can inhibit the competing submerged aquatic plants that require solar energy in their growth and can also exclude algae from water bodies. The ability of duckweed to form a mat over the surface of water is one of its competitive advantages which reduce the production of carbon dioxide from other growing aquatic plants. The water beneath the dense cover of duckweed is highly anaerobic due to limited oxygen transfer into water/wastewater; hence, the lack of surface reaeration, would facilitate denitrification. Anaerobic conditions are advantages because mosquito larva cannot survive, and algae do not have an opportunity to develop. The duckweed mat prevents odor by reducing the diffusion of products of anaerobic fermentation of organic matter. Moreover, evapotranspiration from the duckweed mat is lower than evaporation from an open water surface at the same environmental conditions; solar radiation, air temperature, humidity, and wind speed (Skillicorn et al. 1993; Brix 1995; Brix 1997; Becerra et al. 1995; Iqbal 1999; Ansa et al. 2015).

Zayed et al. (1998) showed in their study, that wetland plants play an important role in the trace elements removal process. They investigated the potential of duckweed (*Lemna minor* L.) to accumulate Cd, Cr, Cu, Ni, Pb, and Se when supplied individually in a nutrient solution at a series of concentrations ranged from 0.1 to 10 mg L^{-1}. Duckweed shows promise for the removal of Cd, Se, and Cu from contaminated wastewater since it accumulates high concentrations of these elements. Further, the growth rates and harvest potential make duckweed a good species for phytoremediation activities. Ferrara et al. (1985) showed that the reliability of wastewater treatment by some aquatic plants including duckweed in adsorption of the heavy metals cadmium and zirconium. Viet et al. (1988) reported that duckweed plants proved to be an excellent bioaccumulator of various heavy metals, which allowed it to treat a variety of wastewaters including industrial and highly polluted wastes. Hammouda et al. (1995) evaluated the efficiency of duckweed aquatic treatment in heavy metals removal in various water systems data obtained suggested a maximum reliability of systems with mixtures containing high ratios of wastewater.

El-Kheir et al. (2007) conducted an experiment to study the efficiency of duckweed (*Lemna gibba* L.) as an alternative cost effective natural biological tool in wastewater treatment in general and eliminating concentrations of both nutrients and soluble primary treated sewage water systems (from the collector tank) for aquatic treatment over an eight day retention time period under local outdoor natural conditions. Samples were taken below duckweed cover after every two days to assess the plant's efficiency in purifying sewage water from different pollutants and to examine its effect on both phytoplankton and total and fecal coliform bacteria. Total suspended solids, biochemical oxygen demand, chemical oxygen demand, nitrate, ammonia, ortho-phosphate, Cu, Pb, Zn, and Cd decreased by: 96.3%, 90.6%, 89.0%, 100%, 82.0%, 64.4%, 100%, 100%, 93.6%, and 66.7%, respectively. Ran et al. (2004) carried out a pilot study on constructed wetlands using duckweed for treatment of domestic primary effluent to be used for reuse purposes. Results indicated that the system efficiently reduced fecal coliform by approximately 95% under average hydraulic residence time of about 4.26 days. Steen et al. (2000) compared fecal coliform (FC) decay in a series of five shallow algal ponds to FC decay in an integrated system of algal and duckweed ponds. In algal ponds, light attenuation by algal matter became rate limiting for FC decay. In the integrated system, the algal concentration in the algal ponds was reduced by the intermediary duckweed ponds. This increases the FC decay in the algal ponds of the integrated system considerably, compared to the FC decay in the algal ponds alone. Falabi et al. (2002) carried out a study to determine the ability of duckweed ponds used to treat domestic wastewater to remove Giardia and Cryptosporidium. The influent and effluent of a pond covered with duckweed with six days retention time was tested for Giardia cysts, Cryptosporidium oocysts, fecal coliforms, and coliphage. Results showed that these structures were reduced by 98 and 89% respectively, total coliforms by 61%, fecal coliforms by 62% and coliphage by 40%. Pandey (2001) showed that bacteriological analysis in influent and treated effluent at Delhi duckweed pond indicated removal of fecal coliform in the range of 99.27% and 99.78% at hyrdraulic retention time of 6.4 to 14.2 days.

2. Duckweed grown on nutrient-rich water has a high concentration of trace minerals, potassium, phosphorus, nitrogen, fat, vitamins, and pigments, particularly carotene and xanthophyll, which make duckweed suitable for animal feed like birds, cows, goats, and sheep. In addition, the plant has a high protein content of 20–40%, and is rich in amino acids. In many villages around the world, duckweed is used in duck-farms, where, ducklings are given small quantities of fresh duckweed in case of an illness or vitamin insufficiency. Compared to most other plants, duckweed contains little fiber, and has little to no indigestible material, and is even used for monogastric animals. Duckweed is a very useful diet for aquatic animals as grass carp, tilapia, catfish, fingerling, prawn, and turtle (Mbagwu and Aeniji 1988; Haustein et al. 1988; Haustein et al. 1992; Haustein et al. 1994; Leng et al. 1995; Anh and Preston 1997).

3. Duckweed is eaten by humans in some parts of Southeast Asia due to its nutritional value, and it provides a rich source of vitamins A and B. It can be eaten in salads or soups, on sandwich or as a component of vegetable spread. If grown for human consumption, careful attention should be paid to the possible presence of bacteria.

4. Duckweed is used in China to treat acne (Fleming 2000). It can be used in the treatment of flatulence, kidney infections, respiratory tract infections, jaundice, and hypothermia. The active ingredients in duckweed are the flavonoids.

5. Duckweed can be used as a reproducible model plant system for the study of host-pathogen interactions in human pathogenesis. It can be used as a fast, inexpensive, and reproducible model plant system for the study of host-pathogen interactions, could serve as an alternative choice for the study of some virulence factors, and could also potentially be used in large-scale screening for the discovery of antimicrobial chemicals (Zhang et al. 2010; Chaudhary and Sharma 2014).

6. Duckweed can be utilized to produce ethanol, butanol, 3 and biogas, which are promising alternative energy sources to minimize dependence on limited crude oil and natural gas. The advantages of this aquatic plant include high rate of nutrient (nitrogen and phosphorus) uptake, high biomass yield, and great potential as an alternative feedstock for the production of fuel ethanol, butanol, and biogas. Where, research is needed in these areas that include duckweed selection, optimization of duckweed biomass production, and enhancement of starch accumulation in duckweeds and use of duckweeds for production of various biofuels (Cui and Cheng 2015).

Case Study

The case study is from Jordan. The country located about 80 kilometers east of the Mediterranean Sea between 29° 11' to 33° 22' north, and 34° 19' to 39° 18' east. Jordan is bordered by Syria to the North, Saudi Arabia to the South, Palestine and

Israel to the West and Saudi Arabia to the East. The area of land mass is approximately 88,778 km², while the area of water bodies is approximately 482 km² that includes the Dead Sea and the Gulf of Aqaba. The altitude ranges from less than –400 m (below mean sea level) at the surface of the Dead Sea up to the 1750 m of Jebel Rum.

Jordan's water resources consist of surface and ground water, with reclaimed wastewater being used at an increasing scale for irrigation. Jordan's surface water is distributed unevenly in 15 basins. The major basins are Yarmouk, Zarqa, Jordan Riverside, Wadi Mujib, Dead Sea, and Hasa. Based on the general drainage direction, these basins can be classified into three major groups:

1. Basins that drain into the Dead Sea,

2. Basins that drain into the Red Sea, and

3. Basins that drain eastward into mudflats of the desert.

Geographically, four main topographical features characterize Jordan:

1. The lowlands (Ghors), which consist of three areas: the **Jordan valley** which starts at Lake Tiberias in the north, the **lowlands** along the Dead Sea, and the **Wadi Araba** which extends in a southerly direction to the northern shores of the Red Sea. The Jordan Valley is part of the Jordan River system or the North Dead Sea Basin which drains an area of about 18,300 km². The Jordan River basin consists of three sub-basins, namely: the upper Jordan River, the Yarmouk River, and the lower Jordan catchments. The Jordan River basin or North Dead Sea Basin is part of the Great Rift Valley that extends from the Horn of Africa as far to north western Syria. Within Jordan, the Jordan Rift Valley extends from south Lake Taberia at an elevation of –207 passing into the Dead Sea with an elevation of –418 m extending then south to the Red Sea. The area between the lake Taberia to north edge of the Dead Sea is known as the Jordan Valley. The Jordan Valley is the most important agricultural areas. It exhibits a very unique environment with fertile soil that allows growing crops and vegetables off-season giving it an advantageous value. The water source of irrigation is the flow from King Talal Dam which is fed from the Zarqa River. The rainfall of Jordan Valley decreases from approximately 300 mm in the north to 102 mm in the south. Annual rainfall of southern Ghor is less than 100 mm. The Jordan Valley below sea level is warm during winter and hot in summer. Wadi Araba region possesses a hot arid climate with average rainfall of 50 mm/year, with limited cultivated areas using underground water.

2. The highlands, these extend from north to south between 600 and 1600 m above sea level. Circumstances as high altitudes and high annual rainfall (350–600 mm) join to create a temperate Mediterranean climate, which ensures the area is home to the vast majority of the population and includes the cities of Amman, Irbid, and Zarqa.

3. The arid plains, these comprise the plains between the Badiah and the highlands. Rainfall ranges between 200 mm in the East and 350 mm in the West.

4. Badiah (Eastern Desert), it is an extension of the Arabian Desert and covers about 90% of the Kingdom. Annual rainfall of Badiah is less than 200 mm.

Jordan is divided into four biogeographic zones, Mediterranean, Irano-Turania, Saharo-Arabian, and Sudania (subtropical penetration). Such regions vary in their geophysical and biological characteristics. With regards to the Jordan flora; Jordan has a diverse range of habitats, ecosystems and biota due to its varied environments between the Jordan Valley, the Mountain Heights Plateau, and the Badia Desert region (Al-Eisawi 1982; Hashemite Kingdom of Jordan 1998; Al-Eisawi et al. 1998; Dutton and Shahbaz 2008). Spring is the high season for Jordanian flora, and from February to May many regions are carpeted with a dazzling array of flowering plants. More than 2000 species of plants grow in Jordan, and the variety of the country's topography and climate is reflected in the diversity of its flora. Most of these species, however, depend heavily on the winter rains. When there is a warm, dry winter—as in 1984—many flowers either fail to appear or are considerably reduced. Accordingly, biodiversity and conservation have received some attention and some serious studies (Amr et al. 2011; Courrier 1992; Barakat and Hegazy 1997; Johnson 1995; Heywood and Watson 1995; AL-Eisawi 1995; Cope and Al-Eisawi 1998; Lemons et al. 2003; Anonymous 2003). The highlands of Jordan host forests of oak and pine, as well as pistachio and cinnabar trees. Olive, eucalyptus, and cedar trees thrive throughout the highlands and the Jordan Valley. Jordan's dry climate is especially conducive to shrub trees, which require less water. Species of shrubs can be found throughout all the geographical regions of Jordan. The Badia of Jordan hosts many small shrub plants, where they are often grazed by the goats of local Bedouin tribes. Several species of acacia trees can be found in the deserts, as well as a variety of sturdy wild flowers and grasses which grow among the rocks in this demanding habitat.

With regards to the Jordan fresh water ecosystems and aquatic plants, most of the running resources have almost natural water quality with minor alteration of its properties. Now-a-days most of the valley systems have polluted water with various types of influxes. Due to this, edible water plants or even ornaments water plants that grow on good quality of fresh water have disappeared totally as *Veronica aquatica* and *Nerium oleander* (Al-Eisawi 2004). Moreover, aquatic plants depend on the seasonal and permanent streams flow through many of wadis, but, scarcity of water is a fact of life in most of the Middle East countries (Gopal 2003). Al-Eisawi (2012) showed in his study "conservation natural ecosystems in Jordan," that different types of natural ecosystems in Jordan subjected to human interference. Mediterranean forest and shrub land ecosystems are severely affected by cutting and fragmentation into farm lands or housing areas, leading to a great loss in biodiversity, especially, rare species, and causing soil erosion and desertification. The second type of system is a unique desert oasis. The importance of this place is its abundance of fresh water supply that for local people as well as major cities of Jordan. However, increased demand of water, especially, the capital Amman, has caused remarkable reduced the size of the oasis dramatically. This phenomenon is worsened due to the presence of hundreds of aquifer wells for the use in the ever increasing desert. Finally, such activities have destroyed this unique site, have made the place dry out, and lose of many rare and unique plants species. The third ecosystem is a marginal land area, classified as steppe land belonging to the Irano-Turanian biogeographic vegetation region subject to cultivation under a little rain water. It caused destruction of vegetation, soil erosion, and weed spreading. The fourth is a fresh water ecosystem also badly altered by mixing of sewage water effluent. It is causing the disappearance of many edible and prominent aquatic plant species.

Thereby, the growth of duckweed in Jordan depends on the availability, quality of water and its nutrients load which support duckweed growth to be abundant in the country.

Location and environmental requirements

In 1982, Al-Eisawi has conducted research and surveys to establish the availability of duckweeds. He recorded the species of *Lemna* that is distributed in water bodies of Jordan (Al-Eisawi 1982). In June 1997, Shammout has conducted extensive field visits and surveys with the goals of locating and identifying water bodies that contain duckweeds. The field surveys and visits were to the following locations (Shammout 1998; Shammout et al. 2008):

a. Water flow of upstream and downstream of the Zarqa River.

b. Jordan valley (Ghors) as North Shuna and South Shuna, and al-Karameh.

c. Dams as King Talal Dam.

The results of field visits and surveys of (Shammout 1998; Shammout et al. 2008), show that the aquatic duckweed plants grow naturally at Sukhnah, and Jerash areas based on the flow of Zarqa River Basin. Duckweed plants grow in a dense green aggregation where their minute leaves or fronds can completely hide the water, thus, giving impression of dry land. The roots of these plants are serving chiefly as an anchor to keep the fronds right side up and to protect colonies in dispersal by water motion. It is noticed that duckweed's distribution is not affected by water depth as it can grow in a shallow water bodies as shown in Fig. 2; the duckweed growth at Jerash area-Jordan. At these locations; surface water of Zarqa River is mixed with treated effluent of Khirbet es-Samra Wastewater Treatment Plant and is used for irrigation purposes. Figure 3 shows the location of duckweed plants along the Jordan's surface resources basins. Moreover, the results of field survey show that duckweed plants grow naturally in the irrigation ponds at Jordan Valley (Ghors). The location of duckweed plants and the water sources of the farm irrigation ponds-Jordan are shown in Fig. 4. It can be seen from Fig. 4 that Zarqa River plays an important role in the environment that encourages the growth of duckweed. The Zarqa River supports these irrigation ponds with preferable nutrients for duckweed plants' growth. This is due to the fact that the river consists of mixed water mainly of its base flow and the flow of the treated effluent of domestic water treatment plants (Shammout et al. 2013). It is also vital to mention that the mixed water of the Zarqa River is held in King Talal Dam before being released to travel a distance via King Abdullah Canal and reach the Jordan Valley. That it is, the King Abdullah Canal is fed by the reservoir of the King Talal Dam, which is fed by the Zarqa River.

The water quality is the main issue for duckweed availability. Table 1 shows the water quality that supports duckweed plants. The ranges of values, in mg/l, are as follows: Ca 120-113, Mg 48-40, K 22-18, Cl 349-340, SO_4 220-188, PO_4 7.6-6.3, Na 212-205, and NO_3 38-31. The average power of hydrogen (pH) value is 7.7, and the

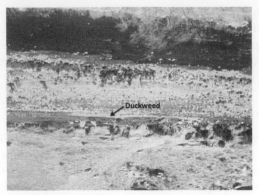

Fig. 2: Duckweed growth at Jerash area-Jordan.

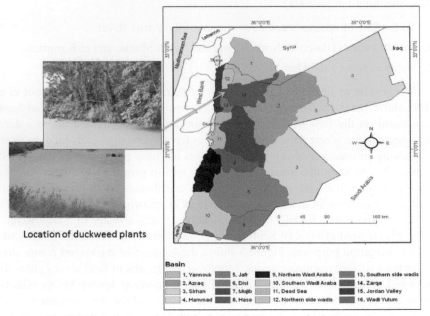

Fig. 3: Location of duckweed plants along the Jordan's surface resources basins.

average electrical conductivity (EC) value is 1.89 mS/cm (Shammout and Zakaria 2015).

Duckweed growth rate varies from one month to another and from one location to another. Its existence is affected by many environmental factors such as:

1. At Sukhnah and Jerash areas; it is observed that during the month of June through October; duckweed plants tend to flourish, become green in color, and of a healthy nature. From the month of October through winter time, the green duckweed disappears. This is due to the fluctuation in water availability, where, Jordan is known for severe water scarcity, as well as, the heavy utilization of many springs along the Zarqa River by farmers for irrigation purposes has seriously affected the River base flow and thus, the availability of duckweed

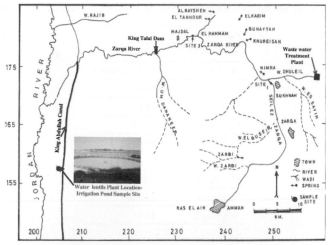

Fig. 4: Location of duckweed plants and the water sources of the irrigation ponds-Jordan.

Fig. 5: Duckweed location at Sukhnah area (dry location)-Jordan.

aquatic plants. Figure 5 shows the duckweed location at Sukhnah area (dry location)-Jordan.

2. At Jordan Valley (Ghors), it is observed that duckweed is available all year around. As long as water is present, duckweed will grow and prosper.

Duckweed efficiency in water bioremediation

Duckweeds can reduce eutrophication effects and provide oxygen from their photosynthesizing activity (Hasan et al. 2009). The basic concept of a duckweed wastewater treatment system is to farm local duckweed on the wastewater requiring treatment. Duckweed wastewater treatment systems have been studied for dairy waste lagoons, raw domestic sewage, secondary effluent, waste stabilization ponds, and fish culture systems (Beeby 1993; Bonomo et al. 1997; Hasan and Chakrabarti 2009). They have been used to remove diverse pollutants, including heavy metals, estrogenic hormones, or for bioremediation of tannery effluent or water contaminated by oil refineries (Mwale and Gwaze 2013). Duckweed waste water treatment systems can remove as much as 99% of the nutrients and dissolved solids contained in waste

Table 1: Water quality that supports duckweed plants (Shammout and Zakaria 2015).

Parameter	Range
pH	7.7 (Average)
EC mS/cm	1.89 (Average)
Ca (mg/l)	120-113
Mg (mg/l)	48-40
K (mg/l)	22-18
Cl (mg/l)	349-340
SO_4 (mg/l)	220-188
PO_4 (mg/l)	7.6-6.3
Na (mg/l)	212-205
NO_3 (mg/l)	38-31

water (Skillicorn et al. 1993). These substances are then removed permanently from the effluent stream following the harvesting of a proportion of the crop. The plants also reduce BOD_5 by reduction of sunlight in lagoons.

Based on the above description, and in order to study the effect of Jordan's duckweed availability in water bioremediation, two reservoirs are being prepared, monitored and analyzed regularly at the University of Jordan (outdoor). These reservoirs are fed by the same water source from the Sukhnah area. One of the reservoirs is considered the control and contains only water source from the Sukhnah area (duckweed free), whereas the second reservoir has duckweed plant. Each of the reservoirs has dimensions of 125 cm length, 75 cm width, and 40 cm depth. The capacity of each reservoir is 0.375 m^3. They are maintained at a water source level by the addition of water source of Sukhnah water to compensate for evaporation losses. Harvesting of duckweed is required every three days when the layer thickness between 1–2 cm. The water of the reservoirs is analyzed to evaluate the water quality in the absence and presence of duckweed from July to October. Table 2 shows the water bioremediation using aquatic plant (duckweed). The ranges of Temperature °C, EC, pH, and the average values of BOD_5, fecal coliform, NO_3, and PO_4 are shown in Table 2. Variations in the parameter values are clearly noticed. These variations are due to the role of duckweed in water bioremediation.

Table 2: Water bioremediation using aquatic plants (duckweed).

Parameter	Control reservoir	Duckweed reservoir
Temperature °C	18.3–23.8	18.1–23.3
EC (mS/cm)	2.51	2.46
pH	7.72	7.87
BOD5 (mg/l)	41	20
Fecal coliform (MPN/100 ml)	2.13×10^2	6.6×10^1
NO_3 (mg/l)	43	19
PO_4 (mg/l)	0.25	0.11

Discussion and Conclusions

With regards to the Jordan's aquatic plants, most of the running water resources that support these plants with nutrients have polluted water with various types of influxes. In addition, aquatic plants depend on the seasonal and permanent streams flow through many of wadis, and scarcity of water is a fact in the Kingdom. Thus, aquatic plants that grow on good quality of fresh water have disappeared totally. Nevertheless, duckweed (*Lemna* sp.) is located in Jordan at different locations as Sukhnah, Jerash, and Jordan Valley (Ghors) areas. At Sukhnah and Jerash areas, duckweed is found during the month of June through October. During this period, duckweed tends to become green in color and shines across the water. In the Jordan Valley (at specific locations), duckweed is found all year round. The water source that supports duckweed with nutrient has been determined. Findings show that water source for the availability of duckweed plants at Sukhnah and Jerash areas is based on the flow of Zarqa River; where the Zarqa River consists mainly of its base flow and the flow of the effluent of domestic water treatment plants, that is, Khirbet es-Samra Wastewater Treatment Plant (Shammout et al. 2013). The main water source for Jordan Valley (Ghors) is based on the King Abdullah Canal. As shown in Table 1, the average values of pH, EC, and the concentration of nutrients of Ca, Mg, K, Cl, SO_4, PO_4, Na, and NO_3 are encouraging the availability of aquatic plants (duckweed) at different locations of the Kingdom.

With regards to the bioremediation using Jordan's aquatic plants (duckweed), Table 2 shows the results of the two monitored reservoirs at the University of Jordan (outdoor). Variations are clear in values of the water Temperature °C, EC, pH, BOD_5, fecal coliform, NO_3, and PO_4. The average range of Temperature from July–October in the absence of duckweed is between 18.3–23.8, whereas, the average range of Temperature in the presence of duckweed is between 18.1–23.3. The Temperature decrease in the presence of duckweed is due to the complete coverage of duckweed which reduces the evaporation and sun penetration. The average value of Electrical Conductivity (EC) in the absence of duckweed is 2.51 mS/cm, whereas, the average value of Electrical Conductivity (EC) in the presence of duckweed is 2.42 mS/cm. This is due to the duckweed's absorbance of nutrients which reduces water conductivity and upgrades the quality water. The average pH value in the reservoir with duckweed is 7.87, but it is 7.72 in the reservoir that has no duckweed. This is due to the reduction in carbon dioxide (CO_2) as a result of the reduction of other growing aquatic plants in the presence of duckweed. As for the reduction in BOD_5 in the duckweed reservoir, it is due to the reduction in microbial activity. The microbial analysis value of fecal Coliform in the reservoir with duckweed is 6.6×10^1 MPN/100 ml but it is 2.13×10^2 MPN/100 ml in the reservoir that has no duckweed. Fecal coliform bacteria decreased by 69%. Here, the duckweed green mat prevents oxygen from entering the water surface, thus, reducing light penetration, and surface re-aeration of the duckweed's reservoir. This results in lowering the photosynthesis activity of other aquatic plants, making the water column largely anaerobic. In addition, duckweed exhibits efficiency in the removal of NO_3 and PO_4. The average NO_3 value in the reservoir with duckweed is 19 mg/l, but it is 43 mg/l in the reservoir that has no duckweed, and regarding the average value of PO_4, it is

0.11 mg/l in the presence of duckweed and 0.25 mg/l in the absence of duckweed. It is vital to mention that the percentages of NO_3 and PO_4 absorbed by duckweed and other nutrients are permanently removed from the water when duckweeds are harvested every three days when the layer thickness between 1–2 cm. The harvesting procedure of duckweed plants from the water surface is easy and not a laborious task. Therefore, duckweed works as a natural water bioremediation agent; purifying water and upgrading the water quality.

In conclusion, bioremediation is an alternative technology that offers the possibility to reduce harmless various contaminants using natural biological activity at low cost. Bioremediation can often be carried out on site at low-technology techniques. Jordan's case study of water bioremediation using aquatic plants (duckweed) proved its efficiency in upgrading the water quality. It provides a good opportunity to examine and present the efficiency of duckweed in water bioremediation. It will establish a baseline in Jordan for further studies to utilize these plants on a large scale as a natural inexpensive agent for water or wastewater treatment. Regular harvesting of duckweed is essential to maintain the continuous healthy growth of duckweed to provide water bioremediation. The nutritive values of duckweed as protein, antioxidants, and pigments, make duckweed attractive for research to be utilized properly.

References

Ahmet, S. and O. Erdal. 2009. The accumulation of Arsenic, Uranium, and Boron in *Lemna gibba* L. exposed to secondary effluents. Ecological Engineering 35: 1564–1567.

Alaerts, G.J., M. Mahbubar-Rahma and P. Kelderman. 1996. Performance analysis of a full-scale duckweed covered lagoon. Water Resources 30: 843–52.

Al-Eisawi, D.M. 1982. List of Jordan Vascular Plants. Mitt. Bot, München.

AL-Eisawi, D.M. 1995. Flora and Vegetation of Azraq Wetland Reserve. Azraq Oasis Conservation Project. RAMSAR and the World Bank 72.

Al-Eisawi, D.M., El-Oqlah, S. Oran and J. Lahham. 1998. Plant biodiversity in Jordan. *In*: Jaradat, A.A. [ed.]. Plant Genetic Resources of Jordan. Proceedings of a National Seminar, 2–4 August, 1994. Amman, Jordan.: IPGRI, West Asia and North Africa Regional Office. Aleppo, Syria. 272.

Al-Eisawi, D.M. 2004. Water scarcity in relation to food security and sustainable use of biodiversity in Jordan. International FORUM on Food Security under Water Scarcity in the Middle East: Problems and Solutions COMO (Italy) November, 24–27.

Al-Eisawi, D.M. 2012. Conservation of natural ecosystems in Jordan. Pak. J. Bot. 44: 95–99.

Amr, Z.S., D. Mordy and M.F. Al-Shudiefat. 2011. Badia, The Living Desert. Al Rai Printing Press. Amman, Jordan. 224.

Anh, N.D. and T.R. Preston. 1997. Evaluation of protein quality in duckweed (*Lemna* sp.) using a duckling growth assay. Livestock Research for Rural Development 9(2). http://www.lrrd.org/lrrd9/2/anh92.htm.

Anonymous. 2003. The Hashemite Kingdom of Jordan, Ministry of Environment. National biodiversity strategy and action plan. Jordan, Amman. 238.

Ansa, E.D.O., E. Awuah, A. Andoh, R. Banu, W.H.K. Dorgbetor, H.J. Lubberding et al. 2015. A review of the mechanisms of fecal coliform removal from algal and duckweed waste stabilization pond systems. American Journal of Environmental Sciences. DOI: 10.3844/ajessp.

Ansal, M.D., A. Dhawan and V.L. Kaur. 2010. Duckweed based bio remediation of village ponds: An ecologically and economically viable integrated approach for rural development through aquaculture. Livestock Research for Rural Development. Volume 22, Article #129: http://www.lrrd.org/lrrd22/7/ansa22129.htm.

Axtell, N.R., S.P.K. Sternberg and K. Claussen. 2003. Lead and nickel removal using microspora and *Lemna minor*. Bioresource Technology 89: 41–48.

Barakat, H.N. and A.K. Hegazy. 1997. Review in ecology desert conservation and development, a festschrift for Prof. M. Kassas on the occasion of his 75th birthday. Cairo, Egypt. 331.

Beeby, A. 1993. Alternative methods of treating sewage effluent. Applying Ecology. 1st edition, London: Chapam and Hall. pp. 350–357.

Becerra, M., T.R. Preston and B. Ogle. 1995. Effect of replacing whole boiled soya beans with duckweed (*Lemna* sp.) in diets of growing ducks. Livestock Research for Rural Development 7, article no 26. Retrived September 19, 2018, from http://www.Irrd.org/Irrd7/3/7.htm.

Bonomo, L., G. Pastorelli and N. Zambon. 1997. Advantages and limitations of duckweed-based wastewater treatment systems. Water Science and Technology 35(5): 239–246.

Bianconi, D., F. Pietrini, A. Massacci and M.A. Iannelli. 2013. Uptake of Cadmium by Lemna minor, a (hyper?-) Accumulator Plant Involved in Phytoremediation Applications. E3S Web of Conferences 1 13002 DOI:http://dx.doi.org/10.1051/e3sconf/20130113002(2013).

Brix, H. 1995. Treatment Wetlands: an overview. pp. 167–176. *In*: Toczylowska, I. and M. Mierzejewski [eds.]. Proceedings Conference on Constructed Wetlands for Wastewater Treatment. Technical University of Gdansk, Poland.

Brix, H. 1997. Do macrophytes play a role in constructed treatment wetlands? Wat. Sci. Tech. 35: 11–17.

Chaudhary, E. and P. Sharma. 2014. Use of duckweed in wastewater treatment. International Journal of Innovative Research in Science, Engineering and Technology 3(6): 13622–13624.

Christian, S.B., R.P.L. Diederik, S.A. Saleh and L.N.L. Piet. 2012. Heavy metal removal in duckweed and algae ponds as a polishing step for textile wastewater treatment. Ecological Engineering 44: 102–110.

Classen, J.J., J. Cheng, B.A. Bergmann and A.M. Stomp. 2000. *Lemna gibba* growth and nutrient uptake in response to different nutrient levels. pp. 40–5. *In*: Animal, Agricultural and Food Processing Waste: Proc. Eighth International Symposium. Des Moines, Iowa. St. Joseph, Mich., ASAE.

Clemens, S., M.G. Palmgren and U. Kramer. 2002. A long way ahead: Understanding and engineering plant metal accumulation. Trends in Plant Science 7(7): 309–314. doi:10.1016/S1360-1385(02)0.

Conserve Energy Future. 2015. What is Bioremediation? http://www.conserve-energy-future.com/what-is-bioremediation.php.

Cope, T.A. and D.M. Al-Eisawi. 1998. Checklist of the flora of badia. pp. 332. *In*: Dutton, W.R., J.I. Clarke and A.M. Battikhi [eds.]. Arid Land Resources and their, Management, Jordan's Desert Margin. Kegan Paul International. London and New York.

Courrier, K. [ed.]. 1992. Global biodiversity strategy, guideline for action plane to save, study and use earth's wealth sustainability and equitably. World Resources Institute (WRI). The World Conservation Union (IUCN), UNITED Nation Environment Programme (UNDP) 244.

Cui, W. and J.J. Cheng. 2015. Growing duckweed for biofuel production: a review. Plant Biol. J. 17: 16–23. doi:10.1111/plb.12216.

Cunningham, S.D. and D.W. Ow. 1996. Promises and prospects of phytoremediation. Plant Physiol. 110: 715–719.

Culley, D.D. and D.A. EPPA. 1973. Use of duckweed for waste treatment and animal feed. Journal of Water Pollution Contra Federation 45: 337–347.

Culley, D.D., E. Rejmankova, J. Kvet and J.B. Frye. 1981. Production, chemical quality and use of duckweeds (Lemnaceae) in aquaculture, waste management, and animal feeds. J. World Mar. Soc. 12(2): 27–49.

Demin, S., N. Drouiche, A. Aouabed, T. Benayad, O. Badache and S. Semsari. 2013. Cadmium and nickel: assessment of the physiological effects and heavy metal removal using a response surface approach by *L. gibba*. Ecological Engineering 61: 426–435.

Dewedar, A. and M. Bahgat. 1995. Fate of fecal coliform bacteria in a wastewater retention reservoir containing *Lemna gibba* L. Water Res. 29: 2598–600.

Donganlar, B.Z., C. Seher and Y. Telat. 2012. Metal uptake and physiological changes in *Lemna gibba* exposed to manganese and nickel. International Journal of Biology 4: 148–151.

Dothan, N.F. 1986. Flora Palaestina. Part Four. The Israel Academy of Sciences and Humanities, Jerusalem.

Dutton, W.R. and Shahbaz. 2008. Jordan's arid Badia: deepening our understanding. Smith-Gordon. Cambridgeshire, Great Britain. 449.

El-Kheir, W., G. Ismail, F. El-Nour, T. Tawfik and D. Hammad. 2007. Assessment of the Efficiency of Duckweed (*Lemna gibba*) in wastewater treatment. Int. J. Agri. Biol. 9(5): 681–687.

Elma, L., J. O'Halloran and A.K.J. Marcel. 2012. Frond development gradients are a determinant of the impact of zinc on photosynthesis in three species of Lemnaceae. Aquatic Botany 101: 55–63.

Falabi, J.A., C.P. Gerba and M.M. Karpiscak. 2002. Giardia and Cryptosporidium removal from waste-water by duckweed (*Lemna gibba* L.) covered pond. Lett. Appl. Microbiol. 34: 384–7.

Ferrara, L., P. Forgione, O. Schettino and V. Rullo. 1985. The use of aquatic plants in wastewater treatment adsorption of zirconium and cadmium. Bollettino Della Societa Italiana Di Biologia Sperimentale 61: 43–8.

Flathman, P.E., D. Jerger and J.E. Exner. 1993. Bioremediation: Field Experience. Lewis, Boca Raton, FL.

Fleming, T. 2000. PDR for Herbal Medicines. 2nd ed. Montvale, NJ: Medical Economics Co.

Gaigher, I.G. and R.V. Short. 1986. An evaluation of duckweed (Lemnaceae) as a candidate for aquaculture in South Africa. Aquaculture 15: 81–90.

Garbisu, C. 2002. Phytoremediation: A technology using green plants to remove contaminants from polluted areas. Reviews on Environmental Health 17(3): 173–188. doi:10.1515/REVEH.2002.17.3.173.

Gopal, B. 2003. Aquatic biodiversity in arid and semi-arid zones of Asia and water management. pp. 199–215. *In*: Lemons, J., R. Victor and D. Schaffer [eds.]. Conserving Biodiversity in Arid Regions, Best Practices in Developing Nations. Kluwer Academic Publishers. Boston, Dordrecht, London.

Hammouda, O., A. Gaber and M.S. Abdel-Hameed. 1995. Assessment of the effectiveness of treatment of waste water-contaminated aquatic systems with *Lemna gibba*. Enzyme Microbiol. Technol. 17: 317–23.

Hasan, M.R. and R. Chakrabarti. 2009. Use of algae and aquatic macrophytes as feed in small-scale aquaculture: A review. FAO Fisheries and Aquaculture technical paper, 531. FAO, Rome, Italy.

Hashemite Kingdom of Jordan. 1998. Jordan country study on biological diversity. General Cooperation for the Environmental Protection (GCEP). Amman, Jordan. 414.

Haustein, A.T., R.H. Gillman, P.W. Skillicorn, V. Vergara, V. Guevara and A. Gastanaduy. 1988. Duckweed, a useful strategy for feeding chickens: Performance of layers fed with sewage-grown Lemnaceae species. Poultry Science 69: 1835–1844.

Haustein, A.T., R.H. Gillman, P.W. Skillicorn, V. Guevara, F. Diaz, V. Vergara et al. 1992. Compensatory growth in broiler chicks fed on *Lemna gibba*. British Journal of Nutrition 68: 329–335.

Haustein, A.T., R.H. Gilman, P.W. Skilicorn, H. Hannan, V. Guevara, V. Verara et al. 1994. Performance of broiler chickens fed diets containing duckweed (*Lemna gibba*). Journal of Agriculture Sciences 122: 285–289.

Heuzé, V. and G. Tran. 2015. Duckweed. Feedipedia.org. A programme by INRA, CIRAD, AFZ and FAO. http://www.feedipedia.org/node/15306.

Heywood, V.H. and R.T. Watson. 1995. Global Biodiversity Assessment. Published for the United Nations Environment Programme (UNEP). Cambridge University Press. Cambridge.

Hillman, W.S. 1976. The Lemnaceae or duckweeds: A review of the descriptive and experimental literature. Bot. Rev. 27: 221–87.

Hillman, W.S. and D.D. Culley. 1978. The uses of duckweed. American Scientist 66: 442–451.

Hinchee, R.E., J.L. Means and D.R. Burrisl. 1995. Bioremediation of Inorganics. Battelle Press, Columbus, OH.

Horvat, T., Z. Vidakovic´-Cifrek, V. Oresˇcˇanin, M. Tkalec and B. Pevalek-Kozlina. 2007. Toxicity assessment of heavy metal mixtures by *Lemna minor* L. Science of the Total Environment 384: 229–238.

Iqbal, S. 1999. Duckweed aquaculture: Potentials, possibilities and limitations for combined waste water treatment and animal feed production in developing countries. SUNDEC Report no. 6/99. Switzerland.

Johnson, N. 1995. Biodiversity in the balance. Approach to Setting Geographic Conservation Priorities. Biodiversity Support Program, A USAID Funded Consortium of Wildlife Fund, Tan Nature Conservation and World Resources Institute. 115.

King, R.B., G.M. Long and J.K. Sheldon. 1997. Practical Environmental Bioremediation: The Field Guide. 2nd ed., Lewis, Boca Raton, FL.

Landolt, E. 1986. Biosystematic investigations in the family of duckweeds (Lemnaceae). Vol. 2. The family of Lemnaceae—A monographic study. Part 1 of the monograph: Morphology; karyology; ecology; geographic distribution; systematic position; nomenclature; descriptions.

Lemons, J., R. Victor and D. Schaffer. 2003. Conserving biodiversity in arid regions, best practices in developing nations. Kluwer Academic Publishers. Boston, Dordrecht, London. 493.

Leng, R.A., J.H. Stambolie and R. Bell. 1995. Duckweed a potential high protein feed resource for domestic animals and fish. Livestock Research for Rural Development 7, no. 1, New England.

McLay, C.L. 1976. The effect of pH on the population growth of three species of duckweed: Spirodela oligorrhiza, *Lemna minor* and *Wolffia arrhiza*. Freshwater Biology, Blackwell, U.K. 6: 125–36.

Mitsch, W.J., C.L. Dorge and J.R. Wiemhoff. 1977. Forested wetlands for water resource management in southern Illinois. Urbana, IL: University of Illinois, Water Resources Center. WRC Research Report.

Mbagwu, I.G. and H.A. Aeniji. 1988. The nutritional content of duckweed in the Kainji Lake Area, Nigeria. Aquat. Bot. 29: 357–366.

Mueller, J.G., C.E. Cerniglia and P.H. Pritchard. 1996. Bioremediation of environments contaminated by polycyclic aromatic hydrocarbons. pp. 125–194. *In*: Bioremediation: Principles and Applications. Cambridge University Press, Cambridge.

Mwale, M. and F.R. Gwaze. 2013. Characteristics of duckweed and its potential as feed source for chickens reared for meat production: a review. Scientific Research and Essays 8(18): 689–697.

Oron, D. 1994. Duckweed culture for wastewater renovation and biomass production. Agric. Wat. Man. 26: 27–40.

Oron, G., A. De Vegt and D. Porath. 1988. Nitrogen removal and conversion by duckweed grown on wastewater. Water Res. 22: 179–84.

Pandey, M. 2001. Duckweed based wastewater treatment. Invention intelligence.

Paz-Alberto, A. and G. Sigua. 2013. Phytoremediation: A green technology to remove environmental pollutants. American Journal of Climate Change 2: 71–86. doi: 10.4236/ajcc.2013.21008.

Ran, N., M. Agami and G. Oron. 2004. A pilot study of constructed wetlands using duckweed (*Lemna gibba* L.) for treatment of domestic primary effluent. Water Res. 38: 2241–8.

Rejmankova, E. 1979. The function of duckweed in fish pond ecosystem. PhD Thesis, Department of Hydrobotany, Terbon, Czechoslovakia.

Salt, D.E., M. Blaylock, I. Chet, S. Dushenkov, B. Ensley, P. Nanda et al. 1995. Phytoremediation: A novel strategy for the removal of toxic metals from the environment using plants. Biotechnology 13(5): 468–474. doi:10.1038/nbt0595-468.

Shammout, M.W. 1998. Propagation of duckweed, Lemna sp. (Lemnaceae) and its application in wastewater treatment in Khirbet As-Samra (Jordan). MSc. Thesis, The University of Jordan, Amman, Jordan.

Shammout, M.W., S. Oran and M. Fayyad. 2008. The application of duckweed (*Lemna* sp.) in wastewater treatment in Jordan. International Journal of Environment and Pollution (IJEP) 33(1): 110–120.

Shammout, M.W., M. Shatanawi and S. Naber. 2013. Participatory optimization scenario for water resources management: A case from Jordan. Water Resour. Manage. 27(7): 1949–1962. doi: 10.1007/s11269-013-0264-9.

Shammout, M.W. and H. Zakaria. 2015. Water lentils (duckweed) in Jordan irrigation ponds as a natural water bioremediation agent and protein source for broilers. Ecological Engineering 83: 71–77. http://dx.doi.org/10.1016/j.ecoleng.2015.05.041.

Skillikorn, P., W. Spira and W. Journey. 1993. Duckweed aquaculture, a new aquatic farming system for developing countries, p. 68. The World Bank. Washington, D.C.

Steen, P. Van. Der, A. Brenner, Y. Shabtai and G. Oron. 2000. Improved fecal coliform decay in integrated duckweed and algal ponds. Water Sci. Technol. 42: 363–70.

Sutton, D.L. and W.H. Ornes. 1975. Phosphorus removal from static sewage effluent using duckweed. J. Environ. Qual. 4: 367–70.

The American Heritage Science Dictionary. 2002. Houghton Mifflin. http://www.dictionary.com/browse/bioremediation.

Tripathi, B.D. and K.W. Misra. 1990. Wastewater treatment by duckweed (*Lemna minor* L.). Agric. Biol. Res. 6: 89–93.

Vidali, M. 2001. Bioremediation. An overview. Pure Appl. Chem. IUPAC. 73(7): 1163–1172. Lecture presented at the 8th International Chemistry Conference in Africa (8th ICCA), 30 July–4 August 2001, Dakar, Sénégal.

Viet, N., P. Warren and S. Hancock. 1988. The *Lemna* technology for wastewater treatment. *In*: SCE/Etal. Environ. Engg. Natl. Conference Vancouver. p. 292.

Zayed, A., S. Gowthaman and N. Terry. 1998. Phytoaccumulation of trace elements by wetland plants: I. Duckweed. J. Environ. Qual. 27: 715–721. doi:10.2134/jeq1998.00472425002700030032x.

Zhang, Y., Yangbo Hu, Baoyu Yang, Fang Ma, Pei Lu, Lamei Li. 2010. Duckweed (Lemna minor) as a model plant system for the study of human microbial pathogenesis. PLoS ONE 5(10): e13527. doi:10.1371/journal.pone.0013527.

4

Heavy Metal Bioremediation by Microalgae
Mechanisms and Applications

Aniefon A. Ibuot,[1] *Sanjay Kumar Gupta,*[2] *Preeti Ansolia*[3]
and *Amit K. Bajhaiya*[4,]*

INTRODUCTION

One of the major consequences of industrialization has been the pollution of both aquatic and terrestrial environments (Chabukdhara et al. 2016a,b; Gupta et al. 2014, 2015). Toxic metals released into the environment as a result of industrial activities can persist for long periods, circulating and eventually accumulating throughout the food chain, thus posing a serious threat to the environment, and to human health. The industrial activities that lead to metal release into the environment includes mining, crude oil exploration, production of agricultural chemicals, metallurgy, tanneries, fly ash dumps, and electronics production. In plants, exposure to high levels of heavy metals may results in reduced rates of photosynthesis, chlorosis, growth inhibition, browning of root tips, decreases in water and nutrient uptake, and in serious cases, plant death, which from an agricultural perspective leads to reduced food productivity and risk of metal contamination of the food chain (Kahle 1993;

[1] Department of Science Technology, Akwa Ibom State Polytechnic, P.M.B. 1200, Ikot Ekpene, Akwa Ibom State, Nigeria.
[2] Environmental Engineering, Department of Civil Engineering, Indian Institute of Technology Delhi, India-110016.
[3] Department of Pharmacy, Radharaman Institute of Technology & Science (RITS), Bhopal, India-462003.
[4] Department of Biochemistry, Chemical Biological Centre (KBC), Umeå University, Umea, Sweden, SE-90187.
* Corresponding author: amitbajhaiya@gmail.com, amit.bajhaiya@umu.se

Di Toppi and Gabbrielli 1999). In animals and humans, chronic exposure to heavy metals can induce a variety of health disorders, including kidney damage (Buchet et al. 1980; Hellstrom et al. 2001), osteomalacia (Nogawa and Kido 1993; Tsuritani et al. 1996), organ developmental defects, and cancer (Desi et al. 1998).

Current methods to remediate metal pollutants include absorption using various types of absorbents, ion exchange, electrochemical treatment, membrane technologies, and chemical precipitation, etc. (Fu and Wang 2011). There are some disadvantages associated with these methods such as these methodologies are costly, relatively inefficient, and environmentally unsustainable due to some of the toxic by-products that are generated in the process (Das et al. 2008). This has therefore led to growing interest in recent decades for the employment of biological mechanisms via the use of microbes, plants, or algae to sustainably bioremediate metal pollution from terrestrial and aquatic ecosystems. The metal bioremediation potential of bacteria, fungi, and higher plants are not the subject of this review. This chapter deals with phycoremediation potential of algae only.

Algae offer an alternative, sustainable, and environmentally friendly metal remediation strategy that may also be economically attractive, due to the efficient metal sequestering properties of these organisms (Balasubramani et al. 2016; Chabukdhara et al. 2017). Algae are often distinguished as the larger, multicellular macroalgae, and the microscopic single-celled microalgae, which include prokaryotic cyanobacteria and many diverse eukaryotic microalgae from distinct taxa (Bhattacharya and Medlin 1998). Microalgae can inhabit freshwater and marine environments, can tolerate wastewater conditions, and are also found on soil and vegetation. Therefore, microalgae have the potential to act as remediation agents in a range of situations (Gupta et al. 2016, 2017a). Though there are some concerns of microalgal harvesting after sequestration of metals from wastewater (Gupta et al. 2017c). Microalgae are especially suitable for bioremediation of heavy metals from wastewater and aquatic environments due to their natural abundance in aquatic environments and metal sequestering properties. These qualities make them a better tool for bioremediation compared to other sources both biological and chemical (Monteiro et al. 2011). This chapter will review the characteristics of metal uptake and accumulation in microalgae, the process of metal bioremediation, and their potential application.

Heavy Metal Removal Capacity of Microalgae

Heavy metal removal by microalgae generally occurs in a course of two-phases involving an initial extracellular binding of metals by the microalgal cell wall followed by a slow sequestration of metal ions inside the cell (Gupta et al. 2017b). The initial binding is rapid as various functional groups present on the algal cell wall, rapidly binds metal ions by electrostatic interactions however this process is reversible. Such type of removal (passive) occurs in both dead and live algal biomass. The binding occurs and involves various reactions such as chelation, complexation, and ion exchange. Moreover, metal ions also get trapped in the structural polysaccharide network, or some of the times easily passed the algal cell wall through diffusion (Muñoz et al. 2006; Sud et al. 2008). Though these are non-metabolic processes, metal removal is fast, while the second phase is basically a metabolic process, which passes the cell membrane barrier through the transport protein across the cell membrane. Within the algal cell, these metal ions get accumulated through binding with intracellular cell organelles and compounds. Such type of metal uptake is generally irreversible, varies slowly, and occurs in living cells only.

Metal uptake through the lipid bilayer plasma membrane surrounding the cell is mediated by specific transport proteins as the metal either in ionic form or as a ligand complex and cannot efficiently diffuse through the membrane (Worms et al. 2006). Depending on the electrochemical gradient across the membrane for the particular metal species, uptake may be mediated by passive diffusion through channel proteins, or by facilitated diffusion or active transport through carriers and pumps (Wang and Chen 2006; Gadd 1988; Mehta and Guar 2005).

Heavy metal adsorption, uptake rates, and accumulation capacities vary depending on various factors like the chemical composition of the microalgae cellular surface, the characteristics of uptake transporters, the internal biological sequestration, tolerance mechanisms, and the chemical characteristics of the polluted environment (Table 1). Therefore, the choice of microalgae for the bioremediation depends in part on the knowledge of the metal pollutants and their chemical speciation in the environment. However, having knowledge of the characteristics and mechanism of metal uptake and tolerance by microalgae can give an understanding of the potential of these organisms as a tool for metal bioremediation.

Metal Toxicity and Tolerance Mechanisms of Microalgae

Due to their accumulation and sequestration properties, species of microalgae have shown their ability to remove relatively high metal concentrations. However, tolerance to metal stress is essential for a microalgal strain to act as a significant bioaccumulator. Indeed many strains are quite insensitive to metal pollutants, and therefore are suitable for use as biological indicators in ecotoxicity assessments (Monteiro et al. 2012). To determine species that show tolerance to excess metals, optimal growth which reflects the proper functioning of various physiological and biochemical processes within the cell (e.g., photosynthesis or nutrient uptake) can be used as a key indicator of metal toxicity. Thus, characteristics such as growth rate, photosynthetic activity, or chlorophyll concentration can easily be monitored in laboratory settings to identify tolerant species (Arunakumara and Xuecheng 2008; Carr et al. 1998; Tripathi and Guar 2006). Inhibition of growth in microalgae is directly related to the amounts of metal ions that are taken up intracellularly (Arunakumara and Xuecheng 2008) besides correlating with the chemical properties of the metal in question (Tripathi and Guar 2006).

The potential of a microalgal cell to resist the metal toxicity may be determined by means of its EC_{50} value, which is the effective concentration of a metal that inhibits 50% of the growth of microalgae (Nyholm 1990). This is unanimously used as an index of toxicity and used for the comparison of resistance/tolerance of various species of microalgal when exposed to a given metal. For example, the effects of Cd on the growth of *Scenedesmus obliquus* and *Desmodesmus pleiomorphus* yielded

Table 1. Comparison of metal removal efficiency of microalgae.

Microalgae as biosorbent	Metal	Initial concentration of metal (mg/l)	Metal removal efficiency (%)	References
Anabaena subcylindrica	Cu	ᵉIE	82	El-Sheekh et al. 2005
Anabaena subcylindrica	Co	IE	34	El-Sheekh et al. 2005
Anabaena subcylindrica	Pb	IE	100	El-Sheekh et al. 2005
Anabaena subcylindrica	Mn	IE	100	El-Sheekh et al. 2005
Chlorella vulgaris	Cu	10	72	Mehta and Gaur 2001a
Chlorella vulgaris	Cu	2.5	80	Mehta and Gaur 2001a
Chlorella vulgaris	Ni	2.5	69	Mehta and Gaur 2001a
Chlorella vulgaris	Cu	10	72	Mehta and Gaur 2001a
	Cd	500	84	Sandau et al. 1996
Chlorella vulgaris (acid-pretreated)	Cu	2.5	96	Mehta and Gaur 2001a
Chlorella vulgaris (HCl-pretreated)	Ni	2.5	93	Mehta and Gaur 2001a
Chlorella vulgaris (Ca-alginate immobilized)	Cu	5.0	> 90	Mehta and Gaur 2001a
Chlorella vulgaris (Ca-alginate immobilized)	Ni	5.0	70	Mehta and Gaur 2001a
Nostoc muscorum	Cu	IE	13	El-Sheekh et al. 2005
Nostoc muscorum	Co	IE	12	El-Sheekh et al. 2005
Nostoc muscorum	Pb	IE	26	El-Sheekh et al. 2005
Nostoc muscorum	Mn	IE	33	El-Sheekh et al. 2005
Phormidium valderianum	Cd	40 mM	90	Karna et al. 1999
Scenedesmus abundans	Cd	10	97	Terry and Stone 2002
Scenedesmus abundans	Cu	10	99	Terry and Stone 2002
Scenedesmus quadricauda	Ni	30	97	Chong et al. 2000
Spirulina platensis	Cd	500	81	Sandau et al. 1996
Spirulina sp.	Al	—	49	Chojnacka et al. 2004
Spirulina sp.	Cd	—	56	Chojnacka et al. 2004
Spirulina sp.	Co	—	21	Chojnacka et al. 2004
Spirulina sp.	Cu	—	48	Chojnacka et al. 2004
Spirulina sp.	Hg	—	60	Chojnacka et al. 2004
Spirulina sp.	Ni	—	22	
Spirulina sp.	Pb	—	95	Chojnacka et al. 2004
Spirulina sp.	Zn	—	67	Chojnacka et al. 2004
Spirulina maxima	Pb	—	92	Gong et al. 2005
Spirulina maxima (CaCl$_2$ pretreated)	Pb		84	Gong et al. 2005
Tetraselmis suecica	Cd	45	60	Perez-Rama et al. 2001

(Adapted from Mehta and Gaur 2005).

EC_{50} values of 0.058 and 1.92 mg/L, respectively for Cd (Monteiro et al. 2012). Studies on growth inhibition of *Tetraselmis suecica* (living cells) after exposure to varying concentrations of cadmium revealed an EC_{50} of 7.9 mg/L by 6-days (Pérez-Rama et al. 2001). These studies have revealed that some species of microalgae are highly tolerant to heavy metals while some are less tolerant (Table 1). Microalgal species thus show marked variation in their metal tolerance, which could be attributed to internal and/or external factors.

A number of general indications have been associated with the metal toxicity to microalgae. The list may include (i) an inhibition of photosynthetic activity and thus the growth, (ii) disruption of membrane integrity due to deterioration of the protein structure, (iii) displacement and substitution of essential ions in biomolecules by toxic ions, and (iv) blockage of functional groups in biologically significant molecules (Kaplan 2004; Arunakumara and Xuecheng 2008). To overcome these adverse effects, a number of intracellular and extracellular metal binding strategies have been adopted by microalgae. These include metal chelation, ion exchange, complexation, and physical adsorption. These mechanisms are effective because they transform the toxic metal into non-toxic forms (Munoz et al. 2006; Sud et al. 2008).

Metal detoxicification can be achieved through a number of strategies (Jjemba 2004; Worms et al. 2006). Metals can bind to the cell wall of the algae and are absorbed by the cell through passive diffusion or mediated transport. When these metals are inside the cell, detoxification is achieved by (a) binding to specific intracellular organelles (where redox reaction takes place) or transport to specific cellular components, for example, vacuoles or polyphosphate bodies and (b) metals can be flushed out into the solution through efflux pump (Fig. 1) (Jjemba 2004; Worms et al. 2006). The intracellular metal detoxification can also be achieved by synthesising class III metallothioneins or phytochelatins (Perez-Rama et al. 2001).

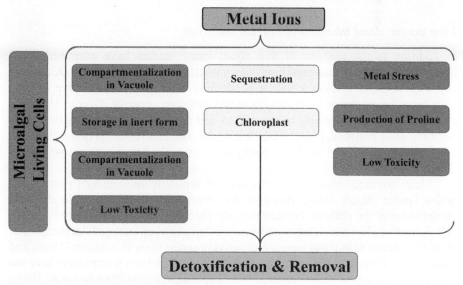

Fig. 1: Mechanism of metal detoxification and removal by microalgae.

Metal Removal through Algal Biosorption

The use of algal biomass for removing metal ions from aqueous solutions has become increasingly interesting in recent years. Despite the abundance of algal species in both aquatic and terrestrial habitats, few species have been explored in depth for toxic metal bioremediation. Heavy metal bioremediation experiments have been conducted using freshwater green algae (e.g., *Chlamydomonas reinhardtii, Cladophora* spp., *Scenedesmus* spp., *Chlorella* spp.) and blue-green algae (like *Oscillatoria* spp. and *Microcystis aeruginosa*). Microalgae are relatively easy to grow in culture and some algal species are being grown commercially in large quantities (Mehta and Guar 2005). Though many species of microalgae are known for their metal sequestering properties, their abilities to bind and accumulate metals differ greatly from species to species. Even among strains of a single species, there are lot of variations in tolerance and accumulation of metals (Table 1). Differences may also be due to environmental and experimental conditions prevalent in different studies. Cells grown under different conditions vary in composition of cell wall, and, hence, in biosorption properties (Chojnacka et al. 2005).

Some algae show a high affinity for binding and accumulating a particular metal ion, whereas others do not show such specificity and may sorb several metal ions (Mehta and Guar 2005). In general, metal ions with greater electronegativity and smaller ionic radii are preferentially sorbed by algal biomass. The available literature suggests that Pb is sorbed maximally compared to other metals in a majority of algal species (Tiem 2002; Davis et al. 2003). In general, the metal sorption capacity of algae is the least for nickel (Ni), although *M. aeruginosa* possesses a very high Ni sorption capability (Pradhan et al. 1998). Species of red and green marine algae were screened for their Cr(VI) adsorbing capacities, and a very high selectivity of Cr(VI) was found for *Pachymeniopsis* sp. (Rhodophyta) (Lee et al. 2000).

Live versus dead biomass for metal sorption

In addition to the analysis of live algae, many studies have examined metal bioremediation using dead algal biomass for metal removal and recovery (Kadukova and Vircikova 2005; Aksu 1998). It is not fully clear as to whether live or dead cells have the greater or have equivalent accumulation capacity (Ozer et al. 2000), although this clearly depends on the species and metal of interest. Using dead cells can have great potential, considering its low cost. Generally, studies reveal that dead cells show greater metal binding capacity than live cells, for reasons not yet known (Mehta and Gaur 2005).

Metal removal using living cells may be achieved through both passive and active uptake (Malik 2004). However, live biomass is sensitive to the chemical composition of the effluent. Furthermore, the potential for metal recovery from live cells is limited when metals are bound intracellularly or when extracellular metabolites form complexes with those metals that actively retain them in solution (Mehta and Guar 2005). Therefore, dead biomass may be preferable when wastewaters have too high concentrations of metals or other toxic impurities present (Sanchez et al. 1999). Conversely, removal of metals by dead biomass entails only a passive process, with metal cations being predominantly adsorbed onto the functional groups on the cell surface. Furthermore, this method of metal binding allows regeneration (and thus reuse) of the biomaterial in multiple sorption/desorption cycles as well as recovery of the metal removed from the solution (Mehta and Guar 2005; Kadukova and Vircikova 2005).

Another factor that may affect metal removal performance by dead biomass is the method used to inactivate microalgal cells. Inactivation by heat may indeed cause partial decay of structural cell components (e.g., protein denaturation), thus decreasing the number of binding sites available for interaction with metal cations relative to those remaining in living cells (Costa and Franca 1998; Terry and Stone 2002; Vannela and Verma 2006). Studies on zinc removal by *Scenedesmus obliquus* and *Desmodesmus pleiomorphus* cells after heat inactivation showed considerably lower amounts of Zn removal than living ones. Uranium sorption by *S. obliquus* was shown to decrease from 13 to 0.6 mg/g dry weight with vacuum drying (Zhang et al. 1997). The corresponding decrease in metal uptake after vacuum drying of the biomass may be due to unavailability of the effective sites for binding metal ions on the cell wall as water evaporates from the cells under vacuum condition (Zhang et al. 1997). Studies reveal that freeze-dried biomass has a higher metal sorption capacity than the live biomass (Winter et al. 1994; Bengtsson et al. 1995). In an experiment to compare sorbing potential of freeze and oven dried biomass, no change in metal sorbing was observed (Neide et al. 2000). Metal sorption capacity of dead cells depends on their pre-treatment and any subsequent change in the structure of their cell wall (Somers 1963; Duddridge and Wainright 1980).

The method of inactivation of cells also influences their metal sorption capacity. Whereas heat killing of biomass may enhance sorption of metals, chemical killing may at times decrease the metal sorption capacity of the biomass (Tobin et al. 1990). Pre-treatment of *Chlorella vulgaris* with dilute HCl considerably increased Ni and copper sorption from aqueous solution (Mehta and Gaur 2001a; Mehta et al. 2002). While the use of inactivated biomass has been preferred, some disadvantages also deserve a mention. Dead cells cannot be used where biological alteration in valency of a metal is sought. Moreover, degradation of organometallic species is not possible with dead biomass. Also, there is no room for biosorption improvement through mutation.

Use of immobilized algae in sorption of metals

Although the use of live algae offers advantages, in practice, where typically the alga biomass is either purchased (as a dried powder) or cultivated in a separate operation prior to use, the method of choice has been to immobilize the biomass by some type of chemical or physical process (Wilde and Benemann 1993). Using algal biomass directly in a standard metal uptake process may be difficult considering its small size, low strength, and density. This may be a threat to the separation of biomass or effluent (Tsezos 1986). The merit of immobilization processes includes the use of high cell densities and column operations (i.e., the adsorption of metal in the packed bed column). The demerit is the diffusion limitations formed (Radovich 1991). This may cause many of the sites on the surface of the biomass partly available to metal ions (Wilde and Benemann 1993).

In most algal immobilization processes, algal cells are firmly fixed in a gelatinous matrix such as calcium alginate and sodium (Fujimura et al. 1989; Jang et al. 1991; Hertzberg and Jensen 1989). These gels along with polyacrylamide and silica gels (Gadd 1990b; Greene and Beldell 1990) are proven to be potentially highly efficient in small-scale systems where diffusions are limited. These gels could be solidified, formed into beads, and used in pack-beds, fluidized-bed, and air-lift bioreactors (Gadd 1990b) that are like the column processes used with ion exchange resins and activated carbon systems.

Immobilized *Chlorella* showed significantly lower efficiency for removing Ni and Cu in all the concentrations used than free cells of the same species (Wong and Park 1992). On the contrary, some studies have shown that free cells show higher efficiency in metal sorption than immobilized cells. Studies on free cells of *Spirulina platensis* revealed high metal accumulation (99 mg/g) of Cd than alginate and silica immobilized cells (71 and 37 mg/g) (Rangsayatorn et al. 2004). Polyacrylamide immobilized cells have the problem of being fragile (Greene and Beldell 1990) but good results have been reported for AlgaSORB, a proprietary material that contains algal cells immobilized in a silica matrix (Greene and Beldell 1990; Darnall and Feller 1991).

Factors Affecting Bio-sorption

Metal speciation and pH

The metal toxicity to the living organisms depends on the concentration of free ions (Sunda and Guillard 1976; Anderson and Morel 1982; Sunda and Fergusson 1983). Heavy metals in free ionic form bind and accumulate in microalgae and hence are more toxic than when complexed with organic or inorganic ligands (Monteiro et al. 2012). The speciation of metals in aquatic environments and the bioavailability of free metal ions in solution are largely influenced by pH (Mehta and Guar 2005).

Considering the importance of pH, few studies have determined pH optima, in order to enhance efficiency of heavy metal removal by algal cells (Rangsayatorn et al. 2002; Gupta and Rastogi 2008). In one study, very little sorption of Cu by *Durvillaea potatorum* at pH below 2 was revealed, but it increased with a rise in pH (Yu and Kaewsarn 1999). Many reports revealed that the sorption of heavy metals by microalgal cells is increased as pH increases (Table 2). For instance, there is an increase in Zn^{2+} and Cd^{2+} removal by *Oedogonium anguistissima* and *Oedogonium* sp., respectively, when pH was increased up to 5 (Ahuja et al. 1999; Gupta and Rastogi 2008). Similarly, there was an increase in Cd^{2+} and Zn^{2+} removal by *Scenedesmus obliquus* when pH increased to above 6 (Monteiro et al. 2009b; Monteiro et al. 2011a). However, it is unclear if this increased uptake is due to altered metal speciation or pH directly affecting transport protein activity.

Table 2: Optimal pH for sorption of heavy metal by microalgae.

	Microalgal species	pH	References
Ag	*Oedogonium* sp.	7.0	Crist et al. 1988
Cd	*Lyngbya taylorii*	3–7	Klimmek et al. 2001
	Stichococcus bacillaris	7.0	Skowronski, 1986
	Chlamydomonas reinharditii	2–7	Tuzan et al. 2005
Cr	*Chlorella vulgaris*	4.0	Aksu and Donmez 2006
	Chlorella vulgaris	2.0	Aksu and Acikel 1999
Cu	*Chlorella vulgaris*	4.0	Aksu and Acikel 1999
	Chlorella vulgaris	3.5	Mehta and Gaur 2001a
	Microcystis aeruginosa	7.0	Pradhan et al. 1998
	Spirulina platensis	6.7	Zhou et al. 1998
Pb	*Chlorella vulgaris*	5.0	Aksu and Kutsal 1991
	Spirulina maxima	5.5	Gong et al. 2005
Ni	*Chlorella vulgaris*	5.5	Mehta and Gaur 2001a
	Microcystis aeruginosa	9.2	Pradhan et al. 1998
Sr	*Spirogyra* sp.	7.0	Crist et al. 1988
	Vaucheria sp.	7.0	Crist et al. 1988

(Adapted from Mehta and Gaur 2005; Moteiro et al. 2012).

Metal concentration

Concentration of metals in the solution plays a major role in the rate at which metals are removed by microalgae. The degree of metal sorption increases with increase in metal concentration, but eventually reaches saturation (Omar 2002; Da Costa and Leite 1991; Aloysius et al. 1999) as predicted by classical adsorption isotherms. An increase in the amount of cadmium (Cd) uptake by *Koliella antarctica* with increasing metal concentration in the growth medium has been reported (La Rocca et al. 2009). The adsorption of Cu^{2+}, Zn^{2+}, and Ni^{2+} by immobilized *Scenedesmus quadricauda* increased as the initial concentration of metal ions increased in the medium, with maximum adsorption capacities of 75.6, 55.2, and 30.4 mg/g, respectively (Bayramoğlu and Arıca 2009).

Temperature

Available reports regarding the effects of temperature on micro-algal sorption has been contradictory. Biomass of *Chlorella vulgaris* increased with temperature, from a maximum of 48.1 at 15°C to 60.2 mg/g at 45°C, and spanning a range of 50–250 mg/l of initial metal concentrations (Aksu 2002). These results suggest that algal metal sorption is an endothermic process. In contrast, some studies have suggested an exothermic nature of metal sorption by algae. For example, studies on cadmium

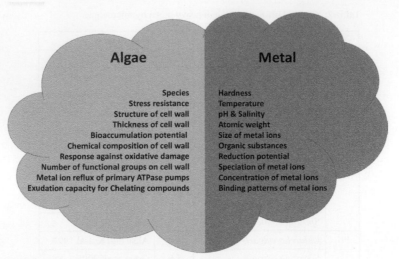

Fig. 2: Factors influencing sequestration of metals by microalgae.

sorption by *Sargassum* sp. showed a slight decrease in metal binding with an increase in ambient temperature (Cruz et al. 2004). Some other studies revealed a similar result (Aksu 2001; Benguell and Benaissa 2002). However, other reports have suggested that temperature does not influence metal sorption (Norris and Kelly 1979; de Rome and Gadd 1987). Studies on binding of Cu and nickel by *C. vulgaris* show that there is no pronounced effect of temperature on metal binding (Mehta and Gaur 2001a; Mehta et al. 2002).

Biomass concentration of microalgae

The amount of metal ion recovered from a solution is directly proportional to the biomass size or concentration. This could be due to the availability of a high number of metal binding sites (Fraile et al. 2005). On the other hand, decrease in the metal removal could be observed at an increased biomass. This could be as a result of partial clustering of biomass that reduces the surface area available for sorption (Munoz et al. 2006). In one study, increasing biomass of dried and grounded cells of *Chlorella* sp. resulted in decreased Cd binding per unit cell mass (Roy et al. 1993). While increasing the biomass concentration from 0.04 to 0.2 g/l, a decrease in metal binding per unit of cell mass was observed (Ahuja et al. 1999). A marked reduction was also reported in Pb^{2+} uptake by *Spirulina maxima* from 121 to 21 mg/g, as the biomass concentration was raised from 0.1 to 20 g/l (Gong et al. 2005).

Biomass Regeneration and Reuse

The basic consideration with metal bioremoval schemes is the generation and reuse of the biomass. Regeneration of algal biosorbent for reuse is essential for the success of microalgae metal bioremediation processes in an industrial perspective as this could make the bioremediation process more economical. Also desorbed metals removed from the liquid phase in a more concentrated and convenient form can be reused afterwards for other industrial processes (Chojnacka et al. 2005). When sorbed on microalgal biomass, heavy metals can be desorbed by a suitable eluent or desorbing solution, which in turn allows the ready reuse of biomass in multiple sorptions/desorption cycles. Nevertheless, the selection of a desorbing agent depends on desorption efficiency and the persistence of the biosorption capacity of adsorbing materials. Also, the desorbing agent should not cause irreversible physical or chemical changes, or damage to the biomass (Monteiro et al. 2011).

Heavy metals can be removed from microalgal biomass by pH change. Reducing the pH of the loaded biomass suspension causes displacement of heavy metal cations back to the solution by protons concomitantly gained by the binding sites. Several organic and inorganic acids, bases, salts, and metal chelators have been found efficient for their metal desorbing ability. Studies revealed that the elution efficiency was maximum in the case of inorganic acids, followed by inorganic salts, chelating agents, and organic acids (in this order), with recoveries above 90%, except for organic acids that could only reach ~ 80% (Vannela and Verma 2006). Desorption of Cr^{3+}, Cd^{2+}, and Cu^{2+} from *Spirulina* sp. using 0.1 M EDTA, 0.1 M HNO_3 or deionized water was investigated and HNO_3 was found to be the most convenient desorbing agent, with efficiencies ranging between 90–98%, and without disrupting the biosorption capacity of the biomaterial (Chojnacka et al. 2005). The reusability of *Spirulina platensis* biomass immobilized on alginate and silica gel, after up to five cycles of adsorption and desorption of Cd^{2+}, using 0.1 M HCl as the desorbing agent was tested. A significant loss of 26% in adsorption capacity resulted after the first cycle. However, the Cd^{2+} adsorption capacity remained essentially constant from the second cycle onward (Rangsayatorn et al. 2004). Although HCl has a high capacity to desorb heavy metals, studies have shown that it decreases the metal sorption ability of biomass when used continuously, likely due to damage to metal binding sites (Chu et al. 1997). The regeneration of biomass over 4 sorption/desorption cycles was also investigated and it was found that there was a decrease in mercury (Hg) removal by *Spirulina platensis* after the second cycle when HCl was used as desorbing agent (Cain et al. 2008). This decrease could be attributed to partial biomass loss, coupled with acid-induced cell damage resulting during the regeneration cycles (Monteiro et al. 2012).

Biomass pre-treatment

Pre-treatment of microalgae has been considered to influence the rate of metal removal from the solution (Mehta and Guar 2005; Romera et al. 2006). A lot of biological tools (like bacteria, fungi) are known to bind to metals to some extents, but the biological tools with high metal binding capability would be a better option for pilot-scale metal removal processes (Mehta and Guar 2005). The most common pre-treatment agents are (i) NaOH, which replaces sodium ions attached to functional groups, thus increasing electrostatic attraction to metal cations and eventually facilitating ion exchange; (ii) HCl, which leads to displacement of light metals by protons released on the biomass surface and can also dissolve polysaccharide compounds of the cell wall to create additional binding sites; (iii) $CaCl_2$, which brings about cross linking of alginate-based polymeric chains and has proven the most economical method for activation (Mehta and Guar 2005; Romera et al. 2006). Gong et al. 2005 reported 84–92% increase in the Pb sorption capacity of *Spirulina maxima* following biomass pre-treatment with $CaCl_2$. Several other chemicals have also been used to enhance the removal capacity of Ni and Cu (Mehta and Gaur 2001). The primary aim of chemical pre-treatment of biomass is obvious; to enhance metal uptake by microalgae but positive results may not always be achieved (Table 1). Also, chemical pre-treatment of micro-algal biomass could be costly, and at times could be high enough not to reconcile with its use considering the fact that metal uptake could be slow (Romera et al. 2006). Finally, it is worthy of note that micro-algal pre-treatment sometimes leads to better physical stability of the microalgal biomass granules, especially in packed bed columns (Matheickal et al. 1999).

Challenges and Prospects

The use of microalgae for bioremediation of heavy metals has recently attracted tremendous attention considering the advantages displayed by these organisms in metal removal compared to other conventional approaches. These advantages include; low cost, its environmental friendly approach, and effective bioremediation due to efficient metal uptake even at low metal contamination levels. The fact that heavy metals can be recovered from algae after adsorption and reuse further provides an economic advantage in addition to reducing the toxicity of the recovered algae. Despite of all these advantages, the technology of using microalgae for metal bioremediation is faced with some challenges. Algal biomass in its natural form cannot be used in the removal of metal ions. Using a continuous system as the algal suspension may block the column; thus, immobilized microalgae should be used for effective metal uptake from wastewaters. Immobilization of algal biomass with chemicals such as alginate may be expensive and hence, not feasible in a large scale setting. Considering this fact, naturally immobilized algae (algal biofilms and mats) could be a great potential tool for the bioremediation of metals (Mehta and Guar 2005; Richards and Mullins 2013). Furthermore, studies on the mechanism of adsorption, such as the relationship between the binding capacity and alginate components, and the biosorption properties of each alginate component could enhance metal bioremediation. This will enhance the effective use of algae and also help in the speedy development of biosorbents with high metal adsorption capacity (Dwivedi 2012).

The vast majority of algal species has not been adequately explored for their metal uptake ability. In these regard, the screening and assessment of algal species with high metal binding capacity could be a great boost to the industry as more research could help perfect the metal binding efficiency of these species. Another approach is to genetically engineer microalgae strains with genes encoding metal transporters or metal binding proteins. The over-expression of such genes could enhance the efficiency and specificity of these strains in metal bioremediation considering the fact that metal transporters have been implicated in the uptake of metals by organisms, in providing tolerance through internal sequestration and chelation (Hanikenne et al. 2005). Manipulation of strains with increased synthesis of phytochelatin and metallothioneins could also lead to an improved capacity in metal uptake (Gardea-Torresdey et al. 1998). Genetic manipulation of a micro-algal cell wall for enhanced synthesis of cell wall polysaccharides to increase metal binding sites could be a great advantage in boasting the metal binding capacity of microalgae (Mehta and Guar 2005).

There are many uncertainties in this field due to the fact that it's relatively new and not yet fully commercialised. With some exceptions, most research for algal-based metal bioremediation systems are laboratory based and have yet to be implemented on a large scale capacity. Therefore, economic assessment of such technology is uncertain and large-scale validation is essential. *Chlorella* and *Spirulina* can currently cost about $20,000 and $10,000 per ton, respectively (Wilde and Benemann 1993). They make microalgae biomass the most expensive of all biomass produced for metal bioremediation (Kuyucak 1990). This is considered a great challenge that could pose lots of limitations in this field. Large scale production of microalgal biomass could reduce the cost depending on current market conditions. Furthermore, cultivation procedures could be made more energy and cost efficient. An alternative could be the use of waste water grown algae, such as algal biomass recovered from industrial effluents or sewage treatment plants (Wilde and Benemann 1993). The production of microalgae biomass specifically for metal bioremediation will therefore likely be expensive, especially if strains produced are to be improved for better performance. Considering the variation in metal bioremediation efficiency of various microalgae, the possible exchange as a result of compromise between cost and performance could be in favour of selected biomass instead of generic biomass. For microalgal technology to be very relevant in large scale bioremediation of metals, more research directed towards industrial effluents and strains (wild) from extremophilic environments should be considered (Monteiro et al. 2012). Metal bioremediation using algal technology may blossom and probably become more efficient than other conventional methods, but more needs to be done to fully explore the metal uptake potentiality of microalgae.

Conclusion

The unique characteristics of microalgae to sequester metals from culture medium make them reliable tools for bioremediation of metals from water and wastewater. Metal biosorption by microalgae has been considered as a very economic and efficient process due to its potential to bind toxic metals even at low concentrations. Another great advantage over the current conventional physicochemical techniques of metal remediation method is that phycoremediation do not generate hazardous sludge. The better understanding of the bioremediation process can subsequently reduce the economics of metal bioremediation from domestic as well as industrial wastewater. The removal of metal can subsequently improve by genetically engineered microalgae. Further research focusing on natural strains from extremophilic environments could also give a boost to this research area. The increase in the metal uptake efficiency of these microalgae could give rise to efficient and commercially viable micro-algal species for metal removal. Heavy metal bioremediation using microalgae is therefore a promising process with significant potential.

References

Ahuja, P., R. Gupta and R.K. Saxena. 1999. Zn^{2+} biosorption by *Oscillatoria anguistissima*. Process Biochem. 34: 77–85.

Aksu, Z. and T. Kutsal. 1991. A bioseparation process for removing lead (II) ions from wastewater by using *C. vulgaris*. J. Chem. Technol. Biotechnol. 52: 109–118.

Aksu, Z. 1998. Biosorption of heavy metals by microalgae in batch and continuous systems. pp. 37–52. *In*: Wong, Y.-S. and N.F.Y. Tam [eds.]. Wastewater Treatment with Algae. Springer-Verlag and Landes Bioscience.

Aksu, Z. and U. Acikel. 1999. A single-staged bioseparation process for simultaneous removal of copper (II) and chromium (VI) by using *C. vulgaris*. Process. Biochem. 34: 589–599.

Aksu, Z. 2001. Biosorption of reactive dyes by dried activated sludge. Equilibrium and kinetic modeling. Biochem. Eng. J. 7: 79–84.

Aksu, Z. 2002. Determination of the equilibrium, kinetic and thermodynamic parameters of the batch biosorption of nickel (II) ions onto *Chlorella vulgaris*. Process Biochem. 38: 89–99.

Aksu, Z. and G. Dönmez. 2006. Binary biosorption of cadmium (II) and nickel (II) onto dried *Chlorella vulgaris*: co-ion effect on mono-component isotherm parameters. Process Biochem. 41: 860–868.

Aloysius, R., M.I.A. Karim and A.B. Arif. 1999. The mechanism of cadmium removal from aqueous solution by non-metabolising free and immobilized live biomass of *Rhizopus oligosporus*. World J. Microbiol. Biotechnol. 15: 571–578.

Anderson, M.A. and F.M.M. Morel. 1982. The influence of aqueous iron chemistry on the uptake of iron by the coastal diatom *Thalassiosira weissflogii*. Limnol. Oceanogr. 27: 789–813.

Arunakumara, K.K.I.U. and Z. Xuecheng. 2008. Heavy metal bioaccumulation and toxicity with special reference to microalgae. J. Ocean Univer. China 7: 60–64.

Balasubramani, R., S.K. Gupta, W.M. Cho, J.K. Kim, S.R. Lee, K.H. Jeong et al. 2016. Microalgae potential and multiple roles—current progress and future prospects—an overview. Sustainability 2015(8): 1–16.

Bayramoğlu, G. and M.Y. Arıca. 2009. Construction a hybrid biosorbent using Scenedesmus quadricauda and Ca-alginate for biosorption of Cu (II), Zn (II) and Ni (II): kinetics and equilibrium studies. Bioresour. Technol. 100(1): 186–193.

Bengtsson, L., B. Johansson, T.J. Hackett, L. Mchale and A.P. Mchale. 1995. Studies on the biosorption of uranium by *Talaromyces emersonii* CBS 814. 70 biomass. Appl. Microbiol. Biotechnol. 42: 807–811.

Benguell, B. and H. Benaissa. 2002. Cadmium removal from aqueous solution by chitin: Kinetic and equilibrium studies. Water Res. 36: 2463–2474.

Bhattacharya, D.L. 1998. Algal phylogeny and the origin of land plants. Plant Physiol. 116: 9–5.

Buchet, J.P., H. Roels, A. Bernard and R. Lauwerys. 1980. Assessment of renal-function of workers exposed to inorganic lead, cadmium, or mercury-vapor. J. Occup. Environ. Med. 22: 741–750.

Cain, A., R. Vannela and L.K. Woo. 2008. Cyanobacteria as a biosorbent for mercuric ion. Bioresor. Technol. 14: 6578–6586.

Carr, H.P., F.A. Carino, M.S. Yang and M.H. Wong. 1998. Characterization of the cadmium-binding capacity of *Chlorella vulgaris*. Bull. Environ. Contam. Toxicol. 60: 433–440.

Chabukdhara, M., A. Munjal, A.K. Nema, S.K. Gupta and R. Kaushal. 2016a. Heavy metal contamination in vegetables grown around peri-urban and urban-industrial clusters in Ghaziabad, India. Human and Ecological Risk Assessment 22(3): 736–752.

Chabukdhara, M., S.K. Gupta and A.K. Nema. 2016b. Assessment of seasonal variation of surface water quality using environmetric and indexing approach. IIOAB Journal, special issue Water, air and soil pollution: Monitoring and remediation. IIOAB Journal 7(11): 16–24.

Chabukdhara, M., S.K. Gupta and M. Gogoi. 2017. Phycoremediation of heavy metals and biofuel production from algal biomass. pp. 163–188. *In*: Gupta, S.K., A. Malik and F. Bux [eds.]. Algal Biofuels: Recent Advances and Future Prospects. Springer International Publishing AG Cham.

Chojnacka, K., A. Chojnacki and H. Gorecka. 2004. Trace metal removal by *Spirulina* sp. from copper smelter and refinery effluent. Hydrometallurgy 73: 147–153.

Chojnacka, K., A. Chojnacki and H. Górecka. 2005. Biosorption of Cr^{3+}, Cd^{2+}, and Cu^{2+} ions by blue-green alga *Spirulina* sp.: kinetics, equilibrium and the mechanism of the process. Chemosphere 59: 75–84.

Chong, A.M.Y., Y.S. Wong and N.Y.F. Tam. 2000. Performance of different microbial species in removing nickel and zinc from industrial wastewater. Chemosphere 41: 251–257.

Chu, A.U., K.H. Hashim, S.M. Phang and V.B. Samuel. 1997. Biosorption of cadmium by algal biomass: adsorption and desorption characteristics. Water Sci. Technol. 35: 115–122.

Costa, A.C.A. and F.P. França. 1998. The behaviour of the microalgae *Tetraselmis chuii* in cadmium contaminated solutions. Aquacult. Int. 6: 57–66.

Costa, A.C.A. and F.P. Franca. 2003. Cadmium interaction with microalgal cells, cyanobacterial cells, and seaweeds; toxicology and biotechnological potential for wastewater treatment. Mar. Biotechnol. 5: 149–156.

Crist, R.H., K. Oberholser, D. Schwartz, J. Marzoff, D. Ryder and D.R. Crist. 1988. Interaction of metals and protons with algae. Environ. Sci. Technol. 22: 755–760.

Cruz, C.C.V., A.C.A. Da Costa, C.A. Henriques and A.S. Luna. 2004. Kinetic modelling and equilibrium studies during cadmium biosorption by dead *Sargassum* sp. Biomass. Bioresour. Technol. 91: 249–257.

Da Costa, A.C.A. and S.G.F. Leite. 1991. Metal biosorption by sodium alginate immobilized *Chlorella homosphaera* cells. Biotechnol. Lett. 13: 359–362.

Darnall, D.W. and H.D. Feiler. 1991. Recovery of heavy metals from contaminated ground waters. Site Assessment/Remediation, pp. 114–116.

Das, N., R. Vimala and P. Karthika. 2008. Biosorption of heavy metals—An overview. Indian J. Biotechnol. 7: 159–169.

Davis, T.A., B. Volesky and A. Mucci. 2003. A review of biochemistry of heavy metal biosorption by brown algae. Water Res. 37: 4311–4330.

De Rome, L. and G.M. Gadd. 1987. Copper adsorption by *Rhizopus arrhizus*, *Cladosporium resinae* and *Penicillium italicum*. Appl. Microbiol. Biotechnol. 26: 84–90.

Desi, I., L. Nagymajtenyi and H. Schulz. 1998. Behavioural and neurotoxicological changes caused by cadmium treatment of rats during development. J. Appl. Toxicol. 18: 63–70.

Di Toppi, L.S. and R. Gabbrielli. 1999. Response to cadmium in higher plants. Environ. Exp. Bot. 41: 105–130.

Duddridge, J.E. and M. Wainwright. 1980. Heavy metal accumulation by aquatic fungi and reduction in viability of *Gammarus pulex* fed Cd^{2+} contaminated mycelium. Water Res. 14: 1605–1611.

Dwivedi, S. 2012. Bioremediation of heavy metal by algae: current and future perspective. J. Adv. Lab. Res. Biol. 3: 229–233.

El-Sheekh, M.M., W.A. El-Shouny, M.F.H. Osman and W.E. El-Gammal. 2005. Growth and heavy metals removal affinity of *Nostoc muscorum* and *Anabaena subcylindrica* in sewage and industrial wastewater effluent. Environ. Toxicol. Pharmacol. 19: 357–365.

Ferguson, C.R., M.R. Peterson and T.H. Jeffers. 1989. Removal of contaminants from waste waters using biomass immobilized in polsulfone beads. pp. 193–200. *In*: Scheiver, B.J. [ed.]. Biotechnology in Minerals and Metal Proc.

Fraile, A., S. Penche, F. González, M.L. Blázquez, J.A. Muñoz and A. Ballester. 2005. Biosorption of copper, zinc, cadmium and nickel by *Chlorella vulgaris*. Chem. Ecol. 21: 61–75.

Fu, F. and Q. Wang. 2011. Removal of heavy metal ions from wastewaters: A review. J. Environ. Man. 92: 407–418.

Fujimura, T., T. Kawai, M. Shiga, T. Kajiwara and A. Hatanak. 1989. Presentation of immobilized living cells from a marine green alga Ulva pertusa Nippon Suisan Gakkaishi. 55: 2211.

Gadd, G.M. 1988. Accumulation of metals by microorganisms and algae. pp. 401–434. *In*: Rehm, H.J. [ed.]. Biotechnology. VCH, Weinheim.

Gadd, G.M. 1990b. Biotechnology. Biosorp. Chem. Ind. 421–426.

Gardea-Torresday, J.L., J.L. Arenas, N.M.C. Francisco, K.J. Tiemann and R. Webb. 1998. Ability of immobilized cyanobacteria to remove metal ions from solution and demonstration of the presence of metallothionein genes in various strains. J. Hazard. Sub. Res. 1: 1–18.

Gong, R., Y. Ding, H. Liu, Q. Chen and Z. Liu. 2005. Lead biosorption and desorption by intact and pre-treated *Spirulina maxima* biomass. Chemosphere 58: 125–130.

Greene, B. and G.W. Beldell. 1990. Algal Gelsor immobilized algae for metal recovery. pp. 109–136. *In*: Akatsuka, I. [ed.]. Introduction to Applied Phycology. Acadamic Publishing. The Hague.

Gupta, S.K., M. Chabukdhara, P.K. Pandey, J. Singh and F. Bux. 2014. Evaluation of potential ecological risk of metal contamination in the Gomti River: a biomonitoring approach. Ecotoxicology and Environmental Safety 110: 49–55.

Gupta, S.K., M. Chabukdhara, J. Singh and F. Bux. 2015. Evaluation and potential health hazard of selected metals in water, sediments, and fish from the Gomti River. Human and Ecological Risk Assessment 21(1): 227–240.

Gupta, S.K., F.A. Ansari, A. Shriwastav, N.K. Sahoo, I. Rawat and F. Bux. 2016. Dual role of chlorella sorokiniana and scenedesmus obliquus for comprehensive wastewater treatment and biomass production for bio-fuels. J. Cleaner Production 115: 255–264.

Gupta, S.K., F.A. Ansari, N. Mahmoud, L. Rawat, N.M. Kumar and F. Bux. 2017a. Cultivation of Chlorella sorokiniana and Scenedesmus obliquus in wastewater: Fuzzy intelligence for evaluation of growth parameters and metabolites extraction. Journal of Cleaner Production 147: 419–430.

Gupta, S.K. A. Sriwastav, F.A. Ansari, M. Nasr and A.K. Nema. 2017b. Phycoremediation: An ecofriendly algal technology for bioremediation and bioenergy production. pp. 431–456. *In*: Bauddh, K., B. Singh and J. Korstad [eds.]. Phytoremediation Potential of Bioenergy Plants. Springer International Publishing AG Cham. Chapter 3.

Gupta, S.K., F.A. Ansari, K. Bauddh, B. Singh, A.K. Nema and K.K. Pant. 2017c. Harvesting of microalgae for biofuels: comprehensive performance evaluation of natural, inorganic, and synthetic flocculants. pp. 131–156. *In*: Ritu Kumar and Sanjeev Kumar [eds.]. Green Technologies and Environmental Sustainability. Springer Nature, Switzerland. Chapter 6.

Gupta, V.K. and Rastogi. 2008. Equilibrium and kinetic modelling of cadmium(II) biosorption by nonliving algal biomass Oedogonium sp. from aqueous phase. J. Haz. Mater. 153: 759–766.

Hanikenne, M., U. Kramer, V. Demoulin and D. Baurain. 2005. A comparative inventory of metal transporters in the green algae *Chlamydomonas reinhardtii* and the red algae *Cyanidioschizon merolae*. Plant Physiol. 137: 428–446.

Hellstrom, L., C.G. Elinder, B. Dahlberg, M. Lundberg, L. Jarup, B. Persson et al. 2001. Cadmium exposure and end-stage renal disease. Am. J. Kidney Dis. 38: 1001–1008.

Hertzberg, S. and A. Jensen. 1989. Studies of alginate—immobilized marine algae. Botanica Marina 32: 266–273.

Jalali-Rad, R., H. Ghalocerian, Y. Asef, S.T. Dalir, M.H. Sahafipour and B.M. Gharanjik. 2004. Biosorption of cesium by native and chemically modified biomass of marine algae: Introduce the new biosorbent for biotechnology application. J. Haz. Mat. 116: 125–134.

Jang, L.K., S.L. Lopez, S.L. Eastman and P. Pryfogle. 1991. Recovery of copper and cobalt by biopolymer gels. Biotechnol. Bioeng. 37: 266–273.

Jeffers, T.H., C.R. Ferguson and D.C. Seidel. 1989. Biosorption of metal contaminants using immobilized biomass. pp. 317–327. *In*: McReady, R. and P. Wwichlacz [eds.]. Pro. Intl. Symp. Biohydeometallurgy Aug. 13–18 Jackson Hole Wy. CANMET special publication.

Jjemba, P.K. 2004. Interaction of metals and metalloids with microorganisms in the environment. pp. 257–270. *In*: Jjemba, P.K. [ed.]. Environmental Microbiology—Principles and Applications. New Hampshire: Science Publishers.

Kadukova, J. and E. Vircı́kova. 2005. Comparison of differences between copper bioaccumulation and biosorption. Environ. Int. 31: 227–232.

Kahle, H. 1993. Response of roots of trees to heavy metals. Environ. Exp. Bot. 33: 99–119.

Kaplan, D. 2004. Water pollution and bioremediation by microalgae—absorption and adsorption of heavy metals by microalgae. pp. 439–44. *In*: Richmond, A. [ed.]. Handbook of Microalgal Culture-Biotechnology and Applied Phycology. Iowa: Blackwell Publishing.

Karna, R.R., L. Uma, G. Subramanian and P.M. Mohan. 1999. Biosorption of toxic metal ions by alkali-extracted biomass of a marine cyanobacterium, *Phormidium valderianum* BDU 30501. World J. Microbiol. Biotechnol. 15: 729–732.

Klimmek, S., H.J. Stan, A. Wilke, G. Bunke and R. Buchholz. 2001. Comparative analysis of the biosorption of cadmium, lead, nickel, and zinc by algae. Environ. Sci. Technol. 35: 4283–4288.

Kuyucak, N. 1990. Feasibility of biosorbents application. In: Volesky, B. [ed.]. Biosorption of Heavy Metals. CRC Press Inc., Boca Raton, Florida 4.3: 372–377.

La Rocca, N., C. Andreoli, G. Giacometti, N. Rascio and I. Moro. 2009. Responses of the Antarctic microalga Koliella Antarctica (Trebouxiophyceae, Chlorophyta) to cadmium concentration. Photosynthetica 47: 471–479.

Lee, D.C., C.J. Park, J.E. Yang, Y.H. Jeong and H.I. Rhee. 2000. Screening of hexavalent chromium biosorbent from marine algae. Appl. Microbiol. Biotechnol. 54: 997–600.

Malik, A. 2004. Metal bioremediation though growing cells. Environ. Int. 30: 261–278.

Matheickal, J.T., Q. Yu and G.M. Woodburn. 1999. Biosorption of cadmium(II) from aqueous solutions by pre-treated biomass of marine alga Durvillaea potatorum. Water Res. 33: 335–342.

Mehta, S.K., B.N. Tripathi and J.P. Gaur. 2000. Influence of pH, temperature, culture age and cations on adsorption and uptake of Ni by *Chlorella vulgaris*. Eur. J. Protistol. 36: 443–450.

Mehta, S.K. and J.P. Gaur. 2001. Characterization and optimization of Ni and Cu sorption from aqueous solution by *Chlorella vulgaris*. Ecol. Eng. 18: 1–13.

Mehta, S.K., A. Singh and J.P. Gaur. 2002. Kinetics of adsorption and uptake of Cu^{2+} by *Chlorella vulgaris*: influence of pH, temperature, culture age, and cations. J. Environ. Sci. Health, Part A 37: 399–414.

Mehta, S.K. and J.P. Gaur. 2005. Use of algae for removing heavy metal ions from wastewater: progress and prospects. Crit. Rev. Biotechnol. 25: 113–152.

Monteiro, C.M., A.P.G.C. Marques, P.M.L. Castro and F.X. Malcata. 2009b. Characterization of *Desmodesmus pleiomorphus* isolated from a heavy metal-contaminated site: biosorption of zinc. Biodegradation 20: 629–641.

Monteiro, C.M., S.C. Fonseca, P.M.L. Castro and F.X. Malcata. 2011. Toxicity of cadmium and zinc on two microalgae, Scenedesmus obliquus and Desmodesmus pleiomorphus, from Northern Portugal. J. Appl. Phycol. 23: 97–103.

Monteiro, C.M., P.M.L. Castro and F.X. Malcata. 2011a. Biosorption of zinc ions from aqueous solution by the microalga *Scenedesmus obliquus*. Environ. Chem. Lett. 9: 169–176.

Monteiro, C.M., P.M.L. Castro and F.X. Malcata. 2012. Metal uptake by microalgae; underlying mechanisms and practical applications. Biotechnol. Prog. 28: 2.

Muñoz, R., M.T. Alvarez, A. Muñoz, E. Terrazas, B. Guieysse and B. Mattiasson. 2006. Sequential removal of heavy metal ions and organic pollutants using an algal-bacterial consortium. Chemosphere 63: 903–911.

Neide, E., E.N. Carrilho and T.R. Gilbert. 2000. Assessing metal sorption on the marine alga *Pilayella littoralis*. J. Environ. Monit. 2: 410–415.

Nogawa, K. and T. Kido. 1993. Biological monitoring of cadmium exposure in itai-itai disease epidemiology. Int. Arch. Occup. Environ. Health 65: S43–S46.

Norris, P.R. and D.P. Kelly. 1979. Accumulation of cadmium and cobalt by Saccharomyces cerevisiae. J. Gen. Microbiol. 99: 317–324.

Nyholm, N. 1990. Expression of results from growth inhibition toxicity tests with algae. Arch. Environ. Contam. Toxicol. 19. 518–522.

O Zer, D., A. O Zer and G. Dursun. 2000. Investigation of zinc (II) adsorption on *Cladophora* crispata in a two-staged reactor. J. Chem. Technol. Biotechnol. 75: 410–416.

Omar, H.H. 2002. Bioremoval of zinc ions by Scenedesmus obliquus and Scenedesmus quadricauda and its effect on growth and metabolism. Int. Biodeterior. Biodegrad. 50: 95–100.

Perez-Rama, M., C.H. Lopez, J.A. Alonso and E.T. Vaamonde. 2001. Class III metalothioneins in response to cadmium toxicity in the marine microalga Tetraselmis suecica (Kylin) Butch. Environ. Toxicol. Chem. 20: 2061–2066.

Perez-Rama, M., J.A. Alonso, C.H. Lopez and E.T. Vaamonde. 2002. Cadmium removal by living cells of the marine microalga Tetraselmis suecica. Bioresour. Technol. 84: 265–270.

Pradhan, S., S. Singh, L.C. Rai and D.L. Parker. 1998. Evaluation of metal biosorption efficiency of laboratory-grown *Microcystis* under various environmental conditions. J. Microb. Biotechnol. 8: 53–60.

Radovich, J.M. 1991. Mass transfer limitations in immobilized cells. Biotech. Adv. 3: 1–2.

Rangsayatorn, N., E.S. Upatham, M. Kruatrachue, P. Pokethitiyook and G.R. Lanza. 2002. Phytoremediation potential of Spirulina (Arthrospira) platensis: biosorption and toxicity studies of cadmium. Environ. Pollut. 119: 45–53.

Rangsayatorn, N., P. Pokethitiyook, E.S. Upatham and G.R. Lanza. 2004. Cadmium biosorption by cells of *Spirulina platensis* TISTR 8217 immobilized in alginate and silica gel. Environ. Int. 30: 57–63.

Richards, R.G. and B.J. Mullins. 2013. Using microalgae for combined lipid production and heavy metal removal from leachate. Ecological Modelling 249: 59–67.

Romera, E., F. González, A. Ballester, M.L. Blázquez and J.A. Muñoz. 2006. Biosorption with algae: a statistical review. Crit. Rev. Biotechnol. 26: 223–235.

Roy, D., P.N. Greenlaw and B.S. Shane. 1993. Adsorption of heavy metals by green algae and ground rice hulls. J. Environ. Sci. Health, Part A 28: 37–50.

Sanchez, A., A. Ballester, M.L. Blazquez, F. Gonzalez, J. Munoz and A. Hammaini. 1999. Biosorption of copper and zinc by Cymodoceanodosa. FEMS Microbiol. Rev. 23: 527–536.

Sandau, E., P. Sandau and O. Pulz. 1996. Heavy metal sorption by microalgae. Acta Biotechnol. 16: 227–235.

Sheng, P.X., L.H. Tan, J.P. Chen and Y.P. Ting. 2004b. Biosorption performance of two brown marine algae for removal of chromium and cadmium. J. Dispersion Sci. Technol. 25: 679–686.

Skowronski, T. 1986. Adsorption of cadmium on green microalga *Stichococcus bacillaris*. Chemosphere 15: 69–76.

Somers, E. 1963. The uptake of copper by fungal cells. Ann. Appl. Biol. 51: 425–437.

Sud, D., G.M.P. Mahajan and M.P. Kaur. 2008. Agricultural waste material as potential adsorbent for sequestering heavy metal ions from aqueous solutions—a review. Bioresour. Technol. 99: 6017–6027.

Sunda, W.G. and R.R.L. Guillard. 1976. The relationship between cupric ion activity and the toxicity of copper to phytoplankton. J. Mar. Res. 34: 511–529.

Sunda, W.G. and R.L. Ferguson. 1983. Sensitivity of natural bacterial communities to addition of copper and to cupric ion activity: A bioassay of copper complexation in seawater. pp. 871–891. *In*: Wong, C.S., E. Boyle, K.W. Bruland, J.D. Burton and E.D. Goldberg [eds.]. Trace Metals in Seawater. Plenum Press.

Terry, P.A. and W. Stone. 2002. Biosorption of cadmium and copper contaminated water by *Scenedesmus abundans*. Chemosphere 47: 249–255.

Tiem, C.J. 2002. Biosorption of metal ions by freshwater algae with different surface characteristics. Proc. Biochem. 38: 605–613.

Tobin, J.M., D.G. Cooper and R.J. Neufeld. 1990. Investigation of the mechanism of metal uptake by denatured *Rhizopus arrhizus* biomass. Enzyme Microb. Technol. 12: 591–595.

Tripathi, B.N. and J.P. Gaur. 2006. Physiological behaviour of Scenedesmus sp. during exposure to elevated levels of Cu and Zn and after withdrawal of metal stress. Protoplasma. 229: 1–9.

Trujillo, E.M., T.H. Jeffers, C. Ferguson and H.Q. Stavenson. 1991. Mathematically modelling the removal of heavy metals from waste water using immobilized biomass. Env. Sci. Technol. 25: 1559–1563.

Tsezos, M. 1986. Adsorption by microbial biomass as a process for removal of ions from process or waste solutions. pp. 201–218. *In*: Eccles, H. and S. Hunt [eds.]. Immobilization of Ions by Bio-Sorption. Ellis Harwood, Chichester, UK.

Tsuritani, I., R. Honda, M. Ishizaki, Y. Yamada and M. Nishijo. 1996. Ultrasonic assessment of calcaneus in inhabitants in a cadmium-polluted area. J. Toxicol. Environ. Health 48: 131–140.

Tüzün, İ., G. Bayramoğlu, E. Yalçın, G. Başaran, G. Çelik and M.Y. Arıca. 2005. Equilibrium and kinetic studies on biosorption of Hg(II), Cd(II) and Pb(II) ions onto microalgae *Chlamydomonas reinhardtii*. J. Environ. Manag. 77: 85–92.

Vannela, R. and S.K. Verma. 2006. Co^{2+}, Cu^{2+}, and Zn^{2+} accumulation by cyanobacterium *Spirulina platensis*. Biotechnol. Prog. 22: 1282–1293.

Wang, J. and C. Chen. 2006. Biosorption of heavy metals by *Saccharomyces cerevisiae*: a review. Biotechnol. Adv. 24: 427–451.

Wilde, E.W. and J.R. Benemann. 1993. Bioremoval of heavy metals by the use of microalgae. Biotechnol. Adv. 11: 781–812.

Winter, C., M. Winter and P. Pohl. 1994. Cadmium adsorption by nonliving biomass of the semimacroscopic brown alga, *Ectocarpus siliculosus*, grown in actinic mass culture and localisation of the adsorbed Cd by transmission electron microscopy. J. Appl. Phycol. 6: 479–487.

Wong, M.H. and D.C.H. Pak. 1992. Removal of copper and nickel by free and immobilized microalgae. Biomed. Environ. Sci. 5: 99–108.

Worms, I., D.F. Simon, C.S. Hassler and K.J. Wilkinson. 2006. Bioavailability of trace metals to aquatic microorganisms: importance of chemical, biological and physical processes on biouptake. Biochimie. 88: 1721–1731.

Yan, G. and T. Viraraghavan. 2000. Effect of pre-treatment on the bioadsorption of heavy metals on Mucor rouxii. Water SA 26: 119–123.

Yu, Q. and P. Kaewsarn. 1999. Fixed-bed study for copper (II) removal from aqueous solutions by marine alga *Durvillaea potatorum*. Environ. Technol. 20: 1005–1008.

Zhang, X., S. Luo, Q. Yang, H. Zhang and J. Li. 1997. Accumulation of uranium at low concentration by green alga *Scenedesmus obliquus*. J. Phycol. 9: 65–71.

Zhou, J.L., P.L. Huang and R.G. Lin. 1998. Sorption and desorption of Cu and Cd by macroalgae and microalgae. Environ. Pollut. 101: 67–75.

5

Contaminant Removal and Energy Recovery in Microbial Fuel Cells

Soumya Pandit,[1,*] *Kuppam Chandrasekhar,*[2]
Dipak Ashok Jadhav,[3] *Makarand Madhao Ghangrekar*[4,*]
and *Debabrata Das*[1,*]

INTRODUCTION

An ever increasing energy demand has induced fossil fuel consumption, and consequently pollution and global warming, driving the world towards an unprecedented high and potentially devastating energy crisis. Therefore, water and energy securities are considered as major concerns in present scenario. Waste/wastewater signifies a potential renewable feedstock to generate various forms of bioenergy aside from the remediation process by regulating the biological process. Bioenergy has gained significant attention as a sustainable and futuristic alternative to fossil fuels. Using waste for bioenergy through its remediation has instigated considerable interest and has further opened a new avenue for the use of renewable and inexhaustible energy sources. Therefore, the field of wastewater management via

[1] Bioprocess Engineering Laboratory, Department of Biotechnology, Indian Institute of Technology Kharagpur, West Bengal, India 721302. Amity Institute of Biotechnology, Amity University, Mumbai, 410206, India.

[2] School of Applied Bioscience, Agriculture Department, Building-1, Kyungpook National University, Dong-Daegu-702701, South Korea.

[3] School of Water Resources, Indian Institute of Technology Kharagpur, West Bengal, India 721302.

[4] Department of Civil Engineering, Indian Institute of Technology Kharagpur, West Bengal, India 721302.

* Corresponding authors: ddas.iitkgp@gmail.com; ghangrekar@civil.iitkgp.ernet.in; sounip@gmail.com

contaminant removal and alternative energy generation are the most unexplored fields of Biotechnology and Science. Microbial fuel cell (MFC) is "gaining popularity as a promising tool for simultaneous waste treatment and electricity generation without polluting environment" (Logan and Regan 2006).

Most of the chemical pollutants have a detrimental effect on human health and cause serious damage in the environment. Contaminants discarded in wastewater and sediments induce different ranges of toxicity to susceptible organisms. Polycyclic aromatic hydrocarbons (PAHs), endocrine disrupting agents, and pesticides are recalcitrant in nature. Recent research on MFC treating recalcitrant compounds with concomitant current generation gained attention in the research community (Pisciotta and Dolceamore Jr. 2016). The breakdown of a wide range of organic substrates to carbon dioxide and water is usually only possible with several enzymatic reaction steps which are easily achieved in MFCs.

MFCs received much attention in recent years due to its advantages over the current technologies applied in biological wastewater treatment like anaerobic digester (AD), activated sludge process (ASP), etc. MFC is capable of direct conversion of waste to energy in terms of electricity in a single step, this ensures high conversion efficiency. Rabaey et al. 2007 showed that the conversion efficiency of bioethanol, biogas and biohydrogen is around 10–25%; 25–38% and 15–30%, respectively. In comparison to this conversion efficiency, MFC has the conversion efficiency as high as 80% (Rabaey and Keller 2008). In anaerobic digestion, methane or hydrogen gas is produced along with carbon dioxide and little amount of hydrogen sulfide. Unlike these technologies, a MFC does not require gas treatment. The energy input is required for aeration in ASP, whereas MFCs do not need the aeration as cathode can be passively aerated. Compare to ASP and AD, MFC produces less amount of sludge. In addition, MFCs stack can be applied as decentralized source of power (Pham et al. 2006).

Basic Principle of MFC

Being a fuel cell, an MFC comprises of an anode, a cathode, and electrolyte(s). The anodic and cathodic chambers may or may not be separated by an ion exchange membrane (IEM). Live microbes in planktonic state or by forming biofilm in the anodic chamber oxidize substrates and as a result produce electrons, protons and other metabolites as end products. The electrons released by the microbes are collected by the anode and pass into the cathode through an external load. On the other hand, the protons percolate via the IEM or simply diffuse to the cathode to get reduced by the arriving electrons, thus completing the circuit (Pandit et al. 2017). The flow of electrons through the external load generates electric current. As the protons get reduced in presence of oxygen at the anode, water is formed as a by-product which makes the process environment-friendly (Fig. 1).

The major components of MFCs are electrodes (i.e., anode, cathode) and membranes/separators. Different carbon materials like graphite plates, rods, carbon paper, cloth, glassy carbon, carbon foam, stainless steel mesh, carbon felt, Pt, Pt black, reticulated vitreous carbon have been applied as anode materials. The anode material should allow bacteria to form biofilm and should be conductive and noncorrosive.

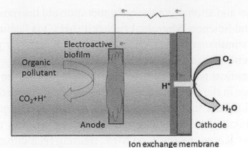

Fig. 1: Schematics of a single chambered MFC.

On a contrary, the cathode usually contains catalyst materials like Pt, Pt black, MnO_2, and polyaniline for improving oxygen reduction reaction (ORR) applied over the base material. Graphite, carbon felt, carbon cloth/paper, carbon brush, glassy carbon, stainless steel, Tt-mesh are among the materials used as a base material for cathode. In dual chambered MFC, ion exchange membrane or separator is used (Logan and Regan 2006). In single chambered MFC, IEM was hot-pressed or chemically fixed to cathode to prepare cathode membrane assembly for passive aeration to provide oxygen directly for ORR. Initially research started with salt-bridge as separator between anodic and cathodic chamber. Gradually it was replaced by IEMs. Both CEM (Nafion, Ultrex) and anion exchange membrane (Ralex AEM, Ultrex) were reported to utilize as membrane (Pandit et al. 2012). However, AEM was found effective in terms of power generation owing to its ability to generate stable voltage and reduced salt deposition.

Microbes Involved for Bioremediation in MFC

MFCs are having potential applications in pollutant removal, and bioremediation. In case of MFCs, both electrochemically active bacteria (EAB) as well as electrochemically inactive ones are hired. EAB also known as exoelectrogens play a vital role in enabling efficient electron transfer and maximizing current densities and energy efficiency. Hence, the prime focus for MFC development involves those strategies that optimize electron transfer abilities of EAB. These current producing bacteria are also called exoelectrogens because of their ability of direct or indirect extracellular electron transfer (EET) to electrodes (Yang et al. 2012). Bacterial growth is supplemented by coupling of energy conservation with transfer of electrons from exoelectrogens leading to cathode/anode respiration.

Various types of EET mechanisms were proposed. Some of the bacterial population present in the inoculum carries out EET through mediators, whereas microbes responsible for biofilm formation perform EET either directly from the cell envelope to the electrode or across the biofilm layer through conductive matrices, nanowires, etc. Kinetic rates would be higher in case of an electron transfer through the latter than the former (Rabaey et al. 2007). Hence, the primary attribute in magnifying MFC power density is the ability of a microbe to adhere to electrode as mono-layered or preferably multi-layered biofilm. In this regards, *Shewanella* and *Geobactor* are the two bacterial species generally known as electronic twins, considered as a model for MFC (Stams et al. 2006).

Microbial growth in the form of planktonic cells or as microbial matrix depends on the metabolic status and environmental factors. For example, *Shewanella oneidensis* can form a biofilm on the anode electrode in lactate-fed MFCs; however, both planktonic cells and biofilms are able to exist together simultaneously and synergistically produce power (Pandit et al. 2014). On contrary to it, *Geobacter sulfurreducens* favors direct contact with the electrode for electron transfer. These organisms can grow on the electrode and build an electrochemically active, multi-layered bacterial colonies, termed as biofilm (Nevin et al. 2009). On the other hand, some EAB can secrete the redox active compounds (mediators/shuttles) to transfer the electrons to the electrode, for which the mechanism is termed as indirect electron transfer (Rabaey et al. 2004). Such mediated electron transfer mechanism can be classified as exogenous (electron transfer mechanism by externally added mediators) or endogenous (electron transfer by mediators secreted by bacteria such as, flavin).

Various electron transfer mechanisms

Electron transfer by c-type cytochromes

The use of c-Type cytochromes (CTCs) has proposed as one of the predominant EET technique for harvesting the current by EABs. C-type cytochromes are widespread heme-containing proteins found in most bacteria and archaea species. A species *S. oneidensis* MR-1 has 42 putative CTCs where 80% of these are located in the outer membrane and occupy 8–34% of the total cell surface (Shi et al. 2008). According to comparative genomic analysis, six *Geobacter* species displayed an average of 79 putative CTCs in each *Geobacter* genome (Nevin et al. 2009). Only 14% of these are found to be conserved in all genomes. It has been suggested that MacA delivers electrons from inner membrane to PpcA in periplasm, and PpcA further transports electrons to the OMCs (outermost cytochrome C) (Fig. 2).

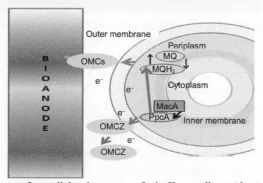

Fig. 2: Proposed pathway of exocellular electron transfer in *Shewanella oneidensis*.

Bacterial nanowire

Bacterial nanowire technique is popularly considered as a novel electron transfer approach. Bacterial nanowire is electrically conductive pili which were discovered during Fe(III) oxide reduction by *G. sulfurreducens* (Richter et al. 2012). These were also found in *S. oneidensis MR-1* and some other bacteria, and many other species signifying a wide environmental network of these bacterial appendages. *S. oneidensis MR-1* nanowires existed in the form of a bunch of conductive pili-like appendages with a size of 3–5 nm diameter under the electron acceptor limiting conditions (Barchinger et al. 2016).

Electron shuttles

Electron shuttles in BES are secreted by mostly gram negative bacteria; the expected properties of electron shuttle are (1) dissolvability, (2) stability, (3) reusability, (4) environment-friendliness, and (5) higher redox potential. Electron shuttle secreted by microbes can be stimulated by power harvesting in MFCs. The most reported electron shuttles in MFCs are endogenously produced flavin by *Shewanella* species, which majorly includes riboflavin as well as flavin mononucleotide (Marsili et al. 2008). Phenazines are intrinsic electron shuttles generated by a diverse bacterial species, mainly *Pseudomonas* species (Pandit et al. 2015). However, on comparing with direct electron transfer, this technique has several shortfalls. These are (1) higher voltage loss, and (2) low diffusion co-efficiency in biofilms. In the case of exoelectrogens, shuttle generation will result due to metabolic pressure and have a competitive drawback, particularly in the case where electron shuttles are degraded or used by other members.

Performance Evaluation in MFC for Contaminant Removal with Simultaneous Power Generation

The performance index of MFC is commonly measured in terms of energetic parameters (volumetric power density, potential difference, current output, internal resistance) and biological wastewater treatment efficiency (COD removal and solids degradation) (Erable et al. 2010). Coulombic efficiency (CE) or current harvesting efficiency express the relation between the amount of electrons contributed to current generation and conversion of the organic substrates over a time period. A polarization curve generator is frequently used in MFC experiments to analyze and characterize the quality of fuel cells in terms of power generation. It can be obtained by varying external resistances using a resistance box or a using programmed liner sweep voltammetry. Electrochemical tool like electrochemical impedance spectroscopy (EIS) is used to evaluate the internal resistance of the different components (charge transfer resistances, solution resistance, and diffusion resistance) in electrochemical systems. Cyclic voltammetry (CV) is a useful method to investigate the mechanisms of oxidation or reduction reactions on the electrode surface. In CV, the external specific range of applied voltage was applied and output current is monitored (Logan 2012).

Oxidation of Organic Waste in MFC

As far as power generation is concerned, MFC cannot compete with a conventional fuel cell, which produces 10^3–10^5 fold higher current compare to MFC. Henceforth, more attention is being paid to applying MFC in wastewater treatment using different kind of organic waste materials (Kelly and He 2014). MFC has great potential for treating recalcitrant wastewater with the help of microorganisms as biocatalysts. A large number of recalcitrant wastes which includes dyes, pesticides, polyalcohol, heterocyclic compounds are produced by the industry. Currently, the discharge of wastewaters having these pollutants is an important environmental hazard, owing to their mutagenicity, recalcitrance, and tendency to accumulate in the environment. Also, their discharge in water bodies causes aesthetic problems and obstructs the light penetration and hence, restrict oxygen transfer into water, thus affecting the aquatic life (Caplan 1993). In MFCs, anodophilic bacteria utilize the biodegradable fraction of wastewater for their energy requirements and release electrons and protons in the anode chamber (Kracke et al. 2015).

Treatment of petroleum hydrocarbon contaminated wastewater in MFC

Most of the chemical pollutants have a detrimental effect on human health and cause serious damage to the environment. Contaminants discarded in wastewater and sediments induce different ranges of toxicity in susceptible organisms. Recent research on MFC treating recalcitrant compounds with concomitant current generation gained the attention of research community (Johnsen et al. 2005).

Aliphatic hydrocarbons (alkanes, alkenes, and cycloalkanes) are straight, branched, or cyclic compounds which may be either saturated or unsaturated. Cycloalkanes share similar physical and chemical properties though cycloalkanes with only one ring are susceptible to nucleophilic attack, and hence less stable and a little more reactive than normal alkanes (Das and Chandran 2011). However, as the number of carbons and ring structure increases, they become more stable. Heterocyclic hydrocarbons are also very reactive and highly susceptible to both chemical and microbial attacks. On the contrary, the benzene, toluene, ethyl benzene, and xylene (BTEX) are aromatic hydrocarbons which are considered as carcinogenic, teratogenic, and neurotoxic. These potentially hazardous compounds are easily soluble in water and therefore, can mix easily in aquifers. Another class of contaminant, PAHs, is bestowed with high toxicity and chemical stability. Moreover, these materials are considered as worldwide environmental contaminants (phenanthrene, fluorene, naphthalene) due to their ability to resist different bio-transformation processes (Chandrasekhar and Venkata Mohan 2012). In industries, mostly skimmers and booms are usually used as a physico-chemical method to remove petroleum contaminated sites; however, this method is incapable of treating or degrading the contaminants to harmless products. For the same reason, the use of the adsorber like organoclay (activated carbon for the removal of a pollutant) is a challenge as disposal of the adsorbant is critical while adopting this technology. Huge costs associated with thermal treatment (pyrolysis, incineration) limits their use for petroleum hydrocarbon-contaminated waste treatment. Application of chemical agents or emulsifying agents followed by photo-degradation creates unwanted by-products (for example: quinones, dihydrodiols, catechols, etc. (Das and Chandran 2011)).

Bioremediation, a microbial based technology was attempted as an alternative approach due to its environmentally favorable and inexpensive nature. The major advantage of bioremediation is the effective utilization of microorganisms to detoxify the pollutant. Moreover, remediation technology can be applied for non-specific contaminant removal. A short description of different biological methods has been provided in Table 1. The application of existing biological methods for remediation became limited and economically nonviable due to several factors. The system parameters like non-availability of substrate, toxicity of contaminant towards microbes, diffusion problem of terminal electron acceptors; the capital costs, expenditure due to maintenance, transportation (*ex situ* remediation), etc. make the processes non-sustainable (Caplan 1993).

Bacteria were utilized widely for the biodegradation of petroleum based hydrocarbon substances. These microorganisms consumes hydrocarbon and "biotransforms" it into an intermediate metabolite; mineralization is the other option in aerobic or anaerobic conditions. In most cases, microorganisms biodegrade complex substrate in co-metabolism process where a primary carbon food source is consumed along with this type of recalcitrant substance. *Pseudomonas, Sphingomonas*, and *Acinetobacter* species are predominant group functioning as hydrocarbon–degraders, apart from microoragnisms like *Micrococcus, Vibrio, Brevibacterium, Corynebacterium, Flavobacterium, Cellulomonas, Alcanivorax, Microbulifer* and some other species of proteobacteria and firmicutes groups (Das and Chandran 2011). Biotransformation of BTEX and PAHs proceeds either by aerobic or anaerobic mechanisms. Aerobic transformation by BTEX and PAHs is achieved by ring oxidation mediated by enzymes (e.g., oxygenase) followed by ring cleavage (Johnsen et al. 2005). Low Molecular Weight (LMW)-PAHs can be bio-transformed to trans-dihydrodiols by bacteria such as *Mycobacterium* sp. using cytochrome P-450 monooxygenase. Utilization of MFC for bioremediation of this kind of recalcitrant product is relatively new. Few studies were performed in the

Table 1: Short description of different existing biological methods for bioremediation.

Biological method	Short description
Bioventing and biosparging	Air at low rates was sparged to increase microbial growth for bioremediation via an injection well; nutrients are also supplemented for the stimulation of microbial growth.
Biopiles	In biopile remediation technology, aeration is done using a perforated pipe in an excavated contaminated area. The height of heaps or pile are around 6 cm.
Land farming	It's a solid phase technology where contaminated soils are dispersed as a thin layer and ploughing is done periodically with nutrient supplementation for better mixing.
Composting	In composting, contaminated waste are mixed with cured compost such as fishbone or bark compost to degrade the target contaminant, it is usually enriched with urea and other carbon sources for the growth of indigenous microbes.
Bioslurry system	This method is widely adopted in industries; contaminated soil/waste is kept in a bioreactor to allow sufficient time to degrade.

anode chamber of MFC to check the possibilities of petroleum based contaminant removal. An 82% reduction in diesel range organic was found when Moris et al. (2009) operated MFC with ex situ method after 21 days of operation, only 32% removal was achieved with same substrate indicates efficacy of bioelectrochemical system for this type of waste treatment. Wang et al. (2012) used *in situ* lab based soil MFC and compared results with non MFC condition; they found enhanced degradation of petroleum hydrocarbon by 120% (within 25 days) compared to the control anaerobic reactor (Wang et al. 2012). These results suggested effectiveness of MFC in treating petroleum based pollutant treatment (Table 2).

Table 2: Treatment of different petroleum derived hydrocarbon contaminated wastewater in MFC.

Pollutant	Microbes type	MFC configuration	Mode of operation	Removal efficiency (%)	References
Furfural	Pretreated mixed culture	Singled chambered	Batch	95	(Luo et al. 2010)
4-nitrophenol	Anaerobic sludge	Two chambered MFC	Batch	100	(Zhu and Ni 2009)
Nitrobenzene	Mixed culture	Two chambered MFC	Continuous	1.2	(Mu et al. 2009b)
Petroleum hydrocarbons	Mixed culture from petroleum contaminated saline soil	Single chamber MFC	Batch	15.2	(Wang et al. 2012)
Toluene	Anaerobic sludge	Two-chamber MFC	Batch	100	(Lin et al. 2014)
Coal Tar containing wastewater	Activated Sludge collected from refinery	Dual-chambered tubular MFC	Batch	88	(Park et al. 2012)
Phenanthrene, Pyrene	Mixed culture from PAH contaminated area	Dual Sediment type MFC	Batch	Pyrene-95; Phenanthrene -99	(Yan et al. 2012)

Xenobiotics/Pollutant as Feedstock for MFC

Xenobiotics are chemicals to which an organism is exposed that are extrinsic to the normal metabolism of that organism. The effective operation of wastewater treatment plants plays an important role in minimizing the discharge of xenobiotic compounds into the aquatic environment. Although the technologies available today can produce high quality water even from wastewater, most wastewater treatment plants are not designed to remove emerging xenobiotic contaminants such as endocrine disrupting compounds (EDCs) and pharmaceutical and personal care products (PPCPs) (Roccaro et al. 2013). Previously, Kaewkannetra et al. (2011)

Table 3: Comparing the performance of MFCs using various wastes and Xenobiotics.

S. No.	Wastewater/Xenobiotics	Concentrations	Electrical output	References
1.	Cyanide containing Cassava wastewater	1.086 g.COD/L	22.2 W/m^3	(Kaewkannetra et al. 2011)
2.	Brewery effluent	0.6 g.COD/L	0.83 W/m^3	(Wen et al. 2010)
3.	Starch processing wastewater	4.852 g.COD/L	0.9 A/m^2	(Lu et al. 2009)
4.	Congo red	0.3 g.COD/L	0.897 A/m^2	(Sun et al. 2009)
5.	Selenium	0.025 g/L	2.9 W/m^2	(Catal et al. 2009)

have utilized agro waste such as cassava wastewater as a substrate for harvesting the current with the application of dual chambered MFC and achieved power output of 1.7 W/m^2 with 88% COD removal. All the azo dyes released during industrial process (i.e., textiles) are distinguished by the presence of the –N = N– bond, which serve as an terminal electron acceptor in cathodic chamber and can be reduced to hydrazo or amine through 2 or 4 electron transfer pathways. Liu et al. (2009) have reported production of electricity in MFC using azo dyes as the catholyte and using pure culture of *Klebsiella pneumonia* strain L17 in the anodic chamber. The other pollutants such as xenobiotics discharged from industrial waste, selenite, nitrate and other pollutants can be effectively used as effective feedstock and efficiently treated in MFC (Kiran Kumar et al. 2012). The treatment of different xenobiotics and pollutants in BES has been summarized in tabular form as below (Table 3).

Dye discoloration in MFC

Usually dyes are coloring chemical substances majorly utilized in the textile, pharmaceutical, food processing, cosmetics and paper manufacturing industries. More than ten thousands dyes and pigments are applied on bulk scale while 7×10^5 tons of synthetic colorants are produced yearly all over world. Nevertheless, it is reported that around 10–15% of these dyes are wasted during application as colorant. The release of unutilized dye in wastewater can cause environmental problems. These are chemically stable substance and therefore degradation of these to harmless by product is hard to achieve. The azo dye containing one or more –N = N– groups are predominantly used as synthetic dye produces toxic and mutagenic by product following degradation. Henceforth, discharge of this kind of dye before adequate treatment is necessary to avoid hazards (Solanki et al. 2013). Different physicochemical techniques like flocculation, coagulation are employed in industries; however these techniques failed to gain popularity due to high volume sludge generation and additional requirement for disposal technology. In spite of its effective removal efficiency, processes ozonization and advanced oxidation process, UV/peroxide application are not used frequently owing to their high cost and energy consumption. The sluggish for biodegradation of the dyes in the existing biological treatment technique like anaerobic reduction process limits their uses. Furthermore, requirement of co-substrates increases the cost of operation and it is responsible for greenhouse gas generation. The enzymatic processes for dye removal were found efficient; still this method is not yet established due to lack of chemical stability

Table 4: Different azo dyes, MFC configuration and inoculum type used in MFC.

Azo dye	Inoculum type	MFC configuration	Power density	References
Congo red	Anaerobic sludge	dual chambered MFC	364.5 mW/m²	(Li et al. 2010a)
Amaranth	Mixed culture	dual chambered MFC	137.37 mW/m²	(Fu et al. 2010)
Methyl orange	Anaerobic sludge	dual chambered MFC	–	(Ding et al. 2010)
Active brilliant red X-3B	Mixture of aerobic and anaerobic sludge	Air cathode single chambered MFC	234 mW/m²	(Sun et al. 2009)
Acid orange 7	Mixed culture	dual chambered MFC	0.31 W/m³	(Mu et al. 2009a)

and product inhibition of enzyme for decolorization of dye wastewater. Recently, the MFC was used for discoloration of different azo dyes in both cathode and anode chamber. The different types of azo dyes being treated in MFCs are Congo red, acid orange 7, methyl orange, amaranth, active brilliant red X-3B, reactive blue 221 and orange I. Basically, dye decolorization reactions obtained in the anodic chamber through biological actions under anoxic conditions for the breakdown of the azo bond of Congo red and in the cathodic compartment abiotically for complete removal of intermediate products (Li et al. 2010a). Decolorization rate of dye in MFC depends on different operational factors like dye concentration, anolyte and catholyte pH, dye structure on decolorization activity, magnitude of external resistance used in MFC, hydraulic retention time (HRT) and presence of co-substrate. Various dyes containing wastewater treatment in MFCs along with simultaneous power generation has been summarized in Table 4.

Sulfur Species Conversion in MFC

Different form of sulfur compound available in nature

Sulfur can exist in many oxidation states and it exists in different chemical forms, which play an important role in aqueous systems. A number of remediation strategies like physicochemical or biotechnological processes for removal of sulfur species have been developed and implemented. Out of the many allotropes that sulfur has, orthorhombic α S_8 is thermodynamically most stable and is the main constituent of commercial sulfur. Elemental sulfur is weakly soluble in water and this property doesn't allow significant reduction in aqueous media. Biologically produced elemental sulfur particles have more hydrophilic properties than chemically produced ones. Sulfides are weak acids in aqueous solutions. The different form of sulfur compound depends on the electrochemical potential and pH for sulfur species in aqueous systems. A Pourbiax diagram visualizes the thermodynamic stability area of different sulfur species. It enables us to see at a glance the stability region of aqueous sulfur species and solid compounds at particular electrochemical potential and pH values in relation to overall sulfur concentration and aqueous sulfur species at equilibrium conditions.

Drawback of existing technologies

One of the major disadvantages of the existing sulfide removal technologies in sewage systems is its high cost. Biotechnological or microbiological sulfide removal processes also involve high aeration costs and if air is used as oxidant, they need stringent control to avoid stripping of sulfides. There is still a need for more cost effective and efficient methods to control sulfur compounds in effluent of sewage treatment plants. Abiotic electrochemical systems can only enable selective aqueous sulfide removal, but a biotic electrochemical system can remove sulfide and organics simultaneously without the expense of energy and with the scope of harvesting solid sulfur in both cases.

Removal of aqueous sulfide in MFC

The operation of MFC is influenced by the conversion of inorganic sulfur in it. Sulfate serve as an alternative electron acceptor during oxidation of organic carbon compounds in BESs. Sulfate is reduced to sulfide by sulfate reducing bacteria (SRB) in anaerobic conditions using organic matter as electron donors. When sulfate reduction and sulfide oxidation occurred simultaneously in anodic chamber, simultaneous removal of organics, sulfide and sulfate from wastewaters can be obtained in MFCs (Rabaey et al. 2004). Although, sulfate reduction can be carried out in a separate vessel, externally while the produced sulfide could be oxidized in an abiotic fuel cell. SRB could also reduce thiosulfate and sulfite to sulfide. In the anode compartment of a MFC, the bacteria appear only to reduce sulfate with fermentable electron donors. Although when acetate was supplied the sulfate reduction was not observed, during both open and closed circuit conditions. Hence, SRB can use only acetate as substrate and it might not develop in an acetate fed MFC environment.

Sulfide oxidation in a biotic cell

Sulfide oxidizing bacteria (SOB) can donate electrons directly to the electrode. Further studies reported that power generation from marine sediments occurs via at least two reactions: (i) oxidation of sediment sulfide—a by-product of microbial heterotrophy; and (ii) microbially catalyzed organics oxidation. In another study, the presence of biofilms did not impact on the sulfide oxidation products and rates in comparison to abiotic systems. The fact that sulfide removal rates do not change due to the presence of a biofilm does not imply that biological sulfide oxidation doesn't happen. In MFC system, the abiotic and biotic processes might continue parallel, but the combined sulfide oxidation rate would remain unaltered in comparison to abiotic oxidation if the reaction rate is restricted by the electrode or operation of MFC (Zhang and Ni 2010). If favorable potential is maintained, elemental sulfur deposited on an anode can be quite rapidly reduced again to sulfide. The formed elemental sulfur would subsequently be oxidized again in anodic chamber. Hence, the deposited elemental sulfur and sulfide would form mediators, or electron shuttles for carbon oxidation. Microbes used electrodeposited sulfur as suitable electron acceptor over the solid anode surface and soluble sulfate, and produced sulfide irrespective of the electrochemical operating conditions.

Role of MFC in Subsurface Remediation

It is difficult to achieve complete biodegradation of certain kinds of contaminants (e.g., chlorinated solvents, PCB) or mixtures of more than one pollutant, which require redox electrochemical conditions. Considering the remediation applications, microbial electrochemical technologies (METs) or Microbial electro-remediation cells (MRCs) have been provide flexible platform to enhance the oxidative degradation of reduced contaminants (e.g., petroleum hydrocarbons, solvents, dyes, etc.) in anode chamber and the reductive transformation of oxidized contaminants (e.g., chlorinated solvents, SO_4^{2-}, nitrate, Cr^{6+}) in cathode chamber. Recently few scientists proposed application of MFC for reduction of chlorinated compounds present in subsurface contaminants (Li et al. 2010b; Shea et al. 2008).

Chromium, copper, Uranium and vanadium are among the metal, which is found to recover via cathode reduction in MRC. Subsurface contamination is of particular interest because oxidized uranium is toxic to aquatic life, soluble and highly mobile. Williams et al. (2010) demonstrated cathodic U(VI) reduction; while Zhang et al. (2010) studied the removal of sulfide and Vanadium (V) in MRCs. Chromium in hexavalent form is widely used in number of industrial applications such as leather tanning, metallurgy, electroplating, and as a wood preservatives. Chromium found in the aqueous solution either as hexavalent or trivalent form, in which Cr(VI) is more toxic due to its mutagenic and carcinogenic properties. Reduction of Cr(VI) can be coupled with electricity generation using MFC and it can be applied in wastewater treatment containing chromium. Perchlorate salts commonly served as terminal electron acceptor by bacteria and detected in soil and ground water. Shea et al. (2008) evaluated the perchlorate reduction in MFC with using denitrifying biocathode (Table 5) (Shea et al. 2008). Subsurface contamination is of particular interest because oxidized uranium is toxic to aquatic life, soluble and highly mobile. Wastewaters containing chromium were effectively treated in MFCs and simultaneous chromium reduction and electricity production were achieved (Li et al. 2010b). Under light irradiation, about 97% Cr(IV) removal was obtained in photo-electrochemical system (PBS) coupled with MFC within 26 h retention time at the initial concentration of 26 mg/L.

MRC has the ability to reduce some the contaminants such as perchlorate, persulfate, chlorinated solvents or uranium in the cathode chamber. The reduced

Table 5: Performance of MFC for remediation.

Contaminants	Bacteria	Cathode potential (vs. SHE)	Degradation rate (µeq/L.day)	References
2,4 dichlorophenol	Mixed culture	–	40.9	Skadberg et al. 1999
Perchlorate	Mixed culture	–	965	Shea et al. 2008
U (IV)	*Geobacter*	– 300 mV	27	Gregory and Lovley 2005
Trichloroethylene	Mixed culture	– 450 mV	2.7	Aulenta et al. 2009
Perchloroethylene	*Geobacter*	– 300 mV	200	Strycharz et al. 2008
Nitrate	*Geobacter*	– 300 mV	670	Gregory et al. 2004

product of the MRC are non-toxic, i.e., do not require any further treatment for environmental disposal. Perchlorate, Nitrate is one of the common types of salts as contaminant found in waters, and causes a variety of serious environmental and health problems detrimental of human society. Perchlorate is utilized as an electron acceptor in MFC and was suggested through reduction pathway as ClO_4^- $\rightarrow ClO_3^- \rightarrow ClO_2 \rightarrow Cl^- + O_2$ by perchlorate reducing bacteria. Butler et al. (2010) used biocathode for denitrification and obtained perchlorate removal rate of about 24 mg/L.day and cathode conversion efficiency of 84%.

Chlorinated hydrocarbons and its derivatives such as perchlorethylene (PCE), trichloroethylene (TCE), cis-dichloroethene (cis-DCE), vinyl chloride (VC) and trichloroethane (TCA) widely utilized as chlorinated organic solvents for degreasing in the dry cleaning, oil industry, industrial manufacturing and machine maintenance industries, are found in groundwater. Over the last few years, various bioelectrochemical approaches have been suggested to promote the reductive dechlorination processes (Fig. 3; Table 5). Aulenta et al. (2009) proposed an electrochemically assisted reductive dechlorination using a polarized glassy carbon cathode as an electron donor for biological reductive dechlorination of TCE to ethane.

Operational Factors Affecting Performance of MFC

Performance of MFC was governed by anodic parameters, cathodic parameters, design aspects, operating conditions. Considering the anodic factors, the selected

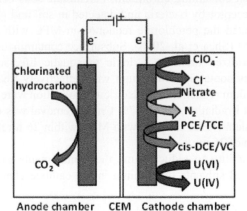

Anode chamber   CEM   Cathode chamber

Fig. 3: Degradation mechanisms for subsurface contaminants in BES.

anode material should have the required biocompatibility, high specific surface area, non-corrosive and non-toxic, highly conductive, chemically stable in wastewater, economical and locally available. Taking into account the cathodic limitations, poor reduction kinetics and slow rate of oxygen reduction reactions at cathode is the major limiting factor which restricts the power output from MFC. Also, essential constituents for MFC construction are bacterial inoculum, wastewater used as anolyte, electrode catalyst and catholyte, which govern the performance of MFC.

Design factors

The performance of MFC depends on the architectural aspects such as shape of MFC, type of MFC, electrode material, electrode spacing, properties and specific surface area of electrode, wastewater flow distribution, etc. Selected electrode material should be biocompatible, conductive, and offer high specific surface area, less resistance and chemically stable in electrolyte solutions (Jabeen and Farooq 2015). Several architectural designs have been proposed to reduce the internal resistance of MFC such as single chambered MFC (Pandit et al. 2014). In such configuration, cathode was placed direct in contact with air instead of submerging it in cathodic chamber (Kadier et al. 2015). In addition, stacking of MFCs by connecting several units in series or parallel connection have been designed to improve the electrical output from MFC.

Physicochemical factors

The growth and metabolic activities of microorganisms affects the electricity generation from MFC. For these reason, MFC system should be operated with optimum operating conditions to favors the microbial growth kinetics. The proton availability at cathode limits the performance of MFC. Therefore, use of NaCl or buffers compensates the retarded transport of protons through membrane. Similarly, temperature also affects the microbial activities, redox reaction rate as well. The microbial component including properties of inoculum, its biofilm forming ability and electron transfer, substrate utilization kinetics plays important role in MFC. Working with complex wastewater, the biological optimization implies selection of suitable bacterial consortia and strategies to adapt bacterial communities to the reactor optimized conditions. Effective utilization of substrate for harvesting the Coulombs and electron transfer mechanism offer the improvement in high current densities at lower anodic overpotential (Du et al. 2007).

Operating conditions

Anodic and cathodic operating conditions are the main factors affecting the performance of MFC. The operating conditions such as external resistance, substrate characteristics and loading rate, temperature, pH and retention time affects the performance of MFC in terms of electricity generation and treatment efficiency. Different varieties of wastewaters ranging from simple sugars to complex wastewater were effectively used as a substrate and efficiently treated in anodic chamber of MFC (Pham et al. 2006). Organic loading rate (OLR) can be expressed as the function of current generation from MFC and substrate degradation (Gil et al. 2003). At higher external resistance, the decrease in anode potential and electron discharge to cathode is limited by external circuit load; however, it is restricted by kinetic or mass transfer limitations at low external resistance. Higher internal resistance resulted into lower power density and current density in MFC.

Bottleneck of MFC in Scaling Up for Wastewater Treatment

There are few bottlenecks which block the function of MFC generally termed as overpotential. Theoretically, 1.1 V is the highest voltage attainable from MFCs, but due to several technical limitations, it has been limited to 0.8 V and around 0.62 V in open circuit mode and during current generation respectively. Hence, the change among measured and theoretical cell potential together indicates the overpotential of the electrodes. This overpotential can be characterized as activation losses, concentration polarization/losses, and ohmic losses (Kim et al. 2007). Several researchers have been investigated the efficiency of pure culture as a biocatalyst for power generation in MFCs. However, mixed cultures are very much suitable when we operating MFCs with complex and/or combined substrates such as wastewater (Yang et al. 2012). Because pure cultures as biocatalyst usually utilize a narrow range of substrates in MFCs. Besides, the type and nature (such as biocompatibility, resistance to corrosion, high surface area and mechanical strength, etc.) of the electrode material influences the biofilm formation, and also the diversity of biocatalyst varies based on the distance between the anode and the membrane (Kim et al. 2007).

Conclusion

The indiscriminate use of fossil fuels lead to not only environmental pollution but also exhausted the limited fuel reserves. In this realm, advent of MFC for green energy source and clean water treatment tool is encouraging. MFC technologies were developed and intensively investigated the performance in last decade. Research on MFC is based on interdisciplinary research which combined microbiology, molecular biology, electrochemistry, material science and engineering platform. Remarkable development was found in MFC research in last decades, the performance also improved significantly. Progress in reactor design, component materials, optimization of operational factors leads to high magnitude of volumetric power output from 1 mW/m³ to 2.87 kW/m³. Treatment capacity of MFC also increased significantly up to 7.1 kg chemical oxygen demand (COD)/m³ reactor volume/day which is comparable to anaerobic digester. Although, there was few scale up studies, most of them were not successful due to water leakage, unwanted by products generation, large variation in power output, influent fluctuation, etc. Henceforth, many aspects need further investigation. Full-scale implementation of such MFC is not easy due to certain microbiological, technological and economic challenges need to be resolved. Molecular studies on the interaction between the electrode and the exoelectrogenic microorganisms is not yet clear due to experimental difficulties in anaerobic attached based growth condition in anode chamber. More information is required to elucidate the detailed mechanism to understand the insight of colonization of EAB and current generating biofilm formation at different operating conditions. Mixed culture may be considered as inoculum in the anode chamber of MFC for practical purpose. The competition for substrate by methanogens and electrogenic bacteria significantly affects MFC performance additional study is required to understand the methanogenesis and electrogenesis. The combination of wastewater treatment along with electricity production helps in compensating the cost of wastewater treatment, making it sustainable. Some technical issues remain unsolved till date. Implementation of such advanced bio-electrochemical wastewater treatment system at field level is not straightforward because several microbiological, electrochemical, technological and economic constraints need to be resolved that have not previously been encountered in any other conventional wastewater treatment system which makes life cycle assessment complicated.

References

Aulenta, F., A. Canosa, P. Reale, S. Rossetti, S. Panero and M. Majone. 2009. Microbial reductive dechlorination of trichloroethene to ethene with electrodes serving as electron donors without the external addition of redox mediators. Biotechnol. Bioeng. 103(1): 85–91. doi: 10.1002/bit.22234.

Barchinger, S.E., S. Pirbadian, C. Sambles, C.S. Baker, K.M. Leung, N.J. Burroughs et al. 2016. Regulation of gene expression in *Shewanella oneidensis* MR-1 during electron acceptor limitation and bacterial nanowire formation. Appl. Environ. Microbiol. 82: 5428–5443. doi: 10.1128/AEM.01615-16.

Butler, C.S., P. Clauwaert, S.J. Green, W. Verstraete and R. Nerenberg. 2010. Bioelectrochemical perchlorate reduction in a microbial fuel cell. Environ. Sci. Technol. 44(12): 4685–4691. doi: 10.1021/es901758z.

Caplan, J.A. 1993. The worldwide bioremediation industry: prospects for profit. Trends Biotechnol. 11: 320–323. doi: 10.1016/0167-7799(93)90153-Z.

Catal, T., H. Bermek and H. Liu. 2009. Removal of selenite from wastewater using microbial fuel cells. Biotechnol. Lett. 31: 1211–1216. doi: 10.1007/s10529-009-9990-8.

Chandrasekhar, K. and S. Venkata Mohan. 2012. Bio-electrochemical remediation of real field petroleum sludge as an electron donor with simultaneous power generation facilitates biotransformation of PAH: Effect of substrate concentration. Bioresour. Technol. 110: 517–525. doi: 10.1016/j.biortech.2012.01.128.

Clauwaert, P., P. Aelterman, T.H. Pham, L.D. Schamphelaire, M. Carballa, K. Rabaey et al. 2008. Minimizing losses in bio-electrochemical systems: the road to applications. Appl. Microbiol. Biotechnol. 79: 901–913. doi: 10.1007/s00253-008-1522-2.

Cucu, A., A. Tiliakos, I. Tanase, C.E. Serban, I. Stamatin, A. Ciocanea et al. 2016. Microbial fuel cell for nitrate reduction. Energy Procedia, EENVIRO-YRC 2015—Bucharest 85: 156–161. doi: 10.1016/j.egypro.2015.12.286.

Das, N. and P. Chandran. 2011. Microbial degradation of petroleum hydrocarbon contaminants: An overview. Biotechnol. Res. Int. 2011: e941810. doi: 10.4061/2011/941810.

Ding, H., Y. Li, A. Lu, S. Jin, C. Quan, C. Wang et al. 2010. Photocatalytically improved azo dye reduction in a microbial fuel cell with rutile-cathode. Bioresour. Technol. 101: 3500–3505. doi: 10.1016/j.biortech.2009.11.107.

Du, Z., H. Li and T. Gu. 2007. A state of the art review on microbial fuel cells: A promising technology for wastewater treatment and bioenergy. Biotechnol. Adv. 25: 464–482. doi: 10.1016/j.biotechadv.2007.05.004.

Erable, B., N.M. Duţeanu, M.M. Ghangrekar, C. Dumas and K. Scott. 2010. Application of electro-active biofilms. Biofouling 26: 57–71. doi: 10.1080/08927010903161281.

Fu, L., S.-J. You, G. Zhang, F.-L. Yang and X. Fang. 2010. Degradation of azo dyes using *in situ* Fenton reaction incorporated into H_2O_2-producing microbial fuel cell. Chem. Eng. J. 160: 164–169. doi: 10.1016/j.cej.2010.03.032.

Gil, G.-C., I.-S. Chang, B.H. Kim, M. Kim, J.-K. Jang, H.S. Park et al. 2003. Operational parameters affecting the performannce of a mediator-less microbial fuel cell. Biosens. Bioelectron. 18: 327–334. doi: 10.1016/S0956-5663(02)00110-0.

Gregory, K.B. and D.R. Lovley. 2005. Remediation and recovery of uranium from contaminated subsurface environments with electrodes. Environ. Sci. Technol. 39(22): 8943–8947.

Jabeen, G. and R. Farooq. 2015. Microbial fuel cells and their applications for cost effective water pollution remediation. Proc. Natl. Acad. Sci. India Sect. B Biol. Sci. 1–11. doi: 10.1007/s40011-015-0683-x.

Johnsen, A.R., L.Y. Wick and H. Harms. 2005. Principles of microbial PAH-degradation in soil. Environ. Pollut. 133: 71–84. doi: 10.1016/j.envpol.2004.04.015.

Kadier, A., Y. Simayi, P. Abdeshahian, N.F. Azman, K. Chandrasekhar and M.S. Kalil. 2015. A comprehensive review of microbial electrolysis cells (MEC) reactor designs and configurations for sustainable hydrogen gas production. Alex. Eng. J. doi: 10.1016/j.aej.2015.10.008.

Kaewkannetra, P., W. Chiwes and T.Y. Chiu. 2011. Treatment of cassava mill wastewater and production of electricity through microbial fuel cell technology. Fuel 90(8): 2746–2750. doi: 10.1016/j.fuel.2011.03.031.

Kelly, P.T. and Z. He. 2014. Nutrients removal and recovery in bioelectrochemical systems: A review. Bioresour. Technol. 153: 351–360. doi: 10.1016/j.biortech.2013.12.046.

Kim, B.H., I.S. Chang and G.M. Gadd. 2007. Challenges in microbial fuel cell development and operation. Appl. Microbiol. Biotechnol. 76: 485–494. doi: 10.1007/s00253-007-1027-4.

Kiran Kumar, A., M. Venkateswar Reddy, K. Chandrasekhar, S. Srikanth and S. Venkata Mohan. 2012. Endocrine disruptive estrogens role in electron transfer: Bio-electrochemical remediation with microbial mediated electrogenesis. Bioresour. Technol. 104: 547–556. doi: 10.1016/j.biortech.2011.10.037.

Kracke, F., I. Vassilev and J.O. Krömer. 2015. Microbial electron transport and energy conservation—the foundation for optimizing bioelectrochemical systems. Front. Microbiol. 6: 575. doi: 10.3389/fmicb.2015.00575.

Li, C., X. Han, F. Cheng, Y. Hu, C. Chen and J. Chen. 2015. Phase and composition controllable synthesis of cobalt manganese spinel nanoparticles towards efficient oxygen electrocatalysis. Nature Comm. 6. doi: 10.1038/ncomms8345.

Li, W.-W., G.-P. Sheng, X.-W. Liu and H.-Q. Yu. 2011. Recent advances in the separators for microbial fuel cells. Bioresour. Technol., Special Issue: Biofuels—II: Algal Biofuels and Microbial Fuel Cells 102: 244–252. doi: 10.1016/j.biortech.2010.03.090.

Li, Z., X. Zhang, J. Lin, S. Han and L. Lei. 2010a. Azo dye treatment with simultaneous electricity production in an anaerobic–aerobic sequential reactor and microbial fuel cell coupled system. Bioresour. Technol. 101: 4440–4445. doi: 10.1016/j.biortech.2010.01.114.

Li, Z., X. Zhang, J. Lin, S. Han and L. Lei. 2010b. Azo dye treatment with simultaneous electricity production in an anaerobic–aerobic sequential reactor and microbial fuel cell coupled system. Bioresour. Technol. 101: 4440–4445. doi: 10.1016/j.biortech.2010.01.114.

Lin, C.-W., C.-H. Wu, Y.-H. Chiu and S.-L. Tsai. 2014. Effects of different mediators on electricity generation and microbial structure of a toluene powered microbial fuel cell. Fuel 125: 30–35. doi: 10.1016/j.fuel.2014.02.018.

Liu, L., F.B. Li, C.H. Feng and X.Z. Li. 2009. Microbial fuel cell with an azo-dye-feeding cathode. Appl. Microbiol. Biotechnol. 85(1): 175–183. doi: 10.1007/s00253-009-2147-2149.

Logan, B.E. and J.M. Regan. 2006. Microbial fuel cells—challenges and applications. Environ. Sci. Technol. 40: 5172–5180. doi: 10.1021/es0627592.

Logan, B.E. 2012. Essential data and techniques for conducting microbial fuel cell and other types of bioelectrochemical system experiments. ChemSusChem. 5: 988–994. doi: 10.1002/cssc.201100604.

Lu, N., S. Zhou, L. Zhuang, J. Zhang and J. Ni. 2009. Electricity generation from starch processing wastewater using microbial fuel cell technology. Biochem. Eng. J. 43: 246–251. doi: 10.1016/j.bej.2008.10.005.

Luo, Y., G. Liu, R. Zhang and C. Zhang. 2010. Power generation from furfural using the microbial fuel cell. J. Power Sources 195: 190–194. doi: 10.1016/j.jpowsour.2009.06.057.

Marsili, E., D.B. Baron, I.D. Shikhare, D. Coursolle, J.A. Gralnick and D.R. Bond. 2008. *Shewanella* secretes flavins that mediate extracellular electron transfer. Proc. Natl. Acad. Sci. 105: 3968–3973. doi: 10.1073/pnas.0710525105.

Morris, J.M., S. Jin, B. Crimi and A. Pruden. 2009. Microbial fuel cell in enhancing anaerobic biodegradation of diesel. Chem. Eng. J. 146: 161–167. doi: 10.1016/j.cej.2008.05.028.

Mu, Y., K. Rabaey, R.A. Rozendal, Z. Yuan and J. Keller. 2009a. Decolorization of azo dyes in bioelectrochemical systems. Environ. Sci. Technol. 43: 5137–5143. doi: 10.1021/es900057f.

Mu, Y., R.A. Rozendal, K. Rabaey and J. Keller. 2009b. Nitrobenzene removal in bioelectrochemical systems. Environ. Sci. Technol. 43: 8690–8695. doi: 10.1021/es9020266.

Nevin, K.P., B.-C. Kim, R.H. Glaven, J.P. Johnson, T.L. Woodard, B.A. Methé et al. 2009. Anode biofilm transcriptomics reveals outer surface components essential for high density current production in *Geobacter sulfurreducens* fuel cells. PLOS ONE 4: e5628. doi: 10.1371/journal. pone.0005628.

Pandit, S., S. Ghosh, M.M. Ghangrekar and D. Das. 2012. Performance of an anion exchange membrane in association with cathodic parameters in a dual chamber microbial fuel cell. Int. J. Hydrog. Energy 37: 9383–9392. doi: 10.1016/j.ijhydene.2012.03.011.

Pandit, S., S. Khilari, S. Roy, D. Pradhan and D. Das. 2014. Improvement of power generation using *Shewanella putrefaciens* mediated bioanode in a single chambered microbial fuel cell: Effect of different anodic operating conditions. Bioresour. Technol. 166: 451–457. doi: 10.1016/j. biortech.2014.05.075.

Pandit, S., S. Khilari, S. Roy, M.M. Ghangrekar, D. Pradhan and D. Das. 2015. Reduction of start-up time through bioaugmentation process in microbial fuel cells using an isolate from dark fermentative spent media fed anode. Water Sci. Technol. J. Int. Assoc. Water Pollut. Res. 72: 106–115. doi: 10.2166/wst.2015.174.

Pandit, S., K. Chandrasekhar, R. Kakarla, A. Kadier and V. Jeevitha. 2017. Basic principles of microbial fuel cell: technical challenges and economic feasibility. pp. 165–188. *In*: Kalia, V.C. and P. Kumar [eds.]. Microbial Applications Vol. 1. Springer International Publishing. doi: 10.1007/978-3-319-52666-9_8.

Park, H.I., C. Wu and L.-S. Lin. 2012. Coal tar wastewater treatment and electricity production using a membrane-less tubular microbial fuel cell. Biotechnol. Bioprocess Eng. 17: 654–660. doi: 10.1007/s12257-011-0374-2.

Pham, T.H., K. Rabaey, P. Aelterman, P. Clauwaert, L. De Schamphelaire, N. Boon et al. 2006. Microbial fuel cells in relation to conventional anaerobic digestion technology. Eng. Life Sci. 6: 285–292. doi: 10.1002/elsc.200620121.

Pisciotta, J.M. and J.J. Dolceamore Jr. 2016. Bioelectrochemical and conventional bioremediation of environmental pollutants. J. Microb. Biochem. Technol. 8. doi: 10.4172/1948-5948.1000306.

Rabaey, K., N. Boon, S.D. Siciliano, M. Verhaege and W. Verstraete. 2004. Biofuel cells select for microbial consortia that self-mediate electron transfer. Appl. Environ. Microb. 70(9): 5373–5382.

Rabaey, K., J. Rodríguez, L.L. Blackall, J. Keller, P. Gross, D. Batstone et al. 2007. Microbial ecology meets electrochemistry: electricity-driven and driving communities. ISME J. 1: 9–18. doi: 10.1038/ismej.2007.4.

Rabaey, K. and J. Keller. 2008. Microbial fuel cell cathodes: from bottleneck to prime opportunity? Water Sci. Technol. 57: 655. doi: 10.2166/wst.2008.103.

Richter, L.V., S.J. Sandler and R.M. Weis. 2012. Two isoforms of *Geobacter sulfurreducens* pila have distinct roles in pilus biogenesis, cytochrome localization, extracellular electron transfer, and biofilm formation. J. Bacteriol. 194: 2551–2563. doi: 10.1128/JB.06366-11.

Rismani-Yazdi, H., S.M. Carver, A.D. Christy and O.H. Tuovinen. 2008. Cathodic limitations in microbial fuel cells: An overview. J. Power Sources 180: 683–694. doi: 10.1016/j. jpowsour.2008.02.074.

Roccaro, P., M. Sgroi and F.G. Vagliasindi. 2013. Removal of xenobiotic compounds from wastewater for environment protection: treatment processes and costs. Chem. Eng. Trans. 32: 505–510.

Shea, C., P. Clauwaert, W. Verstraete and R. Nerenberg. 2008. Adapting a denitrifying biocathode for perchlorate reduction. Water Sci. Technol. 58: 1941–1946. doi: 10.2166/wst.2008.551.

Shi, L., S. Deng, M.J. Marshall, Z. Wang, D.W. Kennedy, A.C. Dohnalkova et al. 2008. Direct involvement of type II secretion system in extracellular translocation of *Shewanella oneidensis* outer membrane cytochromes MtrC and OmcA. J. Bacteriol. 190: 5512–5516. doi: 10.1128/ JB.00514-08.

Skadberg, B., S.L. Geoly-Horn, V. Sangamalli and J.R.V. Flora. 1999. Influence of pH, current and copper on the biological dechlorination of 2,6-dichlorophenol in an electrochemical cell. Water Res. 33(9): 1997–2010. doi: 10.1016/s0043-1354(98)00431-x.

Solanki, K., S. Subramanian and S. Basu. 2013. Microbial fuel cells for azo dye treatment with electricity generation: A review. Bioresour. Technol. 131: 564–571. doi: 10.1016/j.biortech.2012.12.063.

Stams, A.J.M., F.A.M. de Bok, C.M. Plugge, M.H.A. van Eekert, J. Dolfing and G. Schraa. 2006. Exocellular electron transfer in anaerobic microbial communities. Environ. Microbiol. 8: 371–382. doi: 10.1111/j.1462-2920.2006.00989.x.

Sun, J., Y. Hu, Z. Bi and Y. Cao. 2009. Simultaneous decolorization of azo dye and bioelectricity generation using a microfiltration membrane air-cathode single-chamber microbial fuel cell. Bioresour. Technol. 100: 3185–3192. doi: 10.1016/j.biortech.2009.02.002.

Wang, X., Z. Cai, Q. Zhou, Z. Zhang and C. Chen. 2012. Bioelectrochemical stimulation of petroleum hydrocarbon degradation in saline soil using U-tube microbial fuel cells. Biotechnol. Bioeng. 109: 426–433. doi: 10.1002/bit.23351.

Wen, Q., Y. Wu, L. Zhao, Q. Sun and F. Kong. 2010. Electricity generation and brewery wastewater treatment from sequential anode-cathode microbial fuel cell. J. Zhejiang Univ. Sci. B 11: 87–93. doi: 10.1631/jzus.B0900272.

Xafenias, N., Y. Zhang and C.J. Banks. 2015. Evaluating hexavalent chromium reduction and electricity production in microbial fuel cells with alkaline cathodes. Int. J. Environ. Sci. Technol. 12: 2435–2446. doi: 10.1007/s13762-014-0651-7.

Yan, Z., N. Song, H. Cai, J.-H. Tay and H. Jiang. 2012. Enhanced degradation of phenanthrene and pyrene in freshwater sediments by combined employment of sediment microbial fuel cell and amorphous ferric hydroxide. J. Hazard. Mater. 199–200: 217–225. doi: 10.1016/j.jhazmat.2011.10.087.

Yang, Y., M. Xu, J. Guo and G. Sun. 2012. Bacterial extracellular electron transfer in bioelectrochemical systems. Process Biochem. 47: 1707–1714. doi: 10.1016/j.procbio.2012.07.032.

Zhang, B. and J. Ni. 2010. Enhancement of Electricity Generation and Sulfide Removal in Microbial Fuel Cells with Lead Dioxide Catalyzed Cathode. Published in: 2010 4th International Conference on Bioinformatics and Biomedical Engineering. INSPEC Accession Number: 11510248;DOI: 10.1109/ICBBE.2010.5514899, https://ieeexplore.ieee.org/document/5514899/authors#authors.

Zhu, X. and J. Ni. 2009. Simultaneous processes of electricity generation and p-nitrophenol degradation in a microbial fuel cell. Electrochem. Commun. 11: 274–277. doi: 10.1016/j.elecom.2008.11.023.

6

Microbial Removal of Dye Stuffs

T. Selvankumar,* C. Sudhakar and M. Govarthanan*

Chemical Nature of Dyes

Dyes can generally be described as colored substances that have an affinity to the substrates to which they are being applied (Pereira and Alves 2012). They absorb light in the visible spectrum (400–700 nm) with at least one chromophore (colour-bearing group). They have a conjugated system, that is, a structure with alternating double and single bonds, and exhibit resonance of electrons, which is a stabilizing force in organic compounds (Hossain 2014).

The components of the dye consist of chromogen, that is, a chemical compound that is either colored or could be made colored by the attachment of a suitable substituent. The chromophore and the auxochrome(s) are also part of the chromogen (Carmen and Daniela 2012). The chromophore is a chemical group that is responsible for the appearance of color in compounds (the chromogen) where it is located (Fig. 1).

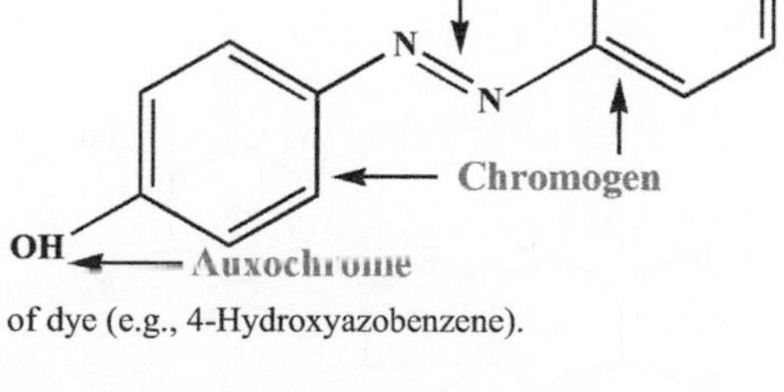

Fig. 1: The components of dye (e.g., 4-Hydroxyazobenzene).

PG and Research Department of Biotechnology, Mahendra Arts and Science College, Kalippatti-637501, Namakkal (District), Tamil Nadu, India.
* Corresponding authors: selvankumar75@gmail.com; gova.muthu@gmail.com

The colorants are sometimes also classified according to their main chromophore (e.g., azo dyes contain the chromophore $-N = N-$) (Iqbal 2008).

Classification of dyes

A dye, as mentioned above, is a colored substance that has an affinity to the substrate to which it is being applied. The dye is generally applied in an aqueous solution, and may require a mordant to improve the fastness of the dye on the fiber. There are two major types of dyes—natural and synthetic dyes. The natural dyes are extracted from natural substances such as plants, animals, or minerals. Synthetic dyes are made in a laboratory. Chemicals are synthesized for making synthetic dyes.

Natural dyes

The natural dyes are obtained from plants, animals, or are of mineral origin with none or very little processing. Some of our most common dyes like yellow, brown, blue and red are still derived from natural sources. The greatest source of dyes has been the plant kingdom, notably roots, berries, barks, leaves, and wood, but only a few have ever been used on a commercial scale. The two old ways to add color to a substance were either to cover it with a pigment (painting), or to colour the whole mass (dyeing). Pigments for painting were usually made from ground up colored rocks and minerals.

Natural dyes are often negatively charged. Positively charged natural dyes do exist, but are not common. In other words, the colored part of the molecule is usually the anion. Although the molecular charge is often shown on a specific atom in structural formulae, it is the whole molecule that is charged. Many, but by no means all, natural dyes require the use of a mordant.

The use of dyes is very ancient. Kermes (natural red 3) is identified in the bible book of Exodus, where references are made to scarlet colored linen. Similar dyes are carmine (natural red 4) and lac (natural red 25). These three dyes are close chemical relatives, obtained from insects of the genus Coccus. All require a mordant. The most commonly used natural dye is undoubtedly hematein (natural black 1), obtained from the heartwood of a tree. This dye also requires a mordant. Saffron (natural yellow 6), is obtained from the stigmata of Crocus sativus, and is used without a mordant, staining as an acid dye. Although its use is very ancient, it is more common now as a coloring and spice for food than for dyeing, due to its expense. Each dye is named according to the pattern: Natural + base colour + number (Fig. 2).

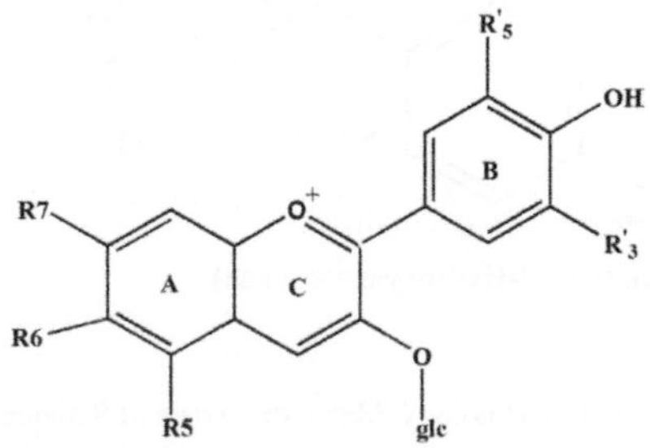

Fig. 2: Dye structure.

Synthetic dye

In 1856, William Henry Perkins accidentally discovered mauveine (Fig. 3), the world's first commercially successful synthetic dye and marked the start of synthetic dye industry. In the last 160 years, several million different colored compounds have been synthesized and approximately 15,000 produced on an industrial scale. Dyes derived from organic or inorganic compound are known as synthetic dyes. Examples of this class of dyes are Direct, Acid, Basic, Reactive, Mordant, Vat, Sulphure, Disperse dye, etc. (Kiernan 2001). Using general dye chemistry as the basis for classification, textile dyestuffs are grouped into 14 categories or classes.

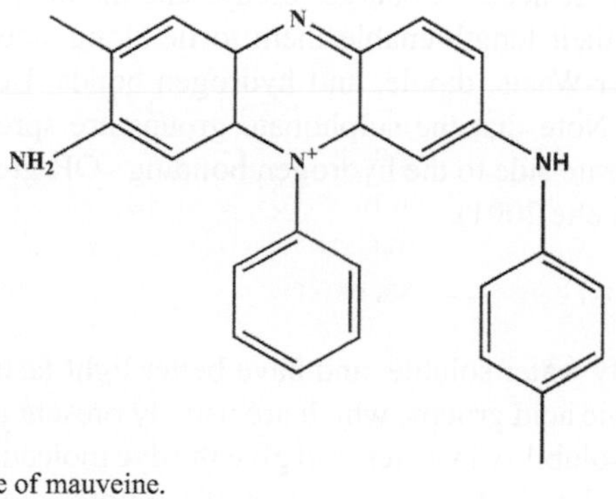

Fig. 3: Chemical structure of mauveine.

Table 1: Some applications of dye.

Group	Application
Direct	Natural fiber (Cotton), Cellulosic, and Blends
Vat dyes	Cotton, Cellulosic, and Blends
Sulphur	Cotton, Cellulosic fibers
Organic pigments	Cotton, Cellulosic, Blended Fabrics, Paper
Reactive	Cellulosic fibers and Fabric
Dispersed dyes	Synthetic fibers
Acid dyes	Wool, Silk, Synthetic fibers, Leather
Azoic	Viscose, Printing inks, and Pigments
Basic	Silk, Wool, Cotton
Oxidation dyes	Hair
Developed dyes	Cellulosic fibers and Fabric
Mordant dyes	Cellulosic fibers and Fabric, Silk, Wool
Optical/Fluorescent brighteners	Synthetic fibers, Leather, Cotton, Sports goods
Solvent dyes	Wood staining, Solvent inks, Waxes, Colouring oils

Direct dyes

Direct or substantive dyeing is normally carried out in a neutral or slightly alkaline dye bath, at or near boiling point, with the addition of either sodium chloride (NaCl) or sodium sulfate (Na_2SO_4) or sodium carbonate (Na_2CO_3). Direct dyes are used on cotton, paper, leather, wool, silk, and nylon. They are also used as pH indicators and as biological stains (Buchanan and Rita 1995).

The name direct dye alludes to the fact that these dyes do not require any form of fixing. They are almost always azo dyes, with some similarities to acid dyes (Seidel et al. 2005). They also have sulphonate functionality, but in this case, it is only to improve solubility, as the negative charges on dye and the fiber will repel each other. Their flat shape and their length enable them to lie along-side cellulose fibers and maximize the Van-der-Waals, dipole, and hydrogen bonds. Below is a diagram of a typical direct dye. Note that the sulphonate groups are spread evenly along the molecule on the opposite side to the hydrogen bonding –OH groups to minimise any repulsive effects (Rouette 2001).

Acid dyes

Acidic dyes are highly water soluble, and have better light fastness than basic dyes. They contain sulphonic acid groups, which are usually present as sodium sulphonate salts. These increase solubility in water, and give the dye molecules a negative charge. In an acidic solution, the $-NH_2$ functionalities of the fibres are protonated to give a positive charge: $-NH^{3+}$ (Rouette 2001). This charge interacts with the negative dye charge, allowing the formation of ionic interactions. Along with this, Van-der-Waals bonds, dipolar bonds, and hydrogen bonds are formed between dye and the fiber. As a group, acid dyes can be divided into two sub-groups: acid-leveling or acid-milling (Buchanan and Rita 1995).

Basic dyes or cationic dyes

Mauveine, the first to be discovered by Perkin, was a basic dye and most of the dyes which followed, including magenta, malachite green, and crystal violet, were of the same type. Basic dyes dye wool and silk from a dye bath containing acid, but they dye cotton fibers only in the presence of a mordant, usually a metallic salt that increases the affinity of the fabric for the dye. Basic dyes include the most brilliant of all the synthetic dyes known, but unfortunately they have very poor light and wash fastness (Pellew 1998).

Fiber-reactive dyes

Dyes that during the process of dyeing form a covalent bond with the substrate are known as reactive dyes. Reactive (fiber-reactive) dyes combine with fiber molecules either by addition or substitution. The color cannot be removed if properly applied. Colors are bright with very good colorfastness but are susceptible to damage by chlorine bleaches. Reactive dyes color cellulosics (cotton, flax, and viscose rayon), silk, wool, and nylon. Reactive dyes are used in conjunction with disperse dyes to dye polyester and cellulosic fiber blends. They were introduced to the industry in 1956 (Warren 1996).

Mordant

Mordant is a Latin word meaning 'to bite'. Mordants act as 'fixing agents' to improve the color fastness of some acid dyes, which have the ability to form complexes with metal ions. Mordants are usually metal salts; alum was commonly used for ancient dyes, but there is a large range of other metallic salt mordants available. Each one gives a different color with any particular dye by forming an insoluble complex with the dye molecules. Chromium salts such as sodium or potassium dichromate are commonly used now for synthetic mordant dyes. The diagrams below show C.I. Mordant Black 1 with and without a chromium (III) ion. Chromium (III) forms 6-coordinate complexes, so two Mordant Black molecules would attach to one ion. Only one is shown below for clarity (Shenai 1997).

Vat dyes

Vat dyes are a good example of the cross-over between dyes and pigments. Large, planar, and often containing multi-ring systems, vat dyes come exclusively from the carbonyl class of dyes (for example, indigo). The ring systems of the vat dyes help to strengthen the Van-der-Waals forces between the dye and fiber.

Vat dyes are insoluble in water, but may become solublised by alkali reduction, for example sodium dithionite (a reducing agent) in the presence of sodium hydroxide. For this reason, they tend not to contain many other functional groups which may be vulnerable to oxidation or reduction. The leuco form produced by alkali reduction is absorbed by the cellulose and, once there, can be oxidised back to its insoluble form. Oxidation is usually performed using hydrogen peroxide, but occasionally with atmospheric oxygen under the correct conditions. Treating the dyed textile with a soap completes the process, since the soap molecules encourage the dye molecules to clump together and become crystalline.

Sulphur

Sulphur dyes are the most commonly used dyes manufactured for cotton in terms of volume. They are cheap, generally have good wash-fastness, and are easy to apply. Sulphur dyes are water insoluble. They have to be treated with a reducing agent and an alkali at temperature of around 80°C where the dye breaks into small particles which then becomes water soluble and hence can be absorbed by the fabric. The dyes are absorbed by cotton from a bath containing sodium sulfide or sodium hydrosulfite and are made insoluble within the fiber by oxidation. During this process these dyes form large complex molecules which are the basis of their good wash-fastness. Their main use is to dye cellulose fibers, cotton and viscose (Chakraborty and Priyadarshi Jaruhar 2014).

Disperse dye

Disperse dyes have low solubility in water, but they can interact with the polyester chains by forming dispersed particles. Their main use is the dyeing of polyesters, and they find minor use dyeing cellulose acetates and polyamides. The general structure of disperse dyes is small, planar, and non-ionic, with attached polar functional groups like $-NO_2$ and $-CN$. The shape makes it easier for the dye to slide between the tightly-packed polymer chains, and the polar groups improve the water solubility, improve the dipolar bonding between dye and polymer and affect the color of the dye. However, their small size means that disperse dyes are quite volatile, and tend to sublime out of the polymer at sufficiently high temperatures (Ramalingam and Vimaladevi 2017).

The dye is generally applied under pressure, at temperatures of about 130°C. At this temperature, thermal agitation causes the polymer's structure to become looser and less crystalline, opening gaps for the dye molecules to enter. The interactions between the dye and polymer are thought to be Van-der-Waals and dipole forces (Sivakumar 2014).

The volatility of the dye can cause loss of color density and staining of other materials at high temperatures. This can be counteracted by using larger molecules or making the dye more polar (or both). This has a drawback, however, in that this new larger, more polar molecule will need more extreme forcing conditions to dye the polymer (Tsatsaroni and Liakopoulou-Kyriakides 1995).

Industrial Applications of Dyes

Organic dyes are used in a wide range of industrial applications in sectors such as textiles (the most significant), cosmetics, food products, pharmaceuticals, and paper printing (Pilarcallao and Soledad Larrechi 2015). Various groups of dyes can be distinguished on the basis of their compound structure: for example, anthraquinones, xanthenes, phtalocyanins, and quinine-imines. The most commonly used dyes in industrial applications are azo dyes because they are cost effective to synthesis.

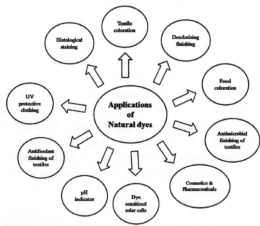

Fig. 4: Applications of natural dyes.

S. No.	Dye	Chemical structure
1	Disperse yellow 3	
2	Basic Brown 1	
3	Direct orange 26	
4	Sulphur red 7	
5	Mordant red 11	
6	Vat blue 4	
7	Acid yellow 36	
8	Reactive blue 5	

Textile Dyeing

Industrial applications of dyes are divided into two groups (Rouette 2001):

I. Textile Dyeing: (Samanta and Agarwal 2009)

These groups include the following:

> ➤ Dyes for cotton, model fibers, and linen: Reactive dyes, direct dyes, sulphur dyes, vat dyes, azoic colorants, and pigments.
> ➤ Dyes for wool, silk, and polyamide: Acid dyes, chrome dyes, metal-complex dyes, and reactive dyes.
> ➤ Dyes for acetate, triacetate, and polyester fibers: disperse dyes.
> ➤ Dyes for acrylic fibers: cationic dyes.

II. Non-Textile Dyeing

Non-textile dyeing includes: Leather dyes, food dyes, fur dyes, hair dyes, photographic dyes, ink dyes, plastic dyes, and indicator dyes.

Food coloration

Natural colors are suitable for a wide range of sugar confectionery products. Hard candies, tablets, gummies, pectin-based candies, panned candies, and gums are perfect applications for natural colors. Natural colors work well in sugar free applications, including in candies sweetened with stevia, and in confections fortified with health ingredients or vitamins (Delgado-Vargas et al. 2000).

There are major three categories of food colors:

i) Natural colors

Annatto and Marigold being non-toxic in nature are widely acceptable food colors for food products that include dairy products like butter, ghee, cheese, ice cream, and margarine as well as some oil and bakery products. They are also used in preparation of cosmetic accessories, pharmaceuticals, crayons, textiles, floor wax, shoe polishes, etc. Yellow color is extracted from Marigold flowers, while butter yellow color is extracted from Annatto seed (Mott MacDonald 2000).

ii) Synthetic colors

Synthetic Food Colors also known as Artificial Food Colours, are manufactured chemically, and are the most commonly used dyes in the food, pharmaceutical, and cosmetic industries. Seven dyes were initially approved under the Pure Food and Drug Act of 1906, but several have been delisted and replacements have been found, such as, Brilliant Blue FCF, Indigotine, Fast green FCF, Allura red AC, Erythrosine, Tartrazine, and Sunset yellow FCF.

Synthetic color was developed by Sir William Henry Perkin and by the turn of the century, unmonitored color additives had spread through Europe and the United States in all sorts of popular foods, including ketchup, mustard, jellies, and wine (Walford 1980). Originally, these were dubbed 'coal-tar' colors because the starting materials were obtained from bituminous coal (Sharma et al. 2011).

Many synthesized dyes were easier and less costly to produce and were superior in coloring properties when compared to naturally derived alternatives. Some synthetic food colorants are diazo dyes. Diazo dyes are prepared by coupling a diazonium compound with a second aromatic hydrocarbon (Klaus Hunger et al. 2005). The resulting compounds contain conjugated systems that efficiently absorb light in the visible parts of the spectrum, that is, they are deeply colored. The attractiveness of the synthetic dyes is that their color, lipophilicity, and other attributes can be engineered by the design of the specific dyestuff. The color of the dyes can be controlled by selecting the number of azo-groups and various substituents. Yellow shades are often achieved by using acetoacetanilide (Konig 2015).

iii) Lakes and dyes

Dyes dissolve in water, but are not soluble in oil. Dyes are manufactured as powders, granules, liquids. or other special purpose forms. They can be used in beverages, dry mixes, baked goods, confections, dairy products, pet foods, and a variety of other products. Dyes also have side effects which lakes do not, including the fact that large amounts of dyes ingested can color stools.

Lakes are made by combining dyes with salts to make insoluble compounds. They tint by dispersion, are not oil soluble, but are oil dispersible, are more stable than dyes,

Table 2: Common color and associated food (Chaitanya Lakshmi et al. 2014).

S. No.	Colour	Chromophore	Plant sources	Nutrients
1	Purple-blue	Anthocyanins	Eggplant, blackberry, purple, cabbage, plum, blueberry, raisins, prunes, purple grapes, figs.	Lutein, zeaxanthin, resveratrol, vitamin C, flavanoid, ellagic acid, quecertin
2	Green	Chlorophyll	Avocado, cucumber, spinach, kale, broccoli, snow pea, zucchini, artichoke, lettuce, kiwi.	Lutein, zeaxanthin, vitamin C, calcium, folate, β-carotene
3	White-tan	Anthoxanthins	Cauliflower, mushrooms, parsnip, potato, ginger, onions, jicama, banana, garlic, onions.	Ancillin, potassium, selenium
4	Red	Lycopene or Anthocyanins	Cranberry, beet, watermelon, tomato, strawberry, pomegranate.	Ellagic acid, quecertin, hesperidin, etc.
5	Yellow-orange	Carotenoids	Papaya, pineapple, apricot, pumkin, peach, carrot, orange, corn.	β-carotene, zeaxanthin, flavanoid, vitamin C, potassium

and are ideal for coloring products containing fats and oils or items lacking sufficient moisture to dissolve dyes. Typical uses include coated tablets, cake and doughnut mixes, hard candies and chewing gums, lipsticks, soaps, shampoos, talc, etc.

Lake pigments find wide usage in areas like foodstuffs, pet foods, drugs and pharmaceuticals, cosmetics, plastics, plastic films, can linings, plastic food containers, and inks and stationery.

Dye Sensitized Solar Cell (DSSC)

A dye-sensitized solar cell is a third generation photovoltaic device for the conversion of visible light into electric energy which provides a technically and economically credible alternative concept to present day p-n junction photovoltaic devices (Gratzel 2003). There have been some interesting explorations of natural dyes in the context of the dye-sensitized solar cell (DSSC) application using pigments obtained from biomaterials. Several natural dyes, such as cyanine, anthocyanins, cyanidins, tannins, chlorophyll and their derivatives, carotenoids, betalains, chalcones, and many others as a cheaper, faster, low-energy requiring and environment-friendly alternative for use in dye-sensitized solar cells (Hao et al. 2006). Although these natural dyes often work poorly in DSSCs, these are expected as low cost and prepared easily comparing to ruthenium (Ru) complex based dyes (Mohammad et al. 2013).

Cosmetics and Pharmaceuticals

Catering to the needs of cosmetic industry, we provide cosmetic colors that are used in the manufacturing of lipsticks, nail polishes, etc. In these colors, mainly Lake dyes are used. These colors are formulated using the best grade raw materials (Dweck 2002).

Colorants or coloring agents are mostly used to impart a distinctive appearance to the pharmaceutical dosage forms. We can also say that the colorants are the cosmetics for the pharmaceutical preparations, because the visual appearance of dosage forms can be improved by using suitable colorants. The main categories of dosage form that are colored are: tablets; hard or soft gelatin capsules (the capsule shell or coated beads); toothpastes; topical creams; oral liquids; ointments; and salves (Krishna and Gannu 2011). The grace and eye appeal of a colored product is valuable, especially for children whom it is often used to treat (that is, in the form of medicinal tablets or capsules and syrups) to avoid injections and allow treatment at home.

Pharmaceutical preparations are colored mainly for the following reasons:

  i) Increases acceptability
 ii) For identification
iii) Standard preparations
 iv) Stability purpose

The food, drug and cosmetic act

The Food Drug and Cosmetic Act of 1938 created three categories of coal tar dyes, of which only the first two are applicable to the manufacture of chewable tablets.

FD and C colors: These are colorants that are certifiable for use in foods, drugs, and cosmetics.

D and C colors: These are dyes and pigments considered safe for use in drugs and cosmetics when in contact with mucous membranes or when ingested.

External D and C colors: These colorants, due to their oral toxicity, are not certifiable for use in products intended for ingestion but are considered safe for use in products applied externally (Lieberman et al. 1989).

Widely used colorants in pharmaceuticals

Betacarotene (Betacarotene; β carotene; β,β carotene): It occurs in the pure state as red crystals when recrystallized from light petroleum. It is capable of producing colors varying from pale yellow to dark orange. It can be used as a color for sugar coated tablets prepared by the ladle process. However, Beta carotene is very unstable to light and air and products containing this material should be securely packaged to minimize degradation. It is particularly unstable when used in spray coating processes, probably owing to atmospheric oxygen attacking the finely dispersed spray droplets.

Brilliant Blue FCF (Erioglaucine; Eriosky blue; Patent Blue AR): It can be combined with Tartrazine to produce various shades of green. It is widely used in mouth washes, shampoos, soaps, and other hygiene and cosmetics applications. It has the capacity for inducing an allergic reaction in individuals with preexisting moderate asthma (Richard et al. 1979).

Indigo Carmine (Indigotine; sodium indigotindisulfonate; soluble Indigo blue): It is a dark blue powder. Aqueous solutions are blue or bluishpurple. The primary use of Indigo carmine is as a pH indicator. Indigo carmine is an indigoid dye used to color oral and topical pharmaceutical preparations and also used with yellow colors to produce green colors (Krishna and Gannu 2011).

It is used as a dye in the manufacturing of capsules. Indigo Carmine is also used to color nylon surgical sutures and is used diagnostically as a 0.8% w/v injection.

Coloring systems for various dosage forms

In selecting a colorant for a given application, prime consideration should be given to the type formulation in which the colorant is to be incorporated. Whatever the form of colorant chosen, it should meet as many characteristics as the ideal colorant.

a) Tablets
   i) Wet granulation
   ii) Direct compression

b) Tablet coating
   i) Sugar coating
   ii) Filmcoating
c) Capsules
   i) Hard gelatin capsules
   ii) Soft gelatin capsules or soft gels
d) Liquid products
e) Ointments and salves
f) Tooth pastes

Source of pollutants

Dye waste waters enter the environment from manufacturers and consumers (i.e., textile, leather and food industries) usually in the form of dispersion or a true solution and often in the presence of other organic compounds originating from operational processes. Color is the first contaminant to be recognized in textile waste water and has to be removed before discharging into water bodies or onto land (Kumar et al. 2005). The presence of small amounts of dyes in water (even < 1 ppm) is highly visible and it affects the aesthetic merit, causes significant loss in luminosity and any increase in the temperature will greatly deplete the dissolved oxygen concentration in waste water. This results in subsequent alteration of the aquatic ecosystem (Moreira et al. 2004).

The removal of color from textile waste water is often more important than the removal of the soluble colorless organic substances which usually contribute to the major fraction of Biochemical Oxygen Demand (BOD). Methods for the removal of BOD from most of these effluents are fairly well established (Pearce et al. 2003). On the other hand, textile waste waters exhibit low BOD to COD ratios (< 0.1) indicating their difficulty to bioremediate or breakdown. On the whole, textile waste water is characterized by unfixed dyes, organic pollutants (much higher than regular domestic waste water), large amounts of COD (organic compounds), high conductivity due to salts, high amounts of sulphide and heavy metals due to chlorinated bleaching agents and halogen, and sulphur or heavy metal dyes (Pagga and Brown 1986).

The ability of microorganisms to degrade textile azo dyes has been studied extensively in both aerobic and anaerobic processes (Pearce et al. 2003). More and more recalcitrant dyes are manufactured with the hope of improving the delivery of color onto fabric, at the expense of becoming increasingly difficult to bioremediate. This has created a need to investigate and understand the actual mechanisms behind the biodegradation of textile waste water. Several enzymes from fungi and bacteria have been identified and used in the breakdown of azo dyes. White rot fungi produce Lignin Modifying Enzymes (LME) namely; laccase, lignin, and manganese peroxidase. These enzymes have been shown to oxidize colored phenolic compounds and aromatic amines via a one-step electron oxidation (Couto et al. 2004). The wood degrading white rot basidiomycete *Phanerodontia chrysosporium* is the only one that has been shown to completely degrade a number of azo dyes (Chivukula and Renganathan 1995). Lignin and manganese peroxidase produced by *P. chrysosporium* appear to initiate the azo degradation while the laccases oxidize aromatic compounds such as aniline and phenols in the presence of oxygen. The substrates are oxidized by one electron to generate corresponding phenoxy radicals, which either polymerize to yield a phenolic polymer or are further oxidized by laccase to produce a quinone (Bollag 1992). The electrons generated from these reactions are transferred onto oxygen which is reduced to water. Although several white rot fungi have shown great potential to degrade azo dyes, they are difficult to apply in the actual bioremediation of textile waste water. The major disadvantage is that they can only be used in lab scale operations where they are cultured in shaking flasks. Their ability to adapt and survive in a sewage waste water treatment plant is minimal.

Earlier research in this field identified bacterial species that are capable of reducing and degrading azo dyes include *Bacillus, Xanthomonas,* and *Pseudomonas* (Zimmermann et al. 1982). Azoreductase enzymes produced by *Pseudomonas* sp. catalyze reductive cleavage of azo dyes in the presence of molecular oxygen, thus making them favorable for the development of azo dye bioremediation systems. These enzymes have high substrate affinity but can only degrade a small fraction of the vast range of azo dyes that are manufactured annually. They are, therefore, not readily applicable to the industrial treatment of textile effluent.

Research on bacterial azo dye reduction has traditionally been focused on the activity of facultative anaerobic bacteria from mammalian intestines. As the removal of dyes from waste water became a cause for concern, bacteria from other origins which included pure cultures, mixed consortia, anaerobic sediments, digester sludge, anaerobic granular sludge, and activated sludge were investigated. Since a large proportion of azo dyes can be reduced by different bacteria under anaerobic conditions, this indicates that azo dye reduction is a non-specific reaction. There are two main enzymatic mechanisms that are thought to be involved in the anaerobic reduction of azo dyes, namely: direct enzymatic azo dye reduction and indirect azo dye reduction.

The implementation of increasingly stringent standards for the discharge of wastes into environment has necessitated the need for the development of alternative processes for waste treatment. A large number of enzymes from different plants and microorganisms have been playing an important role in an array of waste treatment applications (Husain 2006).

These biological methods are founded on enzymes, which are produced by organisms. In the case of enzymatic degradation, laccases, peroxidases, azoreductases, and phenol oxidases have a potential for dye biodegradation, but in certain aspects, each of these enzyme falters in their biodegradation efficiency.

Microbial dye-decolorizing laccases

Laccase (EC 1.10.3.2, *p*-diphenol: dioxygenoxidoreductases) is one of a few enzymes that have been studied since the 19th century. Laccases are classified asphenol oxidases which are able to catalyse one-electron oxidation of the substrate associated with the simultaneous reduction of oxygen in water. These enzymes are produced by a wide spectrum of organisms such as bacteria, filamentous fungi, or plants (Bibi et al. 2011; Zhou and Xiang 2013).

Laccases catalyse the removal of a hydrogen atom from the hydroxyl group via electron oxidation, which produce non-toxic products during synthetic dye biodegradation (Telke et al. 2010; Adnan et al. 2015). Although the biodegradation potential of laccases is mostly evaluated by the change of an absorption spectrum of dye (Chmelova and Ondrejovic 2015), the destruction of chromophores can led to dye degradation (Chen and Ting 2015; Barbora et al. 2016).

Microbial dye-decolorizing peroxidases

Peroxidase (EC 1.11.1.7) is a heme-containing enzyme that is widely distributed in plants, microorganisms, and animals (Duarte-Vazquez et al. 2003). Heme is a complex between an iron ion (Fe^{+3}) and the molecule protoporphyrin IX. Peroxidases are classified into two super families, animal and plant enzymes having a Mr ranging from 30 to 150 kDa (Regalado et al. 2004).

Peroxidases are one of the major H_2O_2 decomposing enzymes and have ability to catalyze the redox reaction of a wide range of phenolic as well as non-phenolic compounds. On the basis of presence or absence of heme, the peroxidases have been classified into heme and nonheme Peroxidases (Passardi et al. 2005). Heme peroxidases have further been assigned into two superfamilies, namely peroxidase-cyclooxygenase (PCOX) superfamily and the peroxidase-catalase (PCAT) superfamily (Yves et al. 2011; Passardi et al. 2007).

Microbial dye-decolorizing azoreductase

Azoreductases (EC 1.7.1.6) require additional co-factors such as $NADH_2$, $NADPH_2$, and $FADH_2$ for the activation of the catalyst. Additionally, azoreductases can degrade only azo dyes via reductive cleavage of azo bonds (Pandey et al. 2007). One of the main disadvantages of this enzyme-catalyzed degradation processes the formation of toxic products. It is known that azoreductases are intracellular enzymes and their use in the pure form without production organisms is problematic due to issues of stability and necessary regeneration of co-factors. Furthermore, the direct application of microorganism produced azoreductases can result in additional problems from the inability of dye diffusion through cell membranes, due to the molecular weight of azo dyes (Robinson et al. 2001).

References

Adnan, L.A., P. Sathishkumar, A.R.M. Yusoff and T. Hadibarata. 2015. Metabolites characterisation of laccase mediated Reactive Black 5 biodegradation by fast growing ascomycete fungus *Trichoderma atroviride* F03. Int. Biodeterior. Biodegradation. 104: 274–282.

Barbora, L., C. Daniela and O. Miroslav. 2016. Degradation of synthetic dyes by laccases—A mini-review. Nova Biotechnologicaet Chimica. 15: 90–106.

Bibi, I., H.N. Bhatti and M. Asgher. 2011. Comparative study of natural and synthetic phenolic compounds as efficient laccase mediators for the transformation of cationic dye. Biochem. Eng. J. 56: 225–231.

Bollag, J.M. 1992. Enzymes catalyzing oxidative coupling reactions of pollutants. Metal Ions Biol. Syst. 28: 205–217.

Buchanan and Rita. 1995. A Dyer's Garden. Loveland, Colorado: Interweave.

Carmen, Z. and S. Daniela. 2012. Textile organic dyes—characteristics, polluting effects and separation/elimination procedures from industrial effluents—a critical overview. pp. 55–86. *In*: Puzyn, T. [ed.]. Organic Pollutants Ten Years After the Stockholm Convention—Environmental and Analytical Update. In Tech Press, Crotia.

Chaitanya Lakshmi, G. 2014. Food coloring: the natural way. Research Journal of Chemical Sciences 4: 87–96.

Chakraborty, J.N. and Priyadarshi Jaruhar. 2014. Dyeing of cotton with sulphur dyes using alkaline catalase as reduction catalyst. Indian Journal of Fibre and Textile Research 39: 303–309.

Chen, S.H. and A.S.Y. Ting. 2015. Biodecolorization and biodegradation potential of recalcitrant triphenylmethane dyes by *Coriolopsis* sp. isolated from compost. J. Environ. Manage. 150: 274–280.

Chivukula, M. and V. Renganathan. 1995. Phenolic azo dye oxidation by laccase from *Pyricularia oryzae*. Appl. Environ. Microbiol. 61: 4374–4377.

Chmelova, D. and M. Ondrejovic. 2015. Effect of metal ions on triphenylmethane dye decolorization by laccase from *Trametes versicolor*. Nova Biotechnol. Chim. 14: 191–200.

Couto, S.R., M.A. Sanaroman, D. Hofer and G.M. Gubitz. 2004. Stainless steel sponge: a novel carrier for the immobilisation of the white–rot fungus *Trametes hirsute* for the decolourisation of textile dyes. Biores. Technol. 1–6.

Delgado-Vargas, F., A.R. Jimenez and O. Paredes-Lopez. 2000. Natural pigments: carotenoids, anthocyanins, and betalains—characteristics, biosynthesis, processing, and stability. Crit. Rev. Food Sci. Nutr. 40: 173–289.

Duarte-vazquez, M.A., J.R. Whitaker, A. Rojo-dominguez, B.E. Garcia-Almendarez and C. Regalado. 2003. Isolation and thermal characterization of an acidic isoperoxidase from turnip roots. J. Agric. Food Chem. 51: 5096–5102.

Dweck, A.C. 2002. Natural ingredients for colouring and styling. Int. J. Cosmet. Sci. 24: 287–302.

Gratzel, M. 2003. Dye-sensitized solar cells. J. Photochem. Photobiol. C. 4: 145–153.

Hao, S., J. Wu, Y. Huang and J. Lin. 2006. Natural dyes as photosensitizers for dye-sensitized solar cell. Sol. Energ. 80: 209–214.

Hossain, I. 2014. Investigation into cotton knit dyeing with reactive dyes to achieve right first time (RFT) shade. Master Thesis. Daffodil International University, Bangladesh.

Husain, Q. 2006. Potential applications of the oxidoreductive enzymes in the decolorization and detoxification of textile and other synthetic dyes from polluted water: a review. Crit. Rev. Biotechnol. 60: 201–221.

Iqbal, M. 2008. Textile dyes. Rahber Publishers, Pakistan.

Kiernan, J.A. 2001. Classification and naming of dyes, stains and fluorochromes. Biotechnic and Histochemistry 76: 261–277.

Klaus Hunger, Peter Mischke, Wolfgang Rieper, RoderichRaue, Klaus Kunde and Aloys Engel. 2005. Azo Dyes in Ullmann's Encyclopedia of Industrial Chemistry. Wiley-VCH, Weinheim.

Konig, J. 2015. Food colour additives of synthetic origin. pp. 35–60. *In*: Scotter, Michael J. [ed.]. Colour Additives for Foods and Beverages. Elsevier.

Krishna, V.A. and P. Gannu. 2011. Colorants the cosmetics for the pharmaceutical dosage forms. International Journal of Pharmacy and Pharmaceutical Sciences 3: 13–21.

Kumar, K., S.S. Devi, K. Krishnamurthi, S. Gampawar, N. Mishra, G.H. Pandya et al. 2005. Decolourisation, biodegradation and detoxification of benzidene based azo dye. Biores. Technol. 97: 407–413.

Lieberman, H.A., L. Lachman and J.B. Schwartz. 1989. Pharmaceutical dosage forms: Tablets. J. Pharm. Sci. 79: 188.

Mohammad, S. Shahid-ul-Islam and M. Faqeer. 2013. Recent advancements in natural dye applications: a review. J. Clean. Prod. 53: 310–331.

Moreira, M.T., C. Viacava and G. Vidal. 2004. Fed-batch decolourisation of poly R-478 by *Trametes versicolor*. Brazilian Arch. Biol. Technol. 47: 179–183.

Mott MacDonald. 2000. Project Profile on Natural Food Colors—Marigold, Annatto, iNDEXTb.

Pagga, U. and D. Brown. 1986. The degradation of dyestuffs: Part II. Behaviour of dyestuffs in aerobic biodegradation tests. Chemosphere 15: 479–491.

Pandey, A., P. Singh and L. Iyengar. 2007. Bacterial decolorization and degradation of azo dyes. Int. Biodeterior. Biodegradation 59: 73–84.

Passardi, F., C. Cosio, C. Penel and C. Dunand. 2005. Peroxidases have more functions than a Swiss army knife. Plant Cell Reports 24: 255–265.

Passardi, F., G. Theiler, M. Zamocky, C. Cosio, N. Rouhier, F. Teixera et al. 2007. PeroxiBase: The peroxidase database. Phytochemistry 68: 605–611.

Pearce, C.I., J.T. Lloyd and J.T. Guthrie. 2003. The removal of colour from textile waste water using whole bacteria cells: A review. Dyes and Pigments 58: 179–196.

Pellew, C. 1998. Dyeing and Printing: Dyes and Dyeing. Abhishek Publications, Chandigarh.

Pereira, L. and M. Alves. 2012. Dyes-environmental impact and remediation. *In*: Malik, A. and E. Grohmann [eds.]. Environmental Protection Strategies For Sustainable Development, Strategies For Sustainability. Springer, New York.

Pilarcallao, M. and M. Soledad Larrechi. 2015. Simultaneous determination of organic dyes using second-order data. Fundamentals and Analytical Applications of Multiway Calibration, Chapter 9: 399–424.

Ramalingam, P. and S. Vimaladevi. 2017. Biodegradation and decolourization of azo dyes using marine bacteria. Int. J. Adv. Biotechnol. Res. 7: 1–12.

Regalado, C., B.E. Garcia-Almendarez and M.A. Duarte-Vazquez. 2004. Biotechnological applications of peroxidases. Phytochem. Rev. 3: 243–256.

Richard, W.W., H. Melvin, A.R.J. Dudley and N. Harold. 1979. Incidence of bronchoconstriction due to aspirin, azo dyes, non-azo dyes, and preservatives in a population of perennial asthmatics. J. Allergy Clin. Immunol. 64: 32–37.

Robinson, T., G. Mcmullan, R. Marchant and P. Nigam. 2001. Remediation of dyes in textile effluent: a critical review on current treatment technologies with a proposed alternative. Bioresour. Technol. 77: 247–255.

Rouette. 2001. Encyclopedia of Textile Finishing, Published by Springer.

Samanta, A.K. and P. Agarwal. 2009. Application of natural dyes on textiles. Indian J. Fibre Text. Res. 34: 384–399.

Seidel, J., S. Grafstrom and L. Eng. 2005. Stimulated emission of surface plasmons at the interface between a silver film and an optically pumped dye solution. Phys. Rev. Lett. 94: 177401.

Sharma, V., T.M. Harold and G.M. Peter. 2011. A global perspective on the history, use, and identification of synthetic food dyes. J. Chem. Educ. 88: 24–28.

Shenai, V.A. 1997. Technology of textile processing (2nd ed.). Mumbai: Sevak.

Sivakumar, T. 2014. Azo dye degradation by fungi—review. Int. J. Adv. Res. Biol. Sci. 1: 224–232.

Telke, A.A., A.A. Kadam, S.S. Jagtap, J.P. Jadhav and S.P. Govindwar. 2010. Biochemical characterization and potential for textile dye degradation of blue laccase from *Aspergillus ochraceus* NCIM-1146. Biotechnol. Bioprocess Eng. 15: 696–703.

Tsatsaroni, E. and M. Liakopoulou-Kyriakides. 1995. Effect of enzymatic treatment on the dyeing of cotton and wool fibres with natural dyes. Dyes and Pigments 29: 203–209.

Walford, J. 1980. Historical development of food colouration. Developments in Food Colours. London: Applied Science Publishers 1: 1–25.

Warren, S.P. 1996. Textile Coloration and Finishing. Durham, N.C. Carolina Academic Press.

Yves, M.E.L., M.S.E.B. Mbassi, W. Mbacham and J.P. Muluh. 2011. Heat stable peroxidases from *Vigna* species (V). Afr. J. Biotechnol. 10: 3168–3175.

Zhou, X. and X. Xiang. 2013. Effect of different plants on azo-dye wastewater biodecolorization. Procedia Environ. Sci. 18: 540–546.

Zimmermann, T., H.G. Kulla and T. Leisinger. 1982. Properties of purified orange II azoreductase, the enzyme initiating azo dye degradation by *Pseudomonas* KF46. Eur. J. Biochem. 129: 197–203.

7

Bioremediation of Organic Xenobiotics from Wastewater
A Review

Mayuri Chabukdhara,[1] *Sanjay Kumar Gupta,*[2,]*
Faiz Ahmad Ansari,[3] *Amit K. Bajhaiya*[4] and
Manish Kumar[5]

INTRODUCTION

Deterioration of water resource quality due to increased anthropogenic activities has raised concern worldwide. Water security is threatened and freshwater systems are increasingly endangered due to release of manufactured and naturally occurring chemical compounds being released or discharged into the environment (Schwarzenbach et al. 2006; Vörösmarty et al. 2010). This key environmental problem results in unprecedented health hazards not having previously existed (Vörösmarty et al. 2010). Recent decades saw increased attention to a set of harmful, human made xenobiotics that are found in both surface and groundwater in very

[1] Department of Environmental Biology and Wildlife Sciences, Cotton University, Guwahati, India 781001.
[2] Environmental Engineering, Department of Civil Engineering, Indian Institute of Technology, Delhi, India 110016.
[3] Institute for Water and Wastewater Technologies, Durban University of Technology, South Africa 4001.
[4] Umeå Plant Science Centre, Umeå University, Umea, Sweden, SE 90187.
[5] Department of Biology, Texas State University, San Marcos, TX 78666.
* Corresponding author: sanjuenv@gmail.com

low, yet relevant concentrations (Gavrilescu 2005; Kreuzinger 2008; El Shahawi et al. 2010) though wastewater may contain higher concentrations (Bester and Schäfer 2009). Among these prominent organic xenobiotics are polycyclic aromatic hydrocarbons (PAHs), phenolic compounds, petrochemicals, chlorinated solvents and dyes, explosives, pesticides, pharmaceutical compounds and personal care products (Bester 2008; Bell et al. 2011; Pal et al. 2010; El Shahawi et al. 2010). Organic xenobiotics are categorized into 4 main types (Colwell and Walker 1977), i.e., saturated hydrocarbons, aromatic hydrocarbons, asphaltenes and resins. These xenobiotics are omnipresent in the global environment (Dórea 2008).

The main concerns with organic xenobiotic compounds are their toxicity, threats to the public health, and the risk the ecosystem. These threats have motivated researchers to develop novel strategies to remove xenobiotics from contaminated environments. The potential health hazards of xenobiotic compounds, including POPs, are functions of their persistence in the environment. These include hydrophobic nature, low solubility in water, and a readily capable binding capacity with surface plankton and other organic particulates often undergoing sedimentation (Wilson et al. 1985). Exposure to extremely low even in nanogram levels of certain xenobiotic and organic compounds such as hydrocarbons containing halogens, polychlorinated biphenyls, dichlorodiphenyltrichloroethane may destabilize the normal metabolic activity of sex hormones (including gonadotropins) of different biota including birds' mammals and fishes (Wu et al. 1999). Food chain contamination through accumulation of various organic compounds (e.g., PCBs, DDT, and total hexachlorocyclohexanes) in seafood can pose serious health threat (Connell et al. 1998). Very high levels of PCBs were reported in fishery products in middle Atlantic Bight of USA (GEASMP 1990).

The development of innovative, cost-effective, and eco-friendly technologies for the removal and remediation of impacted sites is important due to growing release of xenobiotics into the natural water systems. Currently, number of microbial bioremediation techniques are considered as essential, eco-friendly and economically viable solutions for restoration of contaminated ecosystems (Desai et al. 2010). The current scenario of presence of xenobiotics in wastewater, as well as the latest research on bioremediation techniques have been reviewed and are presented in this chapter.

Xenobiotics in Water and Wastewater

In the recent decades, a huge number of xenobiotic compounds have been detected in water bodies (Dordio et al. 2011). Among different water bodies, surface water bodies contain higher concentrations, as well as a wider range of compounds in comparison to groundwater systems (Focazio et al. 2008) due to ease of disposal of wastewater, lesser dilution and self purification capacity in surface water (Barnes et al. 2008). Pharmaceuticals in different levels from ng/L to µg/L are reported in surface waters of the USA, Europe and in African continent (Kolpin et al. 2002; Surujlal-Naicker et al. 2015). A summary of selected organic xenobiotics and their relative occurrence in wastewater and treated effluents have been shown in Table 2.

Table 1: Organic xenobiotic compounds concentration typical concentration in waste water and WWTP effluents.

Xenobiotics name	Application or use	Average typical concentration in wastewater (ng/L)	References
Acetaminophen	Analgesic	11,733[a]; 353[b]; 900[c]	[a,b]Kasprzyk-Hordern et al. 2009; [c]Kosma et al. 2010
Diclofenac	Nonsteroidal anti-inflammatory drug	1366[a]; 179[b]; 1300[c]; 194[d]	[a]Clara et al. 2005; [b]Kasprzyk-Hordern et al. 2009; [c]Kosma et al. 2010; [d]Lishman et al. 2006
Diclofenac	Nonsteroidal anti-inflammatory drug	680	Miège et al. 2009
Oxybenzone		11	Kim et al. 2007
Carbamazepine	Antiepileptic (Poseidon 2005)	471[a]; 405[b]; 713[c]; 596[d]	[a]Snyder et al. 2002; [b]Tixier et al. 2003; [c]Poseidon 2005; [d]Rempharmawater 2003
17E-estradiol	Natural steroid hormone (DEPA 2004)	0.6	DEPA 2003
17E-estradiol	Natural steroid hormone (DEPA 2004)	30	R-Gray et al. 2000
Estrone	Natural steroid hormone (DEPA 2004)	68	R-Gray et al. 2000
17D-ethinyloetradiol	Pharmaceutical steroid hormone (DEPA 2004)	1.1	DEPA 2003
Ethynilestradiol	Pharmaceutical steroid hormone	0.9[a]; 1.3[b]	[a]Miège et al. 2009; [b]Kim et al. 2007

Table 1 contd. ....

...Table 1 contd.

Xenobiotics name	Application or use	Average typical concentration in wastewater (ng/L)	References
Ibuprofen	Anti-inflammatory and analgesic (Stumpf et al. 1999)	63[a]; 190b; 155[c]; 400[d]; 24.5[e]; 318[f]; 489[g]; 263[h]; 1500[i]; 384[j]; 1960[k]; 4130[l]; 250[m]; 335[n]; 530[o]; 527[p]	[a]DEPA 2003; [b]Stumpf et al. 1999; [c]Tixier et al. 2003; [d]Poseidon 2005; [e]Buser et al. 1999; [f]Rempharmawater 2003; [g]Clara et al. 2005; [h]Kasprzyk-Hordern et al. 2009; [i]Kosma et al. 2010; [j]Lishman et al. 2006; [k]Miège et al. 2009; [l]Santos et al. 2009; [m]Yu et al. 2006; [n]Stumpf et al. 1999; [o]Tixier et al. 2003; [p]Rempharmawater 2003
Bisphenol A (BPA)	Intermediate in the production of and a residual in epoxy resins, flame retardants, polycarbonate, etc. (Staples et al. 1998)	511[a]; 571[b]; 30[c]; 200[d]; 270[e]	[a]Snyder et al. 2002; [b]Barber et al. 2000; [c]Kobuke et al. 2002; [d]NOVA 2005; [e]DEPA 2003
Di(2ethylhexyl)-phthalate (DEHP)	Plasticizer: (DEHP-IC 2005)	960[a]; 500[b]; 1800[c]	[a]Fauser et al. 2003; [b]Kobuke et al. 2002; [c]NOVA 2005
Nonylphenol	Metabolite of alkyl phenol ethoxylates widely used in cleaning agents (Lee and Peart 1998)	329	Fauser et al. 2003
Nonylphenol	Metabolite of alkyl phenol ethoxylates widely used in cleaning agents (Lee and Peart 1998)	1060[a]; 100[b]; 3837[c]; 121[d]; 1,165[e]; 16,000[f]; 988[g]	[a]Snyder 2002; [b]Kobuke et al. 2002; [c]Barber et al. 2000; [d]DEPA 2003; [e]Lee and Peart 1998; [f]Rudel et al. 1998; [g]R-Gray et al. 2000
Octylphenol (OP)	Metabolite of alkyl phenol ethoxylates widely used in cleaning agents (Lee and Peart 1998)	15[a]; 200[b]; 150[c]	Snyder 2002[a]; Lee and Peart 1998[b]; Rudel et al. 1998[c]
Lipophilic markers			
1,3,4,6,7,8-Hexahydro-4,6,6,7,8,8-Hexamethyl-cyclopenta(g)-2-benzopyrane (HHCB)	Musk fragrance	1900	Bester 2004; Rüdel et al. 2006

Compound	Description	Concentration	References
7-Acetyl-1,1,3,4,4,6-hexa-methyltetra-hydronaphthalene (AHTN)	Musk fragrance	580	Bester 2004; Rüdel et al. 2006
4,6,7,8-Tetrahydro-4,6,6,7,8,8-hexamethylcyclopenta(g)-2-benzopyrane-1-one (HHCB-lactone)	Metabolite of fragrance from oxidation	200	Bester 2004
1,2,3,4,5,6,7,8-Octahydro-2,3,8,8-tetramethylnaphthalen-2yl ethan-1-one (OTNE)	Fragrance	4000	Bester et al. 2008
5-Chloro-2-(2,4-dichlorophenoxy)-phenol (Triclosan/TCS)	Biocide	700–8000	Adolfsson-Erici et al. 2002; Ying and Kookana 2007; Bester 2003
Triclosan	Biocide	57; 810; 108; 158; 274; 162	Kasprzyk-Hordern et al. 2009; Katz et al.2013; Lishman et al. 2006; Clara et al. 2005; Lishman et al. 2006; Miège et al. 2009
4-Chloro-1-(2,4-dichlorophenoxy)-2- methoxy-benzene (Triclosan-methyl/TCS-ME)	Triclosan-metabolite (methanogenic transformation product)	2–10	Bester 2005
Hydrophilic markers			
Methylthiobenzothiazole (MTB)	Metabolite of biocides and fungicides		Metcalfe et al. 1988; Reemtsma et al. 1995
Tri-n-butylphosphate (TnBP)	Lubricant, plasticiser	16–4600	Bester 2007
Hydrophilic persistent markers			
Tris-(2-chloropropyl)-phosphate	Flame retardant	940–5900	Bester 2007

Table 2: Xenobiotic remediation using different biological methods.

Remediation techniques	Xenobiotics	Organism/plants used	Initial Concentration	Rate constants (k)	Duration of treatment/ Culture	Removal	References
Plants (Phytoremediation)							
	Tebuconazole	*Phragmites australis*	10 µg/l	0.14/day	24 days	96.1%	Lv et al. 2017
	Imazalil	*Phragmites australis*	10 µg/l	0.31/day	24 days	99.8%	Lv et al. 2017
	Iohexol	*Phragmites australis*	10 mg/l	0.06/day	24 days	80%	Zhang et al. 2015
	Iohexol	*Juncus effusus*	10 mg/l	0.007/day	24 days	31%	Zhang et al. 2015
	Ibuprofen	*Typha*	20 µg/l	0.768/day	21 days	99%	Dordio et al. 2011
	Hexachlorobenzene	*Phragmites australis*	300 µg/l				
Microalgae (Phycoremediaiton)							
	Ciprofloxacin (CIP)	*Chlamydomonas mexicana*		0.0121 to 0.079/day	11 days	13 ± 1%	Xiong et al. 2017a
	Polybrominated Diphenyl Ethers (PBDEs)	*Chlorella* sp.				85 ± 3.2%	Deng and Tam 2015

Carbamazepine	*Chlamydomonas raexicana*	1, 10, 25, mg/l	0.0424, 0.0243, 0.0241/day	10 days	37.3 ± 1.3%, 23.6 ± 1%, 23.5 ± 1.4%	Xiong et al. 2016b
Carbamazepine	*Scenedesmus obliquus*	1, 10, 25, mg/l	0.0329, 0.0234, 0.0177/day	10 days	30.4 ± 1.6%, 23.3 ± 0.7%, 17.3 ± 0.3%	Xiong et al. 2016b
Diazinon	*Chlorella vulgaris*	0.5,5,20,40,100 mg/l	0.2304, 0.1613, 0.174, 0.071, 0.0493/day	12 days	100 ± 0.0%,87.67 ± 2.11%, 93.31 ± 0.46%, 61.99 ± 1.52%, 45.22 ± 2.09	Kurade et al. 2016
Levofloxacin (LEV)	*Scenedesmus obliquus*	1 mg/l	0.005/day	11 days	4.53 ± 1.04% (with 0 mM NaCl)	Xiong et al. 2017
Levofloxacin (LEV)	*Scenedesmus obliquus*	1 mg/l	0.289/day	11 days	93.4 ± 0.24% (with 171 mM NaCl)	Xiong et al. 2017b
Fungi (Mycoremediaion)						
Phenols	*Tramates hirsuta IBB 450*				6.9% (removal with glucose)	Batista-García et al. 2017
Phenols	*Pseudogymnoascus sp. TS12*				15.5% (removal with glucose)	Batista-García et al. 2017
PAHs	*Tramates hirsuta IBB 450*				84.9% (removal with glucose)	Batista-García et al. 2017
PAHs	*Pseudogymnoascus sp. TS12*				87.9% (removal without glucose)	Batista-García et al. 2017
Atrazine	*Aspergillus niger*			8 days	72 ± 2%	Marinho et al. 2017
2-Chlorophenol	*Trametes pubescens*	15 mg/L			94.65%	González et al. 2010

Table 2 contd. ....

...Table 2 contd.

Remediation techniques	Xenobiotics	Organism/plants used	Initial concentration	Rate constants (k)	Duration of treatment/ Culture	Removal	References
	2,4-Dichlorophenol	*Trametes pubescens*	15 mg/L			92.41%	González et al. (2010)
	2,4,6-Trichlorophenol	*Trametes pubescens*	15 mg/L			37.88%	González et al. (2010)
Bacterial remediation							
	Phenanthrene	*Bacillus subtilis* (M16K and M19F)	100 mg/l		28 days	> 85%	
	Benzene	*Pseudomonas* sp. YATO411 (suspension)	87.9 mg/L		260 h	50%	Tsai et al. 2013
	Toluene	*Pseudomonas* sp. YATO411 (suspension)	86.7 mg/L		260 h	71%	Tsai et al. 2013

Once the xenobiotics are released into the environment, the biotic and abiotic transformation processes in the environment lead to formation of various derivatives and some of these compounds are more persistent and toxic in comparison to their parent xenobiotics (la Farré et al. 2007). In household survey for products used in Denmark, 900 potential substances was revealed (Erisksson et al. 2002). In another study, grey water derived from bathrooms contained nearly 200 xenobiotic compounds (Erisksson et al. 2003). In addition, knowledge on the toxicity of grey water due to use of laundry products and its treatability is very limited (Knops et al. 2008).

Pesticides such as atrazine, endosulfan, and methyl parathion was reported in groundwater and rainwater samples collected from two municipalities of Mato Grosso state, Brazil (Moreira et al. 2012). Similarly, organochlorine pesticides were reported in aquifers in the Shanxi Province in China (Li et al. 2015). Pesticide contaminations of groundwater, surface water, and drinking water is also reported across India (Dua et al. 1998; Murlidharan 2000; Lari et al. 2014).

Globally, the average annual consumption of pharmaceuticals per person is around 15 g and the value may be higher (50–150 g) in industrialized countries (Alder et al. 2006). Ternes 1998 reported the presence of different drugs and their corresponding metabolites in rivers and streams, and moreover, about 80% selected drugs showed their presence in at least one treated sewage effluent plant. In Europe, wastewater treatment plants effluents contained several pharmaceutical substances and their metabolites and the peak values in surface water exceeded 1 µg/l (Larsen et al. 2004). In the sewage treatment plants, approximately, 40–45% of persistent organic pollutants remains that may get discharged into water systems, and such compound or its derivatives can threaten wildlife and humans (Clara et al. 2007, Zhang et al. 2015). Scheytt et al. 2006 reported that in Germany, medical industry disposes an estimated 16,000 tons of pharmaceuticals annually, and about 60–80% were disposed via toilet or with domestic waste. Due bio-accumulative capacity of some pharmaceuticals, long term exposure to these compounds can cause deleterious effects in aquatic organisms even if the concentrations in wastewater effluents and receiving waters are much lower than therapeutic doses (Cleuvers 2004). In animal feeding operations, veterinary antibiotics are widely used which is of serious concern. Concentrations as high as 650 ng/L of endogenous estrogens E2 and estrone and other hormones were reported in dairy waste lagoons (Kolodziej et al. 2004).

Bioremediation of Organic Xenobiotics from Wastewater

Among a variety of technologies for remediation of xenobiotics, bioremediation is found to be an efficient, economically viable and eco-friendly approach. Plants and microorganisms like fungi, bacteria, and microalgae help in removal of xenobiotics from water and wastewater as well as contaminated tropical sites. Various bioremediation technologies have been discussed separately in this review.

Phytoremediation of Xenobiotics from Wastewater

Phytoremediation is considered to be one of the most economic methods for the remediation of contaminated water and soil (Salt et al. 1998). Concept of plants degrading xenobiotics through metabolism process emerged in the 1940s (Sanderman 1994). This led to development of genomics, proteomics, and metabolic and enhanced remediation through manipulation of the plant metabolism (Eapen et al. 2007). Mechanism of xenobiotic degradation by plants is very complex involving several enzymes (Schwitzguébel and Vanek 2003). The common reed *Phragmites australis* was shown to accumulate organic xenobiotics in their rhizomes (Schröder et al. 2001). *Typha* showed nearly 60% removal of ibuprofen from water within the first 24 hours and over 99% removal by the end of 21 days (Dordio et al. 2011). In a constructed wetland containing Typha, *Phragmites*, *Eleocharis*, *Schoenoplectus*, and *Baume*, over a 3 year period, 50% atrazine was removed (Page et al. 2011). *Leersia oryzoides* and *Typha latifolia* showed significant reduction (45% and 35%, respectively) of atrazine loads when compared to controls in 379 L individual mesocosm (Moore et al. 2013). The contaminant uptake and distribution is dependent upon both physical and chemical properties of the pesticides and biochemical characteristics of the plants (Moore et al. 2013). In addition, pesticide degradation within vegetation may be influenced by abiotic factors including light, heat, and humidity (Garcinuño et al. 2006).

Lv et al. (2016) showed that the removal efficiencies of chiral pesticides such as tebuconazole and imazalil were 25 to 41% and 46 to 96%, respectively by *Typha, Phragmitesis, Iris* and *Juncus* in 24 days. In another study, removal efficiencies of tebuconazole and imazalil by *Phragmites australis* reached 96.1% and 99.8%, respectively, within 24 days (Lv et al. 2017). The transgenic aspen (hybrid) showed tolerance and higher accumulation of TNT (trinitrotoluene) from water and soil (Van Dillewijn et al. 2008). *Rheum rabarbarum* and *Rheum hydrolapatum* could efficiently remove five sulphonated anthraquinones (Aubert and Schwitzguébel 2004). Situation of mixed pollution such as heavy metals co-occurence with organic pollutants may threaten the survival of any given plant species/cultivar (Verkleij et al. 2009). *Phragmites australis* could remove about 96.1% of tebuconazole and 99.8% of imazalil in 24 days from hydroponic solution that had an initial pesticide levels of 10 μg/l (Lv et al. 2017).

Constructed wetlands for remediation of xenobiotics

Constructed wetlands are engineered systems that use wetland vegetation, soils, and microorganisms for treating wastewater (Vymazal et al. 2014). Their role in the treatment of xenobiotics is gaining increasing interests (Davies et al. 2009). Constructed wetlands can mainly be divided into surface flow constructed wetlands, subsurface flow (horizontal or vertical), and hybrid systems (Ávila and García 2015). The hybrid systems consist of vertical flow and horizontal flow systems arranged in number of different possible ways. Due to more aerobic environment of the shallow wetland, ibuprofen removal was found to be 81% in the shallow subsurface flow constructed wetlands and 48% in the deep one planted with *Phragmites* (Matamoros et al. 2005). In two subsurface horizontal flow constructed wetlands (Matamoros and Bayona 2006), efficient removal (> 80%) was found for salicylic acid, caffeine, methyl dihydrojasmonate, and carboxy-ibuprofen and moderate removal (50–80%) for ibuprofen, hydroxy-ibuprofen, and naproxen. The recalcitrant compounds were diclofenac and ketoprofen, while galaxolide and tonalide were eliminated due to hydrophobic interactions (Matamoros and Bayona 2006). In another study, in vertical subsurface flow constructed wetland, > 95% removal was found for salicylic acid, caffeine, CA-ibuprofen, methyl dihydrojasmonate, hydrocinnamic acid, oxybenzone, ibuprofen and OH-ibuprofen were removed; 70 and 90% for naproxen, diclofenac, galaxolide, and tonalide; and < 30% for carbamazepine (Matamoros et al. 2007). In a similar study, in surface flow constructed wetland (Matamoros et al. 2008), more than 90% removal was shown for pharmaceuticals, personal care products and herbicides. The study further suggested that removal of emerging contaminants are due to high HRT (hydraulic retention time) and *via* biodegradation and photodegradation (Matamoros et al. 2008). The attenuation efficiency of restored wetland was compound dependent it ranged from no attenuation to 84% for emerging contaminants (Matamoros et al. 2012). Similarly, in the pond-surface flow constructed wetland consisting of *Phragmites australis* and *Typha latifolia* could remove > 85% emerging contaminants such as galaxolide, ibuprofen, tonalide, naproxen, triclosan, diclofenac, etc. (Matamoros and Salvadó 2012). Further, between 50% and 85% removal were shown for cashmeran, 2-(methylthio)-benzothiazole, tributyl fosphate, methyl dihydrojasmonate, celestollde, diazinon, caffeine, etc. (Matamoros and Salvadó 2012). In vertical subsurface flow wetlands with *Phragmites australis*, 95% removal was obtained for ibuprofen, oxybenzone, bisphenol A and 55 to 90% for tonalide, triclosan, diclofenac (Ávila et al. 2014).

The mean removal of hexachlorobenzene was about 54% in the wetland without vegetation and ~ 69% in the microcosm wetlands planted with *Phragmites australis* and *Typha latifolia* (Zhou et al. 2012). Similarly, 90% of initial hexachlorobenzene concentration was degraded in between 92–192 days in constructed wetland mesocosms containing *Typha latifolia* and *Phragmites australis* and filled with floodplain sediments (Zhou et al. 2013). In a hybrid constructed wetlands vegetated with *Thypha latifolia*, *Iris sibirica* and *Zantedeschia aethiopica* average mass removal of carbamazepine were $62.5 \pm 4.5\%$ and $59.0 \pm 4.5\%$, respectively (Tejeda et al. 2016). Estrone, 17 beta-estradiol and 17 alpha-ethynylestradiol were removed in shallow wetland vegetated with *Phragmites australis* due to aerobic conditions and high root density (Song et al. 2009). Further, monochlorobenzene, 1,4-dichlorobenzene and 1,2-dichlorobenzene removal were observed in a vegetated (*Phragmites australis*) pilot-scale constructed wetlands (Braeckevelt et al. 2011). The average removal efficiency of 3.13–97.3% was observed in estuarine and freshwater constructed wetland systems containing phenolic endocrine disruptors such as nonylphenol diethoxylates and monoethoxylates, nonylphenol (NP), and octylphenol (Hsieh et al. 2013). A study revealed the role of three species of fungi namely *Absidia cylindrospora*, *Cladosporium sphaerospermum*, and *Ulocladium chartarum* in degradation of PAH (Poly aromatic hydrocarbon) particularly fluoranthene in pilot-scale constructed wetlands planted with *Phragmites australis* (Giraud et al. 2001). Efficient removal (> 99% for most samples) of PAHs and phenol was observed in mangrove and non-mangrove (*Canna indica, Acorus calamus* and *Phragmites australis*) constructed wetlands and one of the key removal mechanism is adsorption on the surface of the substrate or organic matter (Polprasert et al. 1996; Leung et al. 2016). Color removal efficiency up to 99% was obtained using a vertical-flow constructed wetlands, planted with *Phragmites* species and further demonstrated cleavage of the azo bond *via* removal of acid orange 7 (AO7) (Davies et al. 2006). In another study, the removal efficiency of approximately 68 ± 8% was obtained for AO7 (average inlet concentration of 748 ± 42 mg/l) by *Phragmites australis* (Davies et al. 2009). In subsurface plants like *Phragmites australis* and *Arudo donax*, 79% and 55% of polycyclic aromatic compounds (PAHs) and linear alkylbenzene sulfonates (LAS) were removed, respectively, while in free water surface constructed wetland, the removal were 68.2% and 30.0%, respectively (Fountoulakis et al. 2009). Absorption in solid media was found to be mainly responsible for removal of xenobiotics in constructed wetlands (Fountoulakis et al. 2009).

Degradation pathways in plants

Plant possesses several pollutant attenuation mechanisms and thus makes the phytoremediation of contaminated soil and water more effective than physicochemical remediations (Huang et al. 2005; Greenberg 2006; Gerhardt et al. 2009). Root uptake, transportation, or translocation of contaminants through various plant tissues is followed by their metabolism, sequestration, and volatilization process (Cunningham et al. 1996; Greenberg et al. 2006; Abhilash 2007; Doty et al. 2007). Several strategies such as phytoextraction, phytodegradation, phytovolatilization, and rhizode-gradation are used for dealing with environmental chemicals (Schnoor 1997). Xenobiotic metabolism in plants can be compared to human hepatic metabolism as it lead to the xenobiotic detoxification and excretion (Shima-bukuro et al. 1979). Availability of glutathione in plants is very important for detoxification and glutathione homeostasis is maintained by its synthesis, bio-degradation, and long-range transport (Noctor et al. 2002).

Bacterial Remediation of Xenobiotics from Wastewater

Microbial bioremediation strategies for restoration of the polluted ecosystems are considered as an essential, eco-friendly and economically feasible solution (Desai et al. 2010). Microbial communities can metabolize recalcitrant xenobiotic anthropogens and get colonized at contaminated sites (Galvao et al. 2005). Microbial inoculation along with nutrient application is a prerequisite for remediation of xenobiotics over a prolonged period (Eapen et al. 2007). Compound such as methyl parathion, p-nitrophenol, 4-nitrocatechol, and 1,2,4-benzenetriol were utilized as sole sources of carbon and energy by *Serratia* sp. strain DS001 (Pakala et al. 2007) Five bacterial genera *viz. Klebsiella, Acinetobacter, Alcaligenes, Flavobacterium*, and *Bacillus* isolated from the sediments of Kor River, Iran were able to degrade endosulfan and produce metabolites of endosulfan diol, lactone, and ether that were less toxic when compared with the original compound (Kafilzadeh et al. 2015). Microcosm's bioaugmented with *Enterobacter, Pandoraea* and *Burkholderia* strains showed 98% removal of total petroleum hydrocarbons (Sarkar et al. 2017). It was reported that isolates of *Pseudomonas aeruginosa* and *stutzeri, Bacillus safensis* and *Staphylococcus arlettae* could degrade 100% of sodium lauryl ether sulfate (SLES) containing 1000 mg/l in the basal salt medium (Karray et al. 2016). However, the growth of these isolates decreased at higher concentrations and concentration as high as 3000 mg/l could be both toxic and lethal for the isolates (Karray et al. 2016). Outstanding breakdown of crude oil (> 94.0%) and phenanthrene (> 85.0%) was observed on 18 d and 28 d post-inoculation, respectively, by M16K and M19F strains of *Bacillus subtilis* (Oyetibo et al. 2017).

The degradation of phenolic compounds and a high COD removal efficiency was reported using the enzyme extracted from *Serratia* sp. AB 90027 (Yao et al. 2006). The amount of TCE (Trichloroethylene) degraded per unit mass of NH_4–N, strongly depended on the initial NH_4–N and TCE concentrations and based on estimated Ty values, it was concluded that nitrifying systems could be efficiently used for biological removal of TCE through co-metabolism (Kocamemi and Çeçen 2010). Wastewater contaminating acetone–butanol–ethanol was successfully treated along with methane production using an anaerobic baffled reactor with four compartments (C1–C4) (Zhang et al. 2015). During community succession from C1 to C4 at stage 3, *Bacteroidetes* and *Methanocorpusculum* became dominant while abundance of *Firmicutes* and ethanoculleus decreased (Zhang et al. 2015). Very high (> 90%) removal efficiency were achieved in moving bed biofilm reactor (MBBR) for diallyl phthalate and diethyl phthalate in optimum conditions (Ahmadi et al. 2015). Among two strains of *Pseudomonas* isolated from an oil refinery wastewater, *Pseudomonas* sp. KA was able not only able to tolerate high concentrations but also could degrade benzene, toluene, and xylene (Martino et al. 2012). A pure strain of *Pseudomonas* sp. YATO411 showed biodegradation of benzene and toluene in an immobilized and a freely suspended system indicating its potential for treating high concentrations of these xenobiotics (Tsai et al. 2013). *Serratia* sp. ISTVKR1 isolated from sewage sludge was investigated to remove complex organic compounds like phosphoric acid triphenyl ester and 4H-1-benzopyran-4-one, 2-(3,4-dimethoxyphenyl)-3, 5-dihydroxy-7-methoxy (Gupta and Thakur 2015). Potentiality of *Serratia liquefaciens* for detoxification of real pulp and paper mill effluent was evaluated and it showed effective reduction of pollution parameters (72% colour, 58% lignin, 85% chemical oxygen demand, and 95% phenol) after 144 h of treatment at 30°C, pH 7.6 and 120 rpm (Haq et al. 2016).

Mycoremediation of Xenobiotics

Fungi can decompose several plant polymers (lignin, cellulose, and hemicelluloses) and can mineralize, release, and accumulate toxic materials (Singh 2006). Yeasts and fungi are extensively used for treatment of wastes and wastewaters (Singh 1991). Among different varieties of fungi, white rot fungi can degrade several xenobiotics, including PAHs, dyes, chlorinated phenols and pharmaceuticals (Vanhulle et al. 2008; Bastos and Magan 2009). The granular bioplastic formulation containing *Phanerochaete chrysosporium* propagules could remove pharmaceuticals such as osel-tamivir (Tamiflu), erythromycin, sulfamethoxazol, and ciprofloxacin from wastewater (Accinelli et al. 2010).

Ganoderma sp. En3 showed greater ability to decolorize Reactive Orange 16 as compared to some other fungal strains (Ma et al. 2014). Young and Yu 1997 examined the decolourisation capabilities of white-rot fungal cultures for several synthetic dyes. *Trametes versicolor* and *Ganoderma lucidum* were used (individually and in combined form) to remove thirteen pharmaceutical pollutants (Vasiliadou et al. 2016). A microcosm of rice husk (lignocellulosic substrate) bioaugmented with the fungus *Trametes versicolor* and a biomixture that contained fungally colonized rice husk was used in the degradation of the insecticide/nematicide carbofuran (Ruíz-Hidalgo et al. 2016). Results suggested that optimization of mixture is essential for efficient remediation and minimizing the residual toxicity and potential chronic effects on aquatic life (Ruíz-Hidalgo et al. 2016). In another study, white-rot fungus *Trametes versicolor* NRRL 66313 showed reduced estrogenic activity of estrogens and BPA (bisphenol A), that are a major endocrine disruptors in domestic wastewater treatment plant effluent (Shreve et al. 2016). *Candida rugosa* and *Yarrowia lipolytic* rM-4A yeast cell cultures, both individually and in combination and their revealed the excellent potential in treating palm oil mill effluent containing high fat and oil by reducing total phenol based compounds, triglycerides and COD level (Theerachat et al. 2017). In another study, *Aspergillus niger* AN 400 was shown to remove atrazine up to 72% (Marinho et al. 2017).

Phycoremediation for Removal of Xenobiotics

Biodegradation studies of naproxen (NP), one of the most prevalent pharmaceuticals, by freshwater algae *Cymbella* sp. showed higher removal efficiency as compared to *Scenedesmus quadricauda* and the mechanism such as hydroxylation, demethylation, decarboxylation, conjunction with tyrosine and glucuronidation caused transformation of naproxen in algal cells (Ding et al. 2017). *Chlorella vulgaris* and three local isolates *of Chlorella* sp. showed rapid and elevated removal of nonylphenol. The highest NP removal was shown by *C. vulgaris* followed by *Chlorella miniata* (Gao et al. 2011a). *Chlamydomonas mexicana* showed the removal of 13 ± 1% of ciprofloxacin (2 mg/l) after 11 days of treatment. However, the addition of 4 g/l of sodium acetate as electron donor, 4 g/l enhanced the removal of ciprofloxacin by > three-fold (56 ± 1.8%) after 11 days (Xiong et al. 2017a). Fungi *Mucor hiemalis* when exposed different concentrations (5–100 ng/l) of acetaminophen (APAP) for 24–48 h, it was able to accumulate 1 mg/g and 2 mg/g on dry weight of fungal biomass, and thus could be used for bioremediation of low concentrations of APAP (Esterhuizen-Londt et al. 2016). *Chlamydomonas mexicana* was found to be more tolerant as compared to *Chlorella vulgaris* for the treatment of bisphenol A (Ji et al. 2014). *Spirulina* strains removed 50% of the polyphosphonate from the solution containing 2.5 mM (Forlani et al. 2011). PBDE (polybrominated diphenyl ethers)-tolerant microalgae from wastewater were isolated and a *Chlorella* isolate, SICh could remove 82–90% of PBDE after 7-days exposure (Deng and Tam 2015). The first step of xenobiotic detoxification in aerobic organisms is hydroxylation that could change the hydrophobic nature of organic pollutants and thus help in the metabolism or excretion (Gao et al. 2011b).

Various species of microalgae have been used for wastewater treatment (Shriwastav et al. 2015; Gupta et al. 2016a,b; 2017a,b; Rawat et al. 2016). Kurade et al. (2016) screened four microalgal species (*Chlamydomonas mexicana, Scenedesmus obliquus, Chlorella vulgaris,* and *Chlamydomonas pitschmannii*) to remove diazinon from the aqueous phase. Among these, *C. vulgaris* could remove (94%) of diazinon at 20 mg/L but above 40 mg/L of diazinon the growth of *C. vulgaris* showed > 30% growth inhibition after 12 days of cultivation (Kurade et al. 2016). With addition of sodium chloride (NaCl) in synthetic saline wastewater containing Levofloxacin (LEV) and its removal by *via* a microalga, *Scenedesmus obliquus* increased significantly from 4.5 to 93.4% (Xiong et al. 2017b). Furthermore, the study showed that biotransformation of levofloxacin in *Scenedesmus obliquus* is possible through cellular biocatalytic reactions, such as decarboxylation, dehydroxylation, demethylation, side chain breakdown, and ring cleavage (Xiong et al. 2017b).

Scenedesmus obliquus and *Chlamydomonas mexicana* and showed 28% and 35% biodegradation of carbamazepine (CBZ) and thus, *C. mexicana* was found to be more tolerant to be used for treatment of CBZ contaminated wastewater (Xiong et al. 2016). Carbamazepine (CBZ) is an antiepileptic agents that has been widely consumed and due to limitation in removal of CBZ, they are frequently detected in drinking water, surface and groundwater, estuaries, wastewater and sludge-amended soil (Zhang et al. 2008). The mechanism and kinetics of a *Chlorella* isolate to remove BDE-47 (Tetrabromodiphenyl ether), one of the most toxic and dominant PBDEs (Polybrominated diphenyl ethers) were investigated and 80% removal of BDE-47 was achieved in the time period between 1 h to 7 days (Deng and Tam 2016). The removal mechanism consisted of (i) a rapid physicochemical adsorption, (ii) slow uptake and (iii) hydroxylation, debromination, and methoxylation wherein adsorption being the dominant one (Deng and Tam 2016). Mishra and Mukherji (2012) suggested that the major adsorption mechanism of hydrophobic organic compounds in microalgae could be due to electrostatic interactions with the negative charges of functional groups. Polybrominated diphenyl ethers (PBDEs) is considered to be the electron donors when reacting with receptors in living cells (Zhao et al. 2008) and organic matter in dissolved state (Nuerla et al. 2013). Enzymatic systems in microalgae, such as cytochrome P450 enzyme families' laccase and glutathione-S-transferases play significant role in the degradation of xenobiotic of organic nature (Semple et al. 1999; Xiong et al. 2016; Kurade et al. 2016). A summary showing xenobiotic remediation using different biological methods is shown in Table 2. As can be seen, both plants and microbial strength have been explored to treat recalcitrant xenobiotics.

Conclusions and Future Prospects

Due to rapid population and industrial growth to meet product demand, the pollution of waterbodies with xenobiotics will continue to rise. Environmental pollution due to xenobiotics is associated with many emerging health problems. Remediation and degradation of xenobiotics from a polluted environment using biological methods has gained worldwide attention due to eco-friendly, non-invasive and cost-effective nature.

In this chapter an overview of recent developments in bioremediation techniques to remove xenobiotics from wastewater is given. Although slow, bioremediation can not only degrade a wide range of recalcitrant compounds but also offers several advantages including cost effectiveness, eco-friendliness. Several studies have been done for remediation of xenobiotics using microbial remediation (using bacteria, fungi, and algae) and phytoremediation and their subtypes. Several scientific studies are in progress to explore bioremediation techniques and their effectiveness in xenobiotics removal. The ample selection of plants and microorganisms capable of treating a wide range of emerging xenobiotics have potential for testing at both the pilot and field scales.

References

Abhilash, P.C. 2007. Phytoremediation: an innovative technique for ecosystem clean up. Our Earth 4: 7–12.

Accinelli, C., M.L. Saccà, I. Batisson, J. Fick, M. Mencarelli and R. Grabic. 2010. Removal of oseltamivir (Tamiflu) and other selected pharmaceuticals from wastewater using a granular bioplastic formulation entrapping propagules of *Phanerochaete chrysosporium*. Chemosphere 81: 436–443.

Adolfsson-Erici, M., M. Pettersson, J. Parkkonen and J. Sturve. 2002. Triclosan, a commonly used bactericide found in human milk and in the aquatic environment in sweden. Chemosphere 46: 1485–1489.

Ahmadi, E., M. Gholami, M. Farzadkia, R. Nabizadeh and A. Azari. 2015. Study of moving bed biofilm reactor in diethyl phthalate and diallyl phthalate removal from synthetic wastewater. Bioresour. Technol. 183: 129–135.

Alder, A.C., A. Bruchet, M. Carballa, M. Clara, A. Joss, D. Löffler et al. 2006. Consumption and occurrence. *In*: Ternes, T.A. and A. Joss [eds.]. Human Pharmaceuticals, Hormones and Fragrances. The Challenge of Micropollutants in Urban Water Management. IWA Publishing.

Aubert, S. and J.P. Schwitzguébel. 2004. Screening of plant species for the phytotreatment of wastewater containing sulphonated anthraquinones. Water Res. 38: 3569–3575.

Ávila Martín, C. and J. García Serrano. 2015. Pharmaceuticals and personal care products (PPCPs) in the environment and their removal from wastewater through constructed wetlands. Comprehensive Anal. Chem. 67: 195–244.

Ávila, C., J. Nivala, L. Olsson, K. Kassa, T. Headley, R.A. Mueller et al. 2014. Emerging organic contaminants in vertical subsurface flow constructed wetlands: Influence of media size, loading frequency and use of active aeration. Sci. Total Env. 494–495: 211–217.

Barber, L.B., G.K. Brown and S.D. Zaugg. 2000. Potential endocrine disrupting organic chemicals in treated municipal wastewater and river water. ACS Symposium Series, American Chemical Society 747: 97–124.

Barnes, K., D. Kolpin, E. Furlong, S. Zaugg, M. Meyer and L. Barber. 2008. A national reconnaissance of pharmaceuticals and other organic wastewater contaminants in the United States I. Groundwater. Sci. Total Environ. 402: 192–200.

Batista-García, R.A., V.V. Kumar, A. Ariste, O.E. Tovar-Herrera, O. Savary, H. Peidro-Guzmán et al. 2017. Simple screening protocol for identification of potential mycoremediation tools for the elimination of polycyclic aromatic hydrocarbons and phenols from hyperalkalophile industrial effluents. J. Environ. Manag. 198: 1–11.

Bell, K.Y., M.J.M. Wells, K.A. Traexler, M.L. Pellegrin, A. Morse and J. Bandy. 2011. Emerging pollutants. Water Environ. Res. 83: 1906–1984.

Bester, K. 2003. Triclosan in sewage plants balances and monitoring data. Water Res. 37: 3891–3896.

Bester, K. 2004. Retention characteristics and balance assessment for two polycyclic musk fragrances (HHCB and AHTN) in a typical German sewage treatment plant. Chemosphere 57: 863–870.

Bester, K. 2005. Fate of triclosan and triclosan-methyl in sewage treatment plants and surface waters. Arch. Environ. Contam. Toxicol. 49: 9–18.

Bester, K. 2007. Personal Care Compounds in the Environment: Pathways, Fate and Methods for Determination. Wiley-VCH, Weinheim.

Bester, K., L. Scholes, C. Wahlberg and C.S. McArdell. 2008. Sources and mass flows of xenobiotics in urban water cycles—an overview on current knowledge and data gaps. Water Air Soil Pollut.: Focus. 8: 407–423.

Bester, K., N. Hüffmeyer, E. Schaub and J. Klasmeier. 2008. Surface water concentrations of the fragrance compound OTNE in Germany a comparison between data from measurements and models. Chemosphere 73: 1366–1372.

Bester, K. and D. Schäfer. 2009. Activated soil filters (bio filters) for the elimination of xenobiotics (micro-pollutants) from storm- and waste waters. Water Res. 43: 2639–2646.

Braeckevelt, M., N. Reiche, S. Trapp, A. Wiessner, H. Paschke, P. Kuschk et al. 2011. Chlorobenzene removal efficiencies and removal processes in a pilot-scale constructed wetland treating contaminated groundwater. Ecol. Eng. 37: 903–913.

Buser, H.R., T. Poiger and M.D. Muller. 1999. Occurrence and environmental behaviour of the chiral pharmaceutical drug ibuprofen in surface waters and in wastewater. Environ. Sci. Technol. 33: 2529–2535.

Clara, M., B. Strenn, O. Gans, E. Martinez, N. Kreuzinger and H. Kroiss. 2005. Removal of selected pharmaceuticals, fragrances and endocrine disrupting compounds in a membrane bioreactor and conventional treatment plants. Water Res. 39: 4797–4807.

Clara, M., S. Scharf, C. Scheffknecht and O. Gans. 2007. Occurrence of selected surfactants in untreated and treated sewage. Water Res. 41: 4339–48.

Cleuvers, M. 2004. Mixture toxicity of the anti-inflammatory drugs diclofenac, ibuprofen, naproxen, and acetylsalicylic acid. Ecotox. Environ. Saf. 59: 309–315.

Colwell, R.R. and J.D. Walker. 1977. Ecological aspects of microbial degradation of petroleum in the marine environment. Crit. Rev. Microbiol. 5(4): 423–45.

Connell, D.W., R.S.S. Wu, B.J. Richardson, K. Leung, P.K.S. Lam and P.A. Connell. 1998. Fate and risk of persistent organic contaminants and related compounds in victoria harbour, Hong Kong. Chemosphere 36: 2019–2030.

Danish Environmental Protection Agency (DEPA). 2003. Endocrine disrupting compounds and pharmaceuticals in wastewater (In Danish: Hormonforstyrrende stoffer og lægemidler i spildevand), Environmental project No. 799, Denmark, 46 pp.

Danish Environmental Protection Agency (DEPA). 2004. Degradation of estrogens in sewage treatment processes. Environmental project No. 899, Denmark, 61 pp.

Davies, L.C., I.S. Pedro, J.M. Novais and S. Martins-Dias. 2006. Aerobic degradation of acid orange 7 in a vertical-flow constructed wetland. Water Res. 40: 2055–2063.

Davies, L.C., G.J.M. Cabrita, R.A. Ferreira, C.C. Carias, J.M. Novais and S. Martins-Dias. 2009. Integrated study of the role of *Phragmites australis* in azo-dye treatment in a constructed wetland: From pilot to molecular scale. Integrated study of the role of *Phragmites australis* in azo-dye treatment in a constructed wetland: From pilot to molecular scale. Ecol. Eng. 35: 961–970.

Deng, D. and N.F.Y. Tam. 2015. Isolation of microalgae tolerant to polybrominated diphenyl ethers (PBDEs) from wastewater treatment plants and their removal ability. Bioresour. Technol. 177: 289–297.

Deng, D. and N.F.Y. Tam. 2016. Adsorption-uptake-metabolism kinetic model on the removal of BDE-47 by a *Chlorella* isolate. Environ. Poll. 212: 290–298.

Desai, C., H. Pathak and D. Madamwar. 2010. Advances in molecular and "-omics" technologies to gauge microbial communities and bioremediation at xenobiotic/anthropogen contaminated sites. Bioresour. Technol. 101: 1558–1569.

Di Martino, C., N.I. López and L.J.R. Iustman. 2012. Isolation and characterization of benzene, toluene and xylene degrading Pseudomonas sp. selected as candidates for bioremediation. Int. Biodeter. Biodegrad. 67: 15–20.

Ding, T., K. Lin, B. Yang, M. Yang, J. Li, W. Li et al. 2017. Biodegradation of naproxen by freshwater algae Cymbella sp. and Scenedesmus quadricauda and the comparative toxicity. Bioresour. Technol. 238: 164–173.

Dordio, A., R. Ferro, D. Teixeira, A.J. Palace, A.P. Pinto and C.M. Dias. 2011. Study on the use of Typha spp. for the phytotreatment of water contaminated with ibuprofen. Int. J. Environ. Anal. Chem. 91: 654–667.

Dórea, J.G. 2008. Persistent, bioaccumulative and toxic substances in fish: Human health considerations. Sci. Total Environ. 400: 93–114.

Doty, S.L., T.Q. Shang, A.M. Wilson, J. Tangen, A.D. Westergreen, L.A. Newman et al. 2007. Enhanced metabolism of halogenated hydrocarbons in transgenic plants containing mammalian cytochrome P450 2E1. Proceedings of the National Academy of Sciences 97(12): 6287–6291.

Dua, K., R. Kumari, R.K. Johari, V.P. Ojha, R.P. Shukla and V.P. Sharma. 1998. Organochlorine insecticide residues in water from five lakes of Nainital (UP), India. Bull. Environ. Contam. Toxicol. 60: 209–215.

Eapen, S., S. Singh and S.F. D'Souza. 2007. Advances in development of transgenic plants for remediation of xenobiotic pollutants. Biotechnol. Adv. 25: 442–51.

El Shahawi, M.S., A. Hamza, A.S. Bashammakh and W.T. Al Saggaf. 2010. An overview on the accumulation, distribution, transformations, toxicity and analytical methods for the monitoring of persistent organic pollutants. Talanta 80: 1587–1597.

Erisksson, E., K. Auffarth, M. Henze and A. Ledin. 2002. Characteristics of grey wastewater. Urban Water 4: 85–104.

Erisksson, E., K. Auffarth, M. Henze and A. Ledin. 2003. Household chemicals and personal care products as sources for xenobiotic organic compounds in grey wastewater. Water SA 29(2): 135–146.

Esterhuizen-Londt, M., K. Schwartz and S. Pflugmacher. 2016. Using aquatic fungi for pharmaceutical bioremediation: Uptake of acetaminophen by Mucor hiemalis does not result in an enzymatic oxidative stress response. Fungal Biol. 120: 1249–1257.

Fauser, P., J. Vikelsøe, P.B. Sørensen and L. Carlsen. 2003. Phthalates, nonylphenols and LAS in an alternately operated wastewater treatment plant—fate modelling based on measured concentrations in wastewater and sludge. Water Resour. 37: 1288–1295.

Focazio, M.J., D.W. Kolpin, K.K. Barnes, E.T. Furlong, M.T. Meyer, S.D. Zaugg et al. 2008. A national reconnaissance for pharmaceuticals and other organic wastewater contaminants in the United States—(ii) Untreated drinking water sources. Sci. Total Env. 402: 201–216.

Forlani, G., V. Prearo, D. Wieczorek, P. Kafarski and J. Lipok. 2011. Phosphonate degradation by spirulina strains: Cyanobacterial biofilters for the removal of anticorrosive polyphosphonates from wastewater. Enzyme Microbial Technol. 48: 299–305.

Fountoulakis, M.S., S. Terzakis, N. Kalogerakis and T. Manios. 2009. Removal of polycyclic aromatic hydrocarbons and linear alkylbenzene sulfonates from domestic wastewater in pilot constructed wetlands and a gravel filter. Ecol. Eng. 35: 1702–1709.

Galvao, T.C., W.W. Mohn and V. Lorenzo. 2005. Exploring the microbial biodegradation and biotransformation gene pool. Trends Biotechnol. 23: 497–506.

Gao, Q.T., Y.S. Wong and N.F.Y. Tam. 2011b. Removal and biodegradation of nonylphenol by different *Chlorella* species. Mar. Poll. Bull. 63: 445–451.

Gao, Q., Y. Wong and N. Tam. 2011a. Removal and biodegradation of nonylphenol by different Chlorella species. Mar. Pollut. Bull. 63: 445–451.

Garcinuño, R.M., P.F. Hernando and C. Cámara. 2006. Removal of carbaryl, linuron, and permethrin by Lupinus angustifolius under hydroponic conditions. J. Agric. Food Chem. 54: 5034–5039.

Gavrilescu, M. 2005. Fate of pesticides in the environment and its bioremediation. Eng. Life Sci. 5: 497–526.

GEASMP. 1990. Joint Group of Experts on the Scientific Aspect of Marine Pollution: the State of the Marine Environment UNEP Regional Seas Reports and Studies. No. 115, UNEP.

Gerhardt, K.E., X.D. Huang, B.R. Glick and B.M. Greenberg. 2009. Phytoremediation and rhizoremediation of organic soil contaminants: potential and challenges. Plant. Sci. 176: 20–30.

Giraud, F., P. Guiraud, M. Kadri, G. Blake and R. Steiman. 2001. Biodegradation of anthracene and fluoranthene by fungi isolated from an experimental constructed wetland for wastewater treatment. Wat. Res. 35: 4126–4136.

González, L.F., V. Sarria and O.F. Sánchez. 2010. Degradation of chlorophenols by sequential biological-advanced oxidative process using Trametes pubescens and TiO$_2$/UV. Bioresour. Technol. 101: 3493–3499.

Greenberg, B.M. 2006. Development and field tests of a multi-process phytoremediation system for decontamination of soils. Can. Reclam. 1: 27–9.

Gupta, A. and I.S. Thakur. 2015. Biodegradation of wastewater organic contaminants using Serratia sp. ISTVKR1 isolated from sewage sludge. Biochem. Eng. J. 102: 115–124.

Gupta, S.K., M. Chabukdhara, J. Singh and F. Bux. 2015. Evaluation and potential health hazard of selected metals in water, sediments, and fish from the gomti river. Human and Ecological Risk Assessment 21(1): 227–240.

Gupta, S.K., N.M. Kumar, R. Mishra, F.A. Ansari, D.D. Dionysios, A. Maity et al. 2016a. Synthesis and performance evaluation of a new polymeric composite for the treatment of textile wastewater. Ind. Eng. Chem. Res. 55(1): 13–20.

Gupta, S.K., F.A. Ansari, A. Shriwastav, N.K. Sahoo, I. Rawat and F. Bux. 2016b. Dual role of *Chlorella sorokiniana* and *Scenedesmus obliquus* for comprehensive wastewater treatment and biomass production for bio-fuels. J. Cleaner Prod. 115: 255–264.

Gupta, S.K., F.A. Ansari, N. Mahmoud, I. Rawat, N.M. Kumar and F. Bux. 2017a. Cultivation of *Chlorella sorokiniana* and *Scenedesmus obliquus* in wastewater: Fuzzy intelligence for evaluation of growth parameters and metabolites extraction. J. Cleaner Prod. 147: 419–430.

Gupta, S.K., A. Sriwastav, F.A. Ansari, M. Nasr and A.K. Nema. 2017b. Phycoremediation: An ecofriendly algal technology for bioremediation and bioenergy production. pp. 431–456. *In*: Bauddh, K., B. Singh and J. Korstad [eds.]. Phytoremediation Potential of Bioenergy Plants. Springer International Publishing AG Cham. Chapter 3. DOI:10.1007/978-981-10-3084-0_18.

Haq, I., S. Kumar, V. Kumari, S.K. Singh and A. Raj. Evaluation of bioremediation potentiality of ligninolytic *Serratia liquefaciens* for detoxification of pulp and paper mill effluent. J. Haz. Mat. 305: 190–199.

Hsieh, C-Y., L. Yang, W-C. Kuo and Y.P. Zen. 2013. Efficiencies of freshwater and estuarine constructed wetlands for phenolic endocrine disruptor removal in Taiwan. Sci. Total Env. 463–464: 182–191.

Huang, X.D., Y.S. El-Alawi, J. Gurska, B.R. Glick and B.M. Greenberg. 2005. A multiprocess phytoremediation system for decontamination of persistent total petroleum hydrocarbons (TPHs) from soils. Microchem. J. 81: 139–47.

Ji, M.-K., A.N. Kabra, J. Choi, J.-H. Hwang, J.R. Kim, R.A.I. Abou-Shanab et al. 2014. Biodegradation of bisphenol A by the freshwater microalgae *Chlamydomonas mexicana* and *Chlorella vulgaris*. Ecol. Eng. 73: 260–269.

Kafilzadeh, F., M. Ebrahimnezhad and Y. Tahery. 2015. Isolation and identification of endosulfan-degrading bacteria and evaluation of their bioremediation in Kor River, Iran. Osong Public Health Res. Perspect. 6(1): 39–46.

Karray, F., M. Mezghani, N. Mhiri, B. Djelassi and S. Sayadi. 2016. Scale-down studies of membrane bioreactor degrading anionic surfactants wastewater: Isolation of new anionic-surfactant degrading bacteria. Int. Biodeter. Biodegrad. 114: 14–23.

Kasprzyk-Hordern, B., R.M. Dinsdale and A.J. Guwy. 2009. The removal of pharmaceuticals, personal care products, endocrine disruptors and illicit drugs during wastewater treatment and its impact on the quality of receiving waters. Water Res. 43: 363–380.

Katz, K., T. Inoue, H. Ietsugu, T. Koba, H. Sasaki, N. Miyaji et al. 2013. Performance of six multi-stage hybrid wetland systems for treating high-content wastewater in the cold climate of Hokkaido, Japan. Ecol. Eng. 51: 256–263.

Kim, S.D., J. Cho, I.S. Kim, B.J. Vanderford and S.A. Snyder. 2007. Occurrence and removal of pharmaceuticals and endocrine disruptors in South Korean surface, drinking and waste waters. Water Res. 41: 1013–1021.

Knops, G., M. Pidou, W. Kadewa, A. Soares, P. Jeffrey and B. Jefferson. Reuse of urban water: impact of product choice. Dangerous pollutants (xenobiotics) in urban water cycle. Springer Publ. 13–22.

Kobuke, Y., H. Tanaka and T. Magara. 2002. Nationwide and regional river monitoring studies as well as bioassays and treatment of EDs in waterworks. IWA World Water Congress Melbourne 2002, Workshop on Endocrine Disruptors Proceedings, Conference Proceedings, 53–62.

Kocamemi, B.A. and F. Çeçen. 2010. Biological removal of the xenobiotic trichloroethylene (TCE) through cometabolism in nitrifying systems. Bioresour. Technol. 101: 430–433.

Kolodziej, E.P., T. Harter and D.L. Sedlak. 2004. Dairy wastewater, aquaculture, and spawning fish a sources of steroid hormones in the aquatic environment. Env. Sci. Technol. 38: 6377–6384.

Kolpin, D.K., E.T. Furlong, M.T. Meyer, M.E. Thurman, S.D. Zaugg, L.B. Barber et al. 2002. Pharmaceuticals, hormones, and other organic wastewater contaminants in US streams, 1999–2000: A national reconnaissance. Env. Sci. Tech. 36: 1202–1211.

Kosma, C.I., D.A. Lambropoulou and T.A. Albanis. 2010. Occurrence and removal of PPCPs in municipal and hospital wastewaters in Greece. J. Hazard. Mat. 179: 804–817.

Kreuzinger, N. 2008 review on the assessment of the removal efficiency of wastewater treatment plants for selected xenobiotics. pp. 227–244. *In*: Hlavinek, P., O. Bonacci, J. Marsalek and I. Mahrikova [eds.]. Dangerous Pollutants (Xenobiotics) in Urban Water Cycle. NATO Science for Peace and Security Series. Springer, Dordrecht.

Kurade, M.B., J.R. Kim, S.P. Govindwar and B.H. Jeon. 2016. Insights into microalgae mediated biodegradation of diazinon by *Chlorella vulgaris*: Microalgal tolerance to xenobiotic pollutants and metabolism. Algal Res. 20: 126–134.

La Farré, M., S. Perez, L. Kantiani and D. Barceló. 2007. Fate and toxicity of emerging pollutants, their metabolites and transformation products in the aquatic environment. Trends Anal. Chem. 27: 991–1007.

Lamoureux, G.L. and D.G. Rusness. 1989. The role of glutathione and glutathione S-transferases in pesticide metabolism, selectivity and mode of action in plants and insects. Glutathione: Chemical, Biochemical, and Medical Aspects 3: 153.

Lari, S.Z., N.A. Khan, K.N. Gandhi, T.S. Meshram and N.P. Thacker. 2014. Comparison of pesticide residues in surface water and ground water of agriculture intensive areas. J. Environ. Health Sci. Eng. 12: 11–19.

Larsen, T.A., J. Lienert, A. Joss and H. Siegrist. 2004. How to avoid pharmaceuticals in the aquatic environment. J. Biotechnol. 113: 295–304.

Lee, H.B. and T.E. Peart. 1998. Occurrence and elimination of nonylphenol ethoxylates and metabolites in municipal wastewater and effluents. Water Quality Res. J. Canada 33: 398–402.

Leung, J.Y.S., Q. CAi and N.F.Y. Tam. 2016. Comparing subsurface flow constructed wetlands with mangrove plants and freshwater wetland plants for removing nutrients and toxic pollutants. Ecol. Eng. 95: 129–137.

Li, J.L., C.X. Zhang, Y.X. Wang, X.P. Liao, L.L. Yao, M. Liu et al. 2015. Pollution characteristics and distribution of polycyclic aromatic hydrocarbons and organochlorine pesticides in groundwater at Xiaodian Sewage Irrigation Area, Taiyuan City. Huan. Jing. Ke. Xue. 36: 172–178.

Lishman, L., S.A. Smyth, K. Sarafin, S. Kleywegt, J. Toito, T. Peart et al. 2006. Occurrence and reductions of pharmaceuticals and personal care products and estrogens by municipal wastewater treatment plants in Ontario, Canada. Sci. Total Environ. 367: 544–558.

Lv, T., Y. Zhang, M.E. Casas, P.N. Carvalho, C.A. Arias, K. Bester et al. 2016. Phytoremediation of imazalil and tebuconazole by four emergent wetland plant species in hydroponic medium. Chemosphere 148: 459–466.

Lv, T., P.N. Carvalho, M.E. Casas, U.E. Bollmann, C.A. Arias, H. Brix et al. 2017. Enantioselective uptake, translocation and degradation of the chiral pesticides tebuconazole and imazalil by *Phragmites australis*. Environ. Poll. 229: 362–370.

Ma, L., R. Zhuo, H. Liu, D. Yu, M. Jiang, X. Zhang et al. 2014. Efficient decolorization and detoxification of the sulfonated azo dye Reactive Orange 16 and simulated textile wastewater containing reactive orange 16 by the white-rot fungus *Ganoderma* sp. En3 isolated from the forest of Tzu-chin Mountain in China. Biochem. Eng. J. 82: 1–9.

Marinho, G., B.C.A. Barbosa, K. Rodrigues, M. Aquino and L. Pereira. 2017. Potential of the filamentous fungus Aspergillus niger AN 400 to degrade atrazine in wastewaters. Biocatalysis Agr. Biotechnol. 9: 162–167.

Matamoros, V., J. Garcia and J.M. Bayona. 2005. Behavior of selected pharmaceuticals in subsurface flow constructed wetlands: a pilot-scale study. Environ. Sci. Technol. 39: 5449–5454.

Matamoros, V. and J.M. Bayona. 2006. Elimination of pharmaceuticals and personal care products in subsurface flow constructed wetlands. Environ. Sci. Technol. 40: 5811–5816.

Matamoros, V., C. Arias, H. Brix and J.M. Bayona. 2007. Removal of pharmaceuticals and personal care products (PPCPs) from urban wastewater in a pilot vertical flow constructed wetland and a sand filter. Environ. Sci. Technol. 41: 8171–8177.

Matamoros, V., J. Garcia and J.M. Bayona. 2008. Organic micropollutant removal in a full-scale surface flow constructed wetland fed with secondary effluent. Water Res. 42: 653–660.

Matamoros, V. and V. Salvadó. 2012. Evaluation of the seasonal performance of a water reclamation pond-constructed wetland system for removing emerging contaminants. Chemosphere 86: 111–117.

Matamoros, V., C.A. Arias, L.X. Nguyen, V. Salvadó and H. Brix. 2012. Occurrence and behavior of emerging contaminants in surface water and a restored wetland. Chemosphere 88: 1083–1089.

Metcalfe, J., M. Fox and J. Carey. 1988. Freshwater leeches (hirudinea) as a screening tool for detecting organic contaminations in the environment. Env. Monit. Assess. 11: 147–169.

Miège, C., J.M. Choubert, L. Ribeiro, M. Eusèbe and M. Coquery. 2009. Fate of pharmaceuticals and personal care products in wastewater treatment plants—conception of a database and first results. Environ. Pollut. 157: 1721–1726.

Mishra, P.K. and S. Mukherji. 2012. Biosorption of diesel and lubricating oil on algal biomass. Biotechnol. 2: 301–310.

Moore, M.T., H.L. Tyler and M.A. Locke. 2013. Aqueous pesticide mitigation efficiency of *Typha latifolia* (L.), *Leersia oryzoides* (L.) Sw., and *Sparganium americanum* Nutt. Chemosphere 92: 1307–1313.

Moreira, J.C., F. Peres, A.C. Simões, W.A. Pignati, E.C. Dores, S.N. Vieira et al. 2012. Groundwater and rainwater contamination by pesticides in an agricultural Region of Mato Grosso State in Central Brazil. Ciência & Saúde Coletiva 17(6): 1557–1568.

Murlidharan, S. 2000. Organochlorine residues in the Waters of Keoladeo National Park. Bull. Environ. Contam. Toxicol. 65: 35–41.

Noctor, G., S. Veljovic-Jovanovic, S. Driscoll, L. Novitskaya and C.H. Foyer. 2002. Drought and oxidative load: a predominant role for photorespiration? Annals of Botany, Ann. Bot. 89(7): 841–850.

NOVA. Point sources. 2003. Revised edition (In Danish: Punktkilder 2003—revideret udgave). Danish Environmental Protection Agency. Orientation paper No. 1, Denmark, 2005, 167 pp.

Nuerla, A., X. Qiao, J. Li, D. Zhao, X. Yang, Q. Xie et al. 2013. Effects of substituent position on the interactions between PBDEs/PCBs and DOM. Chin. Sci. Bull. 58: 884–889.

Oyetibo, G.O., M.-F. Chien, W. Ikeda-Ohtsubo, H. Suzuki, O.S. Obayori, S.A. Adebusoye et al. 2017. Biodegradation of crude oil and phenanthrene by heavy metal resistant *Bacillus subtilis* isolated from a multi-polluted industrial wastewater creek. International Biodeter. Biodegrad. 120: 143–151.

Page, D.W., S.J. Khan and K. Miotliński. 2011. A systematic approach to determine herbicide removals in constructed wetlands using time integrated passive samplers. J. Water Reuse Desalination 1: 11–17.

Pakala, S.B., P. Gorla, A.B. Pinjari, R.K. Krovidi, R. Baru, M. Yanamandra et al. 2007. Biodegradation of methyl parathion and p-nitrophenol: evidence for the presence of a p-nitrophenol 2-hydroxylase in a gram-negative *Serratia* sp. strain DS001. Appl. Microbiol. Biotechnol. 73: 1452–1462.

Pal, A., K.Y. Gin and M. Reinhard. 2010. Impacts of emerging organic contaminants on freshwater resources: Review of recent occurrences, sources, fate and effects. Sci. Total Environ. 408: 6062–6069.

Polprasert, C., N. Dan and N. Thayalakumaran. 1996. Application of constructed wetlands to treat some toxic wastewaters under tropical conditions. Water Sci. Technol. 34: 165–171.

Poseidon, A. 2005. Assessment of technologies for the removal of pharmaceuticals and personal care products in sewage and drinking water facilities to improve the indirect potable water reuse. Final report, EU's Fifth Framework Programme, European Commission 2005: 58.

Rawat, I., S.K. Gupta, A. Srivastav, P. Singh, S. Kumari and F. Bux. 2016. Microalgae applications in wastewater treatment in algae biotechnology: Products and processes. pp. 249–268. *In*: Bux, F. and Y. Chisti [eds.]. Springer International Publishing. DOI: 10.1007/978-3-319-12334-9_13.

Reemtsma, T., O. Fiehn, G. Kalnowski and J. Jekel. 1995. Microbial transformations and biological effects of fungicide-derived benzothiazoles determined in industrial wastewater. Env. Sci. Technol. 29: 478–485.

Rempharmawater. 2003. Ecotoxicological assessments and removal technologies for pharmaceuticals in wastewater. Final report, EU's Fifth Framework Programme, European Commission 2004, 50 pp.

R-Gray, T.P., S. Jobling, S. Morris, C. Kelly, S. Kirby, A. Janbakhsh et al. 2000. Long-term temporal changes in the estrogenic composition of treated sewage effluent and its biological effects on fish. Env. Sci. Technol. 34: 1521–1528.

Rüdel, H., W. Böhmer and C. Schröter-Kermani. 2006. Retrospective monitoring of synthetic musk compounds in aquatic biota from German rivers and coastal areas. J. Environ. Monit. 8: 812–823.

Ruíz-Hidalgo, K., M. Masís-Mora, E. Barbieri, E. Carazo-Rojas and C.E. Rodríguez-Rodríguez. 2016. Ecotoxicological analysis during the removal of carbofuran in fungal bioaugmented matrices. Chemosphere 144: 864–871.

Salt, D.E., R.D. Smith and I. Raskin. 1998. Phytoremediation. Annu. Rev. Plant. Physiol. Plant. Mol. Biol. 49: 643–68.

Sanderman, Jr. H. 1994. Higher plant metabolism of xenobiotics: the 'green liver' concept. Pharmacogenetics 4: 225–41.

Santos, J.L., I. Aparicio, M. Callejón and E. Alonso. 2009. Occurrence of pharmaceutically active compounds during 1-year period in wastewaters from four wastewater treatment plants in seville (Spain). J. Hazard. Mat. 164: 1509–1516.

Sarkar, P., A. Roy, S. Pal, B. Mohapatra, K.K. Sufia, M.K. Maitia and P. Sar. 2017. Enrichment and characterization of hydrocarbon-degrading bacteria from petroleum refinery waste as potent bioaugmentation agent for *in situ* bioremediation. Biores. Technol. 242: 15–27.

Scheytt, T.J., P. Mersmann and T. Heberer. 2006. Mobility of pharmaceuticals carbamazepine, diclofenac, ibuprofen, and propyphenazone in miscible-displacement experiments. J. Contam. Hydrol. 83: 53–69.

Schnoor, J.L. 1997. Phytoremediation. Pittsburgh: National Environmental Technology Applications Center. Technology Evaluation Report TE-97-01.

Schröder, P., C. Scheer and E.J.D. Belford. 2001. Metabolism of organic xenobiotics in plants: conjugating enzymes and metabolic endpoints. Minerva Biotechnol. 13: 85–91.

Schwarzenbach, R.P., B.I. Escher, K. Fenner, T.B. Hofstetter, C.A. Johnson, G.U. Von and B. Wehrli. 2006. The challenge of micropollutants in aquatic systems. Science 313: 1072–1077.

Schwitzguébel, J.P. and T. Vanek. 2003. Some fundamental advances for xenobiotic chemicals. pp. 123–157. *In*: McCutcheon, S. and J. Schnoor [eds.]. Phytoremediation—Transformation and Control of Contaminants. Wiley Interscience, Hoboken, NJ.

Semple, K.T., R.B. Cain and S. Schmidt. 1999. Biodegradation of aromatic compounds by microalgae. FEMS Microbiol. Lett. 170: 291–300.

Shreve, M.J., A. Brockman, M. Hartleb, S. Prebihalo, F.L. Dorman and R.A. Brennan. 2016. The white-rot fungus Trametes versicolor reduces the estrogenic activity of a mixture of emerging contaminants in wastewater treatment plant effluent. Int. Biodeter. Biodegrad. 109: 132–140.

Shriwastav, A., S.K. Gupta, F.A. Ansari, I. Rawat and F. Bux. 2014. Adaptability of growth and nutrient uptake potential of Chlorella sorokiniana with variable nutrient loading. Bioresour. Technol. 174: 60–66.

Singh, H. 1991. Role of yeasts and fungi in wastes and wastewaters treatment. Ph.D. dissertation, Greenwich University, Hilo, HI.

Singh, H. 2006. Mycoremediation: Fungal remediation. John Wiley & Sons, Inc., Publication, New Jersey, pp. 1–20.

Snyder, S. 2002. Endocrine disruptors and pharmaceutically active compounds: US regulations and research. IWA World Water Congress Melbourne 2002, Workshop on Endocrine Disruptors Proceedings, Conference Proceedings, 1–10.

Song, H.-L., K. Nakano, T. Taniguchi, M. Nomura and O. Nishimura. 2009. Estrogen removal from treated municipal effluent in small-scale constructed wetland with different depth. Bioresour. Technol. 100: 2945–2951.

Stumpf, M., T.A. Ternes, R.D. Wilken, S.V. Rodrigues and W. Baumann. 1999. Polar drug residues in sewage and natural waters in the state of Rio de Janeiro, Brazil. Sci. Total Env. 225: 135–141.

Surujlal-Naicker, S., S.K. Gupta and F. Bux. 2015. Evaluating the acute toxicity of the estrogen hormones and South African wastewater effluents using Vibrio fischeri. Human and Ecological Risk Assessment: An International Journal 21.4: 1094–1108.

Tejeda, A., A.X. Torres-Bojorges and F. Zurita. 2016. Carbamazepine removal in three pilot-scale hybrid wetlands planted with ornamental species. Ecol. Eng. http://dx.doi.org/10.1016/j.ecoleng.2016.04.012.

Ternes, T.A. 1998. Occurrence of drugs in German sewage treatment plants and rivers. Water Res. 32: 3245–3260.

Theerachat, M., P. Tanapong and W. Chulalaksananukul. 2017. The culture or co-culture of Candida rugosa and yarrowia lipolytica strain rM-4A, or incubation with their crude extracellular lipase and laccase preparations, for the biodegradation of palm oil mill wastewater. Int. Biodeter. Biodegrad. 121: 11–18.

Tixier, C., H.P. Singer, S. Oellers and S.R. Müller. 2003. Occurrence and fate of carbemazepine, clofibric acid, diclofenac, ibuprofen, ketoprofen, and naproxen in surface waters. Environ. Sci. Technol. 37: 1061–1068.

Tsai, S.L., C.-W. Lin, C.-H. Wu and C.-M. Shen. 2013. Kinetics of xenobiotic biodegradation by the Pseudomonas sp. YATO411 strain in suspension and cell-immobilized beads. J. Taiwan Inst. Chem. Engineers 44: 303–309.

van Dillewijn, P., J.L. Couselo, E. Corredoira, A. Delgado, R.M. Wittich, A. Ballester et al. 2008. Bioremediation of 2, 4, 6-trinitrotoluene by bacterial nitroreductase expressing transgenic aspen. Environ. Sci. Technol. 42: 7405–10.

Vanhulle, S., E. Enaud, M. Trovaslet, L. Billottet, L. Kneipe, J.L.H. Jiwan et al. 2008. Coupling occurs before breakdown during biotransformation of acid blue 62 by white rot fungi. Chemosphere 70: 1097–1107.

Vasiliadou, I.A., R. Sanchez-Vázquez, R. Molina, F. Martínez, J.A. Melero, L.F. Bautista et al. 2016. Biological removal of pharmaceutical compounds using white-rot fungi with concomitant FAME production of the residual biomass. J. Environ. Manag. 180: 228–237.

Verkleij, J.A.C., A. Golan-Goldhirsh, D.M. Antosiewisz, J.-P. Schwitzguébel and P. Schröder. 2009. Dualities in plant tolerance to pollutants and their uptake and translocation to the upper plant parts. Environ. Exp. Bot. 67: 10–22.

Vörösmarty, C.J., P.B. McIntyre, M.O. Gessner, D. Dudgeon and A. Prusevich. 2010. Global threats to human water security and river biodiversity. Nature 467: 555–561.

Vymazal, J. 2014. Constructed wetlands for treatment of industrial wastewaters: A review. Ecol. Eng. 73: 724–751.

Wilson, J.T., J.F. McNabb, J.W. Cochran, T.H. Wang, M.B. Tomson and Bedient, P.B. 1985. Influence of microbial adaptation on the fate of organic pollutants in ground water. Environ. Toxicol. Chem. 4: 721–726.

Wu, R.S.S. 1999. Eutrophication, water borne pathogens and xenobiotic compounds: environmental risks and challenges. Mar. Poll. Bull. 39(1–12): 11–22.

Xiong, J.-Q., M.B. Kurade, R.A.I. Abou-Shanab, M.-K. Ji, J. Choi, J.O. Kim et al. 2016. Biodegradation of carbamazepine using freshwater microalgae *Chlamydomonas mexicana* and *Scenedesmus obliquus* and the determination of its metabolic fate. Biores. Technol. 205: 183–190.

Xiong, J.-Q., M.B. Kurade, J.R. Kim, H.-S. Roh and B.-H. Jeon. 2017a. Ciprofloxacin toxicity and its co-metabolic removal by a freshwater microalga *Chlamydomonas mexicana*. J. Haz. Mat. 323: 212–219.

Xiong, J.-Q., M.B. Kurade, D.V. Patil, M. Jang, K.-J. Paeng and B.-H. Jeon. 2017b. Biodegradation and metabolic fate of levofloxacin via a freshwater green alga, *Scenedesmus obliquus* in synthetic saline wastewater. Algal Res. 25: 54–61.

Yao, R.-S., M. Sun, C.-L. Wang and S.S. Deng. 2006. Degradation of phenolic compounds with hydrogen peroxide catalyzed by enzyme from *Serratia marcescens* AB 90027. Water Res. 40: 3091–3098.

Young, L. and J. Yu. 1997. Ligninase-catalysed decolorization of synthetic dyes. Water Res. 31: 1187–1193.

Yu, J.T., E.J. Bouwer and M. Coelhan. 2006. Occurrence and biodegradability studies of selected pharmaceuticals and personal care products in sewage effluent. Agric. Water Manage. 86: 72–80.

Zhang, Y., T. Lv, P.N. Carvalho, C.A. Arias and Z. Chen. 2015. Removal of the pharmaceuticals ibuprofen and iohexol by four wetland plant species in hydroponic culture: plant uptake and microbial degradation. Environ. Sci. Poll. Res. DOI: 10.1007/s11356-015-5552-x.

Zhang, Y.J., S.U. Geißen and C. Gal. 2008. Carbamazepine and diclofenac: removal in wastewater treatment plants and occurrence in water bodies. Chemosphere 73: 1151–1161.

Zhao, Y., F. Tao and E.Y. Zeng. 2008. Theoretical study on the chemical properties of polybrominated diphenyl ethers. Chemosphere 70: 901–907.

Zhou, Y., S. Trestip, X. Li, M. Truu, J. Truu and U. Mander. 2012. Dechlorination of hexachlorobenzene in treatment microcosm wetlands. Ecol. Eng. 42: 249–255.

Zhou, Y., T. Tegane, X. Li, M. Truu, J. Truu and U. Mander. 2013. Hexachlorobenzene dechlorination in constructed wetland mesocosms. Water Res. 47: 102–110.

8

Fungal Laccase Mediated Bioremediation of Xenobiotic Compounds

Palanivel Sathishkumar,[1,*] *Feng Long Gu,*[1,*] *Fuad Ameen*[2]
and *Thayumanavan Palvannan*[3]

INTRODUCTION

The release of multiple xenobiotic compounds from agricultural, industrial, and domestic usage in wastewater systems is an increasing environmental concern worldwide. Approximately 90% of the used hazardous xenobiotic compounds are disposed into the environment without proper treatment (Reddy and Mathew 2001). The widespread detection of xenobiotic compounds such as pharmaceutically active compounds (PhACs), personal care products (PCP), phenolic compounds, pesticides, herbicides, polycyclic aromatic hydrocarbons (PHAs), and industrial chemicals in the aquatic environment has raised concerns over their potential adverse effects on the eco-system. In addition, these xenobiotic compounds have been considered as emerging contaminants (ECs) due to their persistence in the environment and

[1] Key Laboratory of Theoretical Chemistry of Environment, Ministry of Education; School of Chemistry and Environment, South China Normal University, Guangzhou, 510006, P. R. China.
[2] Department of Botany and Microbiology, Faculty of Science, King Saud University, Riyadh, Saudi Arabia.
  Email: fhasan@ksu.edu
[3] Laboratory of Bioprocess and Engineering, Department of Biochemistry, Periyar University, Salem-636011, Tamil Nadu, India.
  Email: pal2912@periyaruniversity.ac.in
* Corresponding authors: salemsathishkumar@gmail.com; gu@scnu.edu.cn

negative impact on human health (Deblonde et al. 2011; Sathishkumar et al. 2014a). Thus, the effective remediation of xenobiotic compounds is highly desirable for public health and environmental security (Sathishkumar et al. 2015; Adnan et al. 2017). Several conventional approaches have been reported for the xenobiotic compounds remediation (Sathishkumar et al. 2012a). Recent studies highlight that the enzyme based biotechnological approach has been considered as an effective tool for remediation of xenobiotic Compounds due to their simple, eco-friendly, and cost-effective processes. Enzymatic remediation comes under traditional categories of both chemical and biological processes, due to the performance of chemical reactions by the action of biological catalysts. The enzymatic transformation of xenobiotic compounds into less toxic or even harmless products is an alternative to their complete removal. In this regard, a number of microbial enzymes (manganese peroxidase (EC 1.11.1.13), lignin peroxidase (EC 1.11.1.14), laccase (EC 1.10.3.2), and cytochrome P450 (EC 1.14.14.1)) have been reported. Interestingly, fungal laccases have proved their potential in targeting different types of xenobiotic compounds and also achieved maximum detoxification.

Laccases: Source, Structure, Mechanism, and Production

Laccase (p-diphenol:dioxygen oxidoreductase) is a member of the multicopper oxidoreductase enzyme. They use molecular oxygen from air as an electron acceptor, and further generate water as a by-product (Morozova et al. 2007). It catalyzes a wide range of substrates, including aromatic compounds, metal ions, organometallic compounds, organic redox compounds, and iodide anion (Mate and Alcalde 2015). Furthermore, the catalytic activity of laccase expanded to non-phenolic substrates in the presence of redox mediators.

Sources

Laccase was first extracted from the Japanese lacquer tree *Toxicodendron vernicifluum* in the late nineteenth century by Yoshida (1883). It is widely distributed in higher plants, bacteria, fungi, lichens, and insects (Mayer and Staples 2002; Laufer et al. 2009; Santhanam et al. 2011). However, laccases are particularly more abundant in fungi. The redox potentials of fungal laccases is more efficient than plant and insect laccases. The stability of fungal laccases is higher at acidic pH. The fungal laccases have a wide variety of applications, including biosensor development, chemical synthesis, bioleaching of paper pulp, textile finishing, pollutant remediation, and wine stabilization (Kalia et al. 2014; Khomutov et al. 2016; Verrastro et al. 2016; Karaki et al. 2017). The efficient laccase producing fungal strains are *Aspergillus flavus, Bjerkandera adusta, Botrytis cinerea, Cerrena unicolor, Coriolopsis gallica, Coriolopsis polyzona, Coriolus versicolor, Ganoderma lucidum, Lentinula edodes, Marasmius quercophilus, Myceliophthora thermophila, Paraconiothyrium variabile, Phaenerochaaete chrysosporium, Pichia pastoris, Pleurotus eryngii, Pleurotus florida, Pleurotus ostreatus, Pleurotus pulmonarius, Pycnoporus cinnabarinus, Pycnoporus sanguineus, Schizophyllum commune, Streptomyces cyaneus, Trametes hirsuta, Trametes pubescens*, and *Trametes versicolor* (Palvannan and Sathishkumar 2010; Adnan et al. 2015; Mate and Alcalde 2015).

Structure

Generally, laccases are extracellular glycoproteins, which found as monomers or homodimers having three domains in their structure, as shown in Fig. 1. The molecular mass of laccase was approximately 50–70 kDa with 15–20% of carbohydrates. The laccases are mainly characterized by the presence of one type-1 (T1) copper (Cu) with three additional Cu ions, one type-2 (T2) and two type-3 (T3) Cu ions, arranged in a trinuclear cluster (Davies and Ducros 2006; Mot and Silaghi-Dumitrescu 2012). The laccase absorption wavelength peak is close to 600 nm. However, some of the laccases have unusual spectral properties, since they do not present the absorption spectrum feature of Cu T1. These enzymes are referred to as "yellow" or "white" laccases.

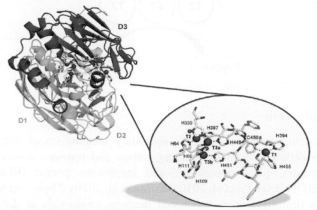

Fig. 1: General structure and details of the active site of *T. trogii* laccase (PDB ID: 2HRG). D1, D2, and D3 are cupredoxin-like domains (Mate and Alcalde 2015).

Mechanism

The catalytic mechanism of laccases is mediated by different Cu centers present in their structure, as shown in Fig. 2. In this catalytic performance, T1 Cu-center oxidizes the reducing substrate and transfers the generated electrons to T2 and T3 Cu-centers. The binding of a reducing substrate to the T1 pocket and subsequent reduction of Cu(II) to Cu(I) occurred at the T1 site during the catalysis. Subsequently, an internal electron transfer occurs from T1 to the T2/T3 cluster due to a strongly conserved His–Cys–His tripeptide motif; the binding and subsequent reduction of oxygen to water also takes place at the T2/T3 cluster. The catalytic activity of laccases is mainly depends on the one electron redox potential difference between the laccase T1 Cu-center and its substrates (phenolic compounds). The first electron transfer from the substrate to the laccase was governed by the "outer-sphere" mechanism (Xu 1996; Méndez et al. 2017).

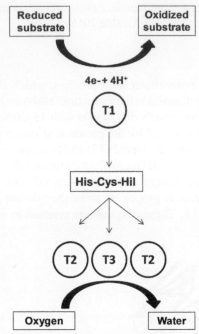

Fig. 2: Mechanism of laccase oxidation reaction.

Laccase production

Laccase production occurs during the secondary metabolism of fungi, which is influenced by different factors, including carbon and nitrogen sources, inducers, agitation, aeration, pH, temperature, and incubation period (Elisashvili and Kachlishvili 2009; Brijwani et al. 2010; Majeau et al. 2010). These factors are highly considered as limiting factors for fungal laccases production at the large-scale; however, fungal species are the most critical factors. In the case of carbon sources, glucose, fructose, mannitol, and cellobiose showed higher laccase production (Mikiashvili et al. 2006). Readily available nitrogen sources such as tryptone, peptone, and asparagine are highly suitable for laccase production (Dong et al. 2005; Sathishkumar et al. 2010). In general, limited use of nitrogen is favored for laccase production; however, some of the reports mentioned that a high concentration of nitrogen in the medium was favorable for laccase production (Majeau et al. 2010). Efficient laccase production occurred at pH 5 to 5.6 and 25 to 30°C. Copper and 2,5-xylidine are the most effective inducers used for the laccase production enhancement (Palmieri et al. 2000; Minussi et al. 2007; Sathishkumar et al. 2013a).

Laccase Immobilization

Although, laccase finds a wide range of applications, its *in vivo* applications have been impeded due to their low stability nature (Poonkuzhali et al. 2011). Interestingly, the laccase immobilized on nanomaterials offer long-term operational stability as well as reusability in continuous cyclic applications. Recent research showed laccases can be immobilized on to nylon nanocomposite membrane (Jasni et al. 2017), cellulose nanofiber (Sathishkumar et al. 2014b), poly(lactic-co-glycolic acid) nanofiber (Sathishkumar et al. 2012b), magnetic nanoparticles (Fortes et al. 2017), graphene nanosheets (Skoronski et al. 2017), carbon nanomaterials (Pang et al. 2015). After immobilization, the thermostability, pH stability, and inhibitors resistance of laccases has been enhanced considerably. These enzyme stability improvements might be due to the enhanced electrical conductivity and biocompatible microenvironment provided by the support materials. In addition, the laccases' immobilized on such eminent supports restricts conformational changes, thus greatly increasing their operational steadiness and lifetime properties. Initially, the enzyme activity of immobilized laccase has slightly decreased, but after certain point the immobilized enzyme is more stable and maintains their stability for a longer time than a free enzyme.

Laccase-mediator System

The redox potential of laccase activity is found between 0.5–0.8 V, which is not enough for the high oxidation of complex xenobiotic compounds. The small molecules can act as redox mediators to extend the laccase enzymatic reactivity towards several substrates (Sathishkumar et al. 2013b). Thus, laccase-mediator system is an interesting approach for wide range of xenobiotic compounds remediation (Bourbonnais and Paice 1990). Redox mediators act as electron shuttles between laccase and target compounds to enhance the catalysis process. The mechanism of laccase-mediator system performance is illustrated in Fig. 3.

In this system, oxidized laccase directly takes electrons from phenolic substrates along with the substrate oxidation to transfer the electrons into molecular oxygen which in turn restores the oxidized laccase state. Further, the specific laccase substrates (mediator) react in the oxidized laccase form with other phenolic and non-phenolic compounds to oxidize through electrons uptake. This restores the mediator function for continuous cycles of laccase-mediator electron transfer chain reactions. Particularly, the oxidized phenolic and non-phenolic compounds can be reactive and might also undergo further chemical reactions. A number of synthetic (2,2'-azino-bis(3-ethylbenz-thiazoline-6-sulfonic acid) (ABTS), 1-hydroxybenztriazole (HBT), N-hydroxyanthranilate (NHA), 2,2,6,6-tetramethylpiperidin-1-yloxy (TEMPO)), and natural (acetosyringone, acetovanillone, ferulic acid, para-coumaric acid, syringaldehyde, vanillin) mediators have been used for the reactions. The chemical structure of mediators is shown in Fig. 4. The influence of different synthetic and natural mediators on the laccase based oxidation system was investigated by Lloret et al. (2010). Among the available mediators, HBT (synthetic) and syringaldehyde (natural) greatly enhanced the action of the laccase enzyme for the remediation of xenobiotic compounds.

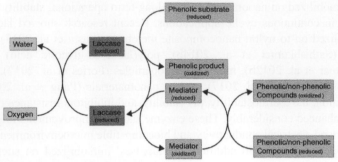

Fig. 3: Mechanism of laccase-mediator system.

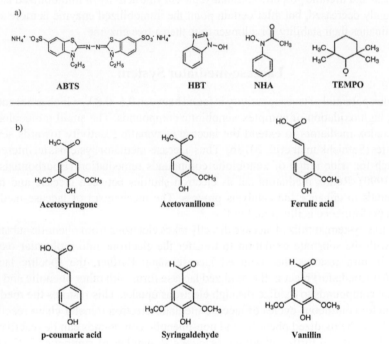

Fig. 4: The chemical structure of (a) synthetic and (b) natural mediators.

Xenobiotic compounds bioremediation by fungal laccases

PhACs

PhACs contamination in the aquatic environment has become an issue of international concern due to their persistence and high biological activity even at low concentrations. The critical ecological risk posed PhACs includes endocrine disruption, increased drug resistance of microorganisms, decreased plant nutrient uptake, and bioaccumulation in the food chain. These risks are associated with a loss of biodiversity, the development of antibiotic-resistant, human infertility, and cancer. Some of the antibiotics, anti-inflammatory drugs, and hormones are currently classified as ECs because of their frequent detection in the environment, the

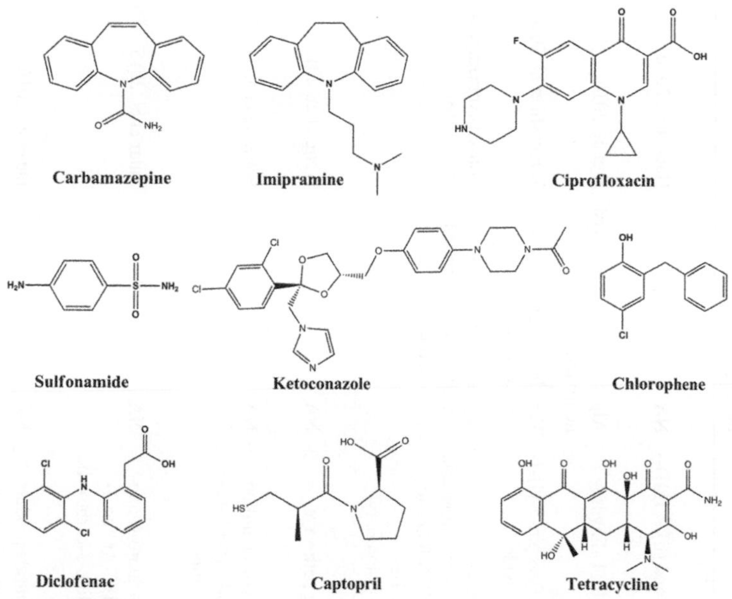

Fig. 5: Structure of PhACs.

uncertainty of their ecological impact, and the legislation concerning their discharge and disposition. Some of the laccase degrading PhACs structures are given in Fig. 5. Table 1 illustrates the previous reports of PhACs and PCPs degradation by fungal laccases.

Carbamazepine. Carbamazepine is used as an antiepileptic drug. It is a highly persistent PhAC and routinely detected in the aquatic environment. Laccase mediated biodegradation provided a promising approach for the elimination of carbamazepine. The repeated treatment of carbamazepine with *T. versicolor* laccase in the presence of a redox mediator HBT showed the effective bioremediation, and two degradation products were indentified namely, 10,11-dihydro-10,11-epoxycarbamazepine, and 9(10H)-acridone by Hata et al. (2010). In 2016, Ji et al. performed the biodegradation of carbamazepine with immobilized *T. versicolor* laccase-mediator (*p*-coumaric acid) membrane hybrid reactor and identified 10,11-dihydro-10,11-dihydroxy-CBZ, 10,11-dihydro-10,11-epoxy-CBZ, and acridone as the major metabolites. Further, the algal toxicity tests proved the laccase effectively removed the toxicity of carbamazepine.

Imipramine. Imipramine is broadly used in the treatment of mental depression. However, it has a toxic effect on the cardiovascular and central nervous systems, which make it a serious threat to life. Tahmasbi et al. (2016) found that *P. variabile* laccase-catalyzed biotransformation of imipramine led to 3-(2,7-dihydroxy-10,11-dihydro-5Hdibenzo[b,f]azepin-5-yl)-1-hydroxy-N,N-dimethyl-3-oxopropan-1-amine oxide as the main metabolite. The metabolite showed very less cytotoxicity on the Caco-2 cell line, which proves laccase is the efficient catalyst for imipramine detoxification.

Table 1: PhACs and PCPs degradation by laccases.

Xenobiotics	Laccase source	Mediators	Degradation metabolites	Toxicity assessment after laccase treatment	References
PhACs					
Carbamazepine	*T. versicolor* laccase	HBT	(i) 10,11-dihydro-10,11-epoxy carbamazepine and (ii) 9(10H)-acridone	NA	Hata et al. 2010
	T. versicolor laccase	*p*-coumaric acid	(i) 10,11-dihydro-10,11-dihydroxy-CBZ, (ii) 10,11-dihydro-10,11-epoxy-CBZ and (iii) acridone	Algal toxicity tests proved no toxicity	Ji et al. 2016
Imipramine	*P. variabile* laccase	-	3-(2,7-dihydroxy-10,11-dihydro-5Hdibenzo[b,f]azepin-5-yl)-1-hydroxy-N,N-dimethyl-3-Oxopropan-1-amine oxide	Very less cytotoxicity on the Caco-2 cell line	Tahmasbi et al. 2016
Ketoconazole	*T. versicolor* laccase	-	(i) 1-(4-{4-[2-(2,4-dichloro-phenyl)-2-imidazol-1-ylmethyl-[1,3]dioxolan-4-ylmethoxy]-phenyl}-4-oxy-piperazin-1-yl)-ethanone, which correspond to C26H28Cl2N4O5 and (ii) 1-(4-{4-[2-(2,4-dichloro-phenyl)-2-imidazol-1-ylmethyl-[1,3]dioxolan-4-ylmethoxy]-phenyl}-3-hydroxy-piperazin-1-yl)-ethanone	A yeast-based micro-toxicity study confirmed decreased toxicity	Yousefi-Ahmadipour et al. 2016
Sulfonamide	*Perenniporia* sp.	ABT and violuric acid	Degraded via deaniline and oxidative coupling pathways	NA	Weng et al. 2012
	T. versicolor laccase	-	(i) Aniline, (ii) aminopyridine, 4-(2-Imino-1-pyridyl)aniline, (iii) N-[4-(2-Imino-1-pyridyl) phenyl]formamide, (iv) N-(3-Pyridyl) pyridine-3-sulfonamide, and (v) 4-(6-Imino-2,4-dimethoxy pyrimidin-1-yl)aniline	NA	Schwarz et al. 2010
Ciprofloxacin	Commercial laccase	-	(i) 7-((2-Aminoethyl) amino)-1-cyclopropyl-6-fluoro-4-oxo-1,4-dihydroquinoline-3-carboxylic acid and (ii) 7-Amino-1-cyclopropyl-6-fluoro-4-oxo-1,4-dihydroquinoline-3-carboxylic acid	NA	Sutar et al. 2015
Captopril	*T. versicolor* laccase	-	(i) Adducts of captopril with quinones, (ii) disulfide product due to auto-dimerization, (iii) dithiolated adducts, (iv) para-quinones, and (v) ortho-quinones	NA	Du et al. 2016

Compound	Laccase source	Mediator	Products/Mechanism	Toxicity	Reference
Diclofenac	*P. sanguineus* laccase cocktail	-	ND	NA	Rodriguez-Delgado et al. 2016
	A. oryzae laccase	-	ND	ToxScreen3 assay revealed no significant toxicity	Nguyen et al. 2014
	P. florida laccase	-	ND	Less toxicity on mouse fibroblast 3T3-L1 preadipocytes	Sathishkumar et al. 2014a
Chlorophene	*T. versicolor* laccase	-	(i) Dimer and (ii) nucleophilic substitution of chlorine	No toxicity on *S. obliquus*	Shi et al. 2016
Tetracycline	*P. ostreatus* laccase	HBT	(i) Oxytetracycline, (ii) hydroxylated, (iii) quinone like structure, (iv) tetracycline-dehydration, (v) tetracycline-demethylation and (vi) tetracycline-deamination	No inhibition of *E. coli*	Sun et al. 2017
	C. unicolor laccase	-	Letone intermediate	Toxicity against *E.coli* and *B. licheniformis* reduced	Yang et al. 2017
PCPs					
Triclosan	*C. polyzona* laccase	-	(i) Dimers, (ii) trimers and (iii) tetramers included in homomolecular oligomers	NA	Cabana et al. 2007
	G. lucidum laccases	HBT	Triclosan oligomers formation	Non-toxic to *E. coli* and *Sphingomonas* sp.	Murugesan et al. 2010
	T. versicolor laccase	-	Dechlorination of Triclosan	NA	Cabana et al. 2011
	T. versicolor laccase	-	Dehalogenated products	92% of toxicity was eliminated	Dai et al. 2015

ND – Not detected; NA – Not assessed.

Ketoconazole. Ketoconazole is a commonly used antifungal drug in both human and veterinary medicine. However, this drug residual cause adverse effects on the ecological system. In 2016, Yousefi-Ahmadipour et al. described the biotransformation and detoxification of ketoconazole using *T. versicolor* laccase. Two major metabolites were identified, namely 1-(4-{4-[2-(2,4-dichloro-phenyl)-2-imidazol-1-ylmethyl-[1,3]dioxolan-4-ylmethoxy]-phenyl}-4-oxy-piperazin-1-yl)-ethanone, which correspond to C26H28Cl2N4O5 and 1-(4-{4-[2-(2,4-dichloro-phenyl)-2-imidazol-1-ylmethyl-[1,3]dioxolan-4-ylmethoxy]-phenyl}-3-hydroxy-piperazin-1-yl)-ethanone with 53% and 44% yield, respectively. A yeast-based micro-toxicity study with *Pseudokirchneriella subcapitata*, *Candida albicans*, *Cryptococcus neoformans*, and *Saccharomyces cerevisiae* confirmed a decreased toxicity of laccase-treated ketoconazole and its degradation metabolites.

Sulfonamide. Sulfonamide is a broad-spectrum antibiotic, which is commonly used in veterinary medicine to control the diseases of aquaculture animals. Its residues appeared in the effluents and groundwater nearby the livestock and aquaculture farms. Weng et al. (2012) reported the biodegradation of sulfonamide by *Perenniporia* strain TFRI 707 laccase-mediator system through deaniline and oxidative coupling pathways. In this laccase-mediator system, ABT and violuric acid showed the fastest biotransformation. In 2010, Schwarz et al. identified aniline, aminopyridine, 4-(2-Imino-1-pyridyl)aniline, N-[4-(2-Imino-1-pyridyl)phenyl]formamide, N-(3-Pyridyl)pyridine-3-sulfonamide, and 4-(6-Imino-2,4-dimethoxypyrimidin-1-yl) aniline as *T. versicolor* laccase treated sulfonamide biodegradation products. Rahmani et al. (2015) performed a microtoxicity study of the inhibition of bacterial growth and reported that the toxicity of *T. versicolor* laccase-treated sulfonamide was decreased.

Ciprofloxacin. Ciprofloxacin is a fluoroquinolone compound, which is mainly used to treat different types of bacterial infections in human. However, this drug is also encountered in non-targeted points. Sutar and Rathod (2015) reported a novel technique of laccase catalyzed degradation of ciprofloxacin in an ultrasound assisted process and identified 7-((2-Aminoethyl) amino)-1-cyclopropyl-6-fluoro-4-oxo-1,4-dihydroquinoline-3-carboxylic acid and 7-Amino-1-cyclopropyl-6-fluoro-4-oxo-1,4-dihydroquinoline-3-carboxylic acid as degradation products. In addition, the study mentioned laccase catalyzed degradation under ultrasound assisted process not only increases ciprofloxacin hydrochloride degradation, but also reduces the biodegradation time compared to other conventional methods.

Captopril. Captopril is a thiol drug used to treat hypertension, heart failure, and kidney disorders. Around 40% of used captopril is excreted and enters into the environment. In 2016, Du et al. used *T. versicolor* laccase for captopril biodegradation, and detected adducts of captopril with quinones, disulfide product due to auto-dimerization, dithiolated adducts, para-quinones, and ortho-quinones as intermediates at different reaction conditions.

Diclofenac. Diclofenac is an anti-inflammatory drug, which is most commonly detected PhACs in aquatic environment. It causes various adverse effects to living system. The laccase mediated diclofenac degradation has been investigated intensively. Laccase immobilized on poly(lactic-*co*-glycolic acid) nanofiber was used for diclofenac biotransformation (Sathishkumar et al. 2012b). Further, the study proved, the immobilized laccase completely transformed the diclofenac up to three cycles, which was extended to six cycles in the presence of syringaldehyde. A novel laccase membrane reactor was proposed for the continuous removal of diclofenac by a commercially available *A. oryzae* laccase. Nguyen et al. (2014) noticed that the toxicity of diclofenac was significantly reduced after laccase treatment. Later, Sathishkumar et al. (2014a) investigated the cytotoxicity assessment of diclofenac and *P. florida* laccase treated diclofenac on mouse fibroblast 3T3-L1 preadipocytes. The study showed that a maximum of 67.9% cell death occurred at 72 h treatment with diclofenac (200 ppm), while the cells treated with laccase catalyzed diclofenac showed less toxicity on the cells. Recently, Rodríguez-Delgado et al. (2016) achieved the efficient biotransformation of diclofeanc using a laccase cocktail from *P. sanguineus* in a real groundwater sample from northwestern Mexico. Nevertheless, none of the studies provide data on the completed biotransformation pathway for diclofenac.

Chlorophene. Chlorophene are emerging broad-spectrum antimicrobial agent, which is mainly used in hospitals. It is likely accumulated in aquatic organisms and sediments after being released to the aquatic environment. Chlorophene could be effectively transformed and eliminated by *T. versicolor* laccase-catalyzed reaction processes through second-order reaction kinetics in water. The laccase-catalysed degradation was occurred as follows: (i) dimer species formation and (ii) nucleophilic substitution of chlorine by the hydroxyl group, followed by oxidation. The toxicity of laccase catalysed chlorophene was effectively eliminated, which was confirmed by toxicity evaluation tests using *Scenedesmus obliquus* (Shi et al. 2016).

Tetracycline. Tetracycline is a broad spectrum polyketide antibiotic. Its accumulation in water sources is a major environmental problem, which leads to increased antibiotic resistance. In 2017, Sun et al. identified *P. ostreatus* laccase-HBT mediated tetracycline intermediate products such as oxytetracycline, hydroxylated, and a quinone-like compound, as well as tetracycline-dehydration, tetracycline-demethylation, and tetracycline-deamination products due to the possible reaction of electron transfer, hydroxylation, dehydrogenation, oxidation, radical reaction, decomposition, and covalent binding. Further, the growth inhibition of *Escherichia coli* confirmed that the antimicrobial activity of tetracycline was remarkably reduced after laccase treatment. Yang et al. (2017) proposed a mechanism of *C. unicolor* laccase mediated tetracycline degradation based on their metabolites. Tetracycline was initially oxidized at position 5 to the corresponding ketone by laccase, and the amino group at position 4 was bi-demethylated and oxidized. Further, water elimination at position 6 and dehydrogenation at position 12 resulted in a degradation product; however, the name of the degradation products was not mentioned. The toxicity of laccase catalyzed tetracycline biodegradation products is considerably reduced against *E. coli* and *Bacillus licheniformis*.

Synthetic Hormone. 17-α-ethinylestradiol (EE2) is a synthetic hormone, which is most frequently encountered in wastewater. It has estrogenic activities and affects endocrine system. Nicotra et al. (2004) reported four homomolecular dimers generated during the laccase-catalyzed biodegradation of EE2. In addition, *L. edodes*, *M. thermophila*, and *T. versicolor* laccases efficiently eliminated the estrogenic activities of this endocrine disrupting hormone (Lloret et al. 2012; Marysková et al. 2016; Eldridge et al. 2017). However, there is no detailed report for the laccase mediated degradation metabolites and their estrogenic nature.

PCPs

PCPs such as toothpaste, facial scrubs, detergents, and soaps are an important contributor of secondary pollutant. Triclosan and some of the synthetic fragrances are the most widely used chemical substances in the PCPs. The structure of triclosan is shown in Fig. 6. The fungal laccase catalyzed remediation of triclosan and some of the synthetic fragrances were reported.

Triclosan. Triclosan is an antimicrobial agent, which is mainly used in toothpaste, soaps, detergents, and surgical cleaning treatments. It inhibits the nitrification process and also acts as an endocrine disruptor. Cabana et al. (2007a) reported that *C. polyzona* laccase treatment of triclosan led to the formation of polymers such as dimers, trimers, and tetramers included in homomolecular oligomers. Murugesan et al. (2010) used *G. lucidum* laccase for the degradation of TCS with (or) without mediator (HBT), and noticed the formation of triclosan oligomers. Further, toxicity of oligomeric products was assessed using microbial inhibition studies with *E. coli* and *Sphingomonas* sp., and confirmed the non-toxic character of laccase catalyzed triclosan. Later, Cabana et al. (2011) proved the direct dechlorination of triclosan through the oxidation process by *T. versicolor* laccase conjugated with chitosan. In 2012, Torres-Duarte et al. highlighted the efficient elimination of triclosan estrogenic activity after *C. gallica* laccase treatment.

Bokare et al. (2010) developed a redox hybrid system for triclosan remediation through (i) a rapid reductive dechlorination under anaerobic conditions, and (ii) laccase-mediated oxidation to form dimers and trimers as degradation products. Similarly, in 2015, Dai et al. studied the triclosan biodegradation using sequential tri-metal reduction and *T. versicolor* laccase-catalytic oxidation, and identified dehalogenated products as metabolites. In this sequential degradation system, around 92.7% of triclosan acute toxicity was eliminated, which was confirmed by the

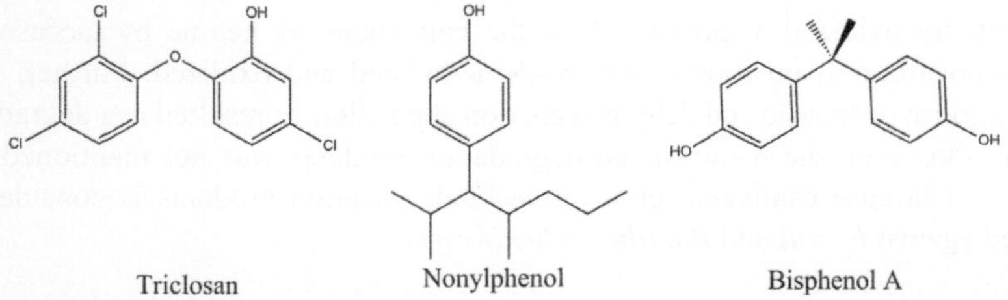

| Triclosan | Nonylphenol | Bisphenol A |

Fig. 6: Structure of PCPs and phenolic compounds.

ecotoxicity analysis of *E. coli* containing green fluorescent protein (GFP). Recently, Melo et al. (2016) also reported the completely reduced toxicity of laccase treated triclosan metabolites against *E. coli*.

Synthetic Fragrances. Synthetic fragrances are used as additives in a wide variety of PCPs. They are considered as ECs, which have been of increasing interest to scientists in recent years. Most recently, Vallecillos et al. (2017) studied the biotransformation of the synthetic fragrances 1-(1,2,3,4,5,6,7,8-octahydro-2,3,8,8,-tetramethyl-2-naphthyl)ethan-1-one (Iso-E-Super, OTNE), 1,3,4,6,7,8,-hexahydro-4,6,6,7,8,8-hexamethylcyclopenta-[g]-2-benzopyran (Galaxolide, HHCB), 7-acetyl-1,1,3,4,4,6-hexamethyl-1,2,3,4-tetrahydronaphtalene (Tonalide, AHTN) and the transformation product of HHCB, and 1,3,4,6,7,8-hexahydro-4,6,6,7,8,8-hexamethylcyclopenta-[g]-2-benzopyran-1-one (Galaxolidone, HHCB-lactone) using *T. versicolor* laccase with redox mediator ABTS in water. Finally, Vallecillos et al. (2017) confirmed that the studied fragrances could be effectively degraded by laccase more than 70% except AHTN. Especially, HHCB-lactone enantiomers showed more than 90% biodegradation by laccase.

Phenolic compounds

Industrial processes generate wastewater that frequently contains toxic phenolic compounds. Phenol and its derivatives can cause adverse health effects in human even at low concentrations. Thus, the presence of phenolic compounds in the aquatic environment has been considered as serious ecological issues (Ahmed et al. 2011). The fungal laccase catalyzed bioremediation of most important phenolic compounds such as nonylphenol and bisphenol A (BPA) were studied extensively. The structure of nonylphenol and BPA is presented in Fig. 6. Table 2 shows the previous reports of phenolic compounds degradation by laccases.

Nonylphenol. Nonylphenol is mainly used to prepare antioxidants, lubricating oil additives, laundry and dish detergents, emulsifiers, and solubilizers. It is highly persistent in the aquatic environment, and also considered as an endocrine disrupting compound. Tsutsumi et al. (2001) performed the degradation of nonylphenol with *T. versicolor* and *P. chrysosporium* laccases and detected some oligomeric reaction products. Later, Cabana et al. (2007) identified dimers, trimers, tetramers, and pentamers as the degradation metabolites of nonylphenol catalyzed by *C. polyzona* laccase. Further, the study confirmed using YES test that 95% of nonylphenol estrogenic activity was removed after laccase treatment. However, the complete structure of the detected oligomeric products is not identified.

BPA. BPA (2, 2-bis (4-hydroxyphenyl) propane) is commonly used in industrial and domestic applications. BPA is one of the endocrine disrupting chemicals and reproductive toxicant. It is unstable under heat or acidic conditions, and can easily enter into the aquatic environment (Kitahara et al. 2010). In 2001, Fukuda et al. observed the formation of polymers of BPA, after laccase (*T. villosa*) treatment. The laccase catalyzed BPA converted into phenol and 4-isopropenylphenol due to oxidative condensation. Further, the study confirmed that the estrogenic activity laccase catalyzed BPA was completely eliminated. Later, Uchida et al. (2001) carried

Table 2: Phenolic compounds degradation by laccases.

Phenolic compounds	Laccase source	Mediators	Degradation metabolites	Toxicity assessment after laccase treatment	References
Nonylphenol	*T. versicolor* and *P. chrysosporium* laccases	-	Oligomeric products	YES test confirmed 95% of Nonylphenol estrogenic activity has removed.	Tsutsumi et al. 2001
	C. polyzona laccase	-	(i) Dimers, (ii) trimers, (iii) tetramers and (iv) pentamers	NA	Cabana et al. 2007
Bisphenol A	*T. villosa* laccases	-	(i) Phenol and (ii) 4-isopropenylphenol	NA	Fukuda et al. 2001
	T. villosa laccase	-	5,5'-bis-[1-(4-hydroxy-phenyl)-1-methyl-ethyl]-biphenyl-2, 2'-diol	NA	Uchida et al. 2001
	C. polyzona laccase	-	(i) Dimers, (ii) trimers and (iii) tetramers	NA	Cabana et al. 2007
	T. villosa laccase	-	4-isopropenylphenol	NA	Fukuda et al. 2004
	C. gallica laccase	HBT	Tartaric acid	NA	Daassi et al. 2016
		-	*b*-Hydroxybutyric acid		

NA – Not assessed.

out the polymerization of BPA using *T. villosa* laccase and identified high molecular weight dimer product 5.5′-bis-[1-(4-hydroxy-phenyl)-1-methyl-ethyl]-biphenyl-2, 2′-diol due to the formation of C-C bond of two monomer units of BPA. Few years later, Fukuda et al. (2004) proposed the degradation mechanism of bisphenol A by a recombinant *T. villosa* laccase. During the degradation dimers and oligomers initially formed by coupling between the C-C or C-O via phenolic groups, and further oligomers were formed to lead the formation of 4-isopropenylphenol. Cabana et al. (2007) reported the formation of dimers, trimers, and tetramers as BPA degradation products by *C. polyzona* laccase treatment. In 2016, Daassi et al. studied three different fungal (*C. gallica*, *B. adusta*, and *T. versicolor*) laccases as biocatalysts for the biotransformation of bisphenol A (BPA). BPA is more rapidly oxidized by *C. gallica* laccase compared to other laccases. Carboxylic acid derivatives such as tartaric acid was found as BPA degradation products in the presence of HBT, while form b-hydroxybutyric acid without HBT. The tartaric and b-hydroxybutyric acids formation due the oxidation of the methyl groups on the propane moiety of BPA by the action of laccase.

Pesticides and herbicides

Pesticides and herbicides are widely used in the modern agricultural field. It is estimated that around 5% of the used pesticides reach their target organism. Unfortunately the majority reach non-targeted ends in the environment (Torres-Duarte et al. 2009). Most of the pesticides have been considered as serious environmental contaminants due to their overuse in the agriculture area. Recent reports highlight that laccase catalyzed the degradation of pesticides which considered are as ECs. The structure of fungal laccase degrading pesticides and herbicides are shown in Fig. 7a. Table 3 demonstrates the previously reported fungal laccase mediated pesticides and herbicides degradation. In 2016, Jin et al. studied a group of pesticides (chlorpyrifos, chlorothalonil, pyrimethanil, atrazine, and isoproturon) degradation by *T. versicolor* laccase with different mediators. The study found best mediators were violuric acid for pyrimethanil and isoproturon, vanillin for chlorpyrifos, and acetosyringone and HBT for chlorothalonil and atrazine, respectively. The degradation rates of pyrimethanil and isoproturon were significantly faster than chlorpyrifos, chlorothalonil, and atrazine. Huang et al. (2016) identified a group of novel genes encoding laccases (at least 22 differentially expressed laccase genes) which are involved in the catabolism/detoxification of atrazine and isoproturon residues in rice. Further, the study confirmed the laccase role in the degradation process by the detection of two laccase catalyzed metabolites, namely hydroxy-dehydrogenated atrazine (HDHA) and 2-OH-isopropyl-IPU in the laccase gene cloned eukaryotic cells.

Recently, Das et al. (2017) studied chlorpyrifos degradation using *T. versicolor* laccase immobilized on magnetic iron nanoparticles and degradation metabolites were identified namely 2,4-bis(1,1 dimethylethyl) phenol and 1,2 benzenedicarboxylic acid, bis(2-methyl propyl) ester. Yang et al. (2015) reported the methoxychlor degradation by host/guest-type immobilized laccase on magnetic tubular mesoporous silica and identified 1,1-diphenylethylene as final degradation products. Zeng et al. (2017) investigated the degradation of the herbicide isoproturon using *T. versicolor*

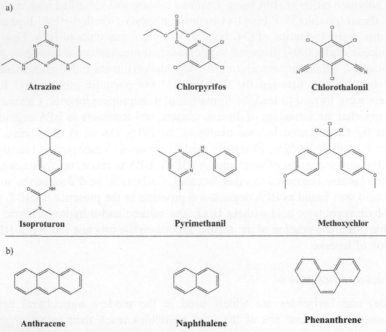

Fig. 7: Structure of (a) pesticides and herbicides and (b) PAHs.

laccase and its laccase-mediator systems. Three isoproturon degradation metabolites namely (i) benzotriazole, (ii) trimer, and (iii) dimer of isoproturon were identified in the laccase-HBT system. Further, the study confirmed lower ecotoxicity of laccase catalyzed isoproturon degradation products against green algae *P. subcapitata*.

PAHs

PAHs are considered as toxic pollutants. It has been widely distributed in the aquatic and terrestrial environment due to incomplete combustion of fossil fuels during industrial processes and vehicle utilization (Collins et al. 1996). In general, PAHs are highly recalcitrant to the degradation process due to their low solubility (Gullotto et al. 2008). Naphthalene, anthracene, and phenanthrene are most important PHAs (Fig. 7b). Table 3 shows the previously reported fungal laccase mediated PAHs degradation. Majcherczyk et al. (1998) performed *T. versicolor* laccase catalyzed oxidation of PAHs, and identified quinines are the major degradation products.

Interestingly, anthracene degradation profiles of laccases from *T. versicolor* and *P. sanguineus* were assessed by Xuan-Zhen et al. (2014). The study indicated that 9,10-anthraquinone was the main metabolites of anthracene catalyzed by *T. versicolor* laccase. Further, the microtox test confirmed that the laccase catalysed degradation products are non-toxic. In 2009, Hu et al. investigated *T. versicolor* laccase mediated oxidations of anthracene in the presence of HBT, and identified anthraquinone as main degradation product. The degradation product showed very less significant genotoxicity to keratinocyte HaCaT and lymphocyte A3 cells.

Table 3: Pesticides, herbicides, and PAHs degradation by laccases.

Xenobiotics	Laccase source	Mediators	Degradation metabolites	Toxicity assessment after laccase treatment	References
Pesticides and herbicides					
Atrazine	*T. versicolor* laccase	HBT	ND	NA	Jin et al. 2016
Chlorpyrifos		Vanillin			
Chlorothalonil		Acetosyringone			
Isoproturon		Violuric acid			
Pyrimethanil		violuric acid			
Chlorpyrifos	*T. versicolor* laccase	-	(i) 2,4-bis(1,1 dimethylethyl) phenol, (ii) 1,2 benzene dicarboxylic acid, (iii) bis(2-methyl propyl) ester	NA	Das et al. 2017
Methoxychlor	*P. ostreatus* laccase	-	1,1-diphenylethylene	NA	Yang et al. 2015
Isoproturon	*T. versicolor* laccase	-	(i) Benzotriazole, (ii) trimer and (iii) dimer	Lower ecotoxicity against green algae *P. subcapitata*	Zeng et al. 2017
PAHs					
Anthracene	*T. versicolor* laccase	HBT	Anthraquinone	No genotoxicity to keratinocyte HaCaT and lymphocyte A3 cells	Hu et al. 2009
	T. versicolor laccase	-	9,10-anthraquinone	Microtox test proved no toxicity	Xuan-Zhen et al. 2014
Naphthalene	*P. ostreatus* laccases	-	Polymeric products	NA	Gullotto 2008
Phenanthrene	*T. hirsuta* laccase	HBT	(i) Phenanthrene-9,10-quinone and (ii) 2,2'-diphenic acid	NA	Bohmer et al. 1998

ND – Not detected; NA – Not assessed.

Gullotto et al. (2008) used *P. ostreatus* laccase for naphthalene degradation, and reported the formation of polymeric degradation products through the oxidation of methylcatechols by laccase. Phenanthrene was efficiently oxidized by *T. hirsuta* laccase in the presence of HBT, and generated phenanthrene-9,10-quinone and 2,2'-diphenic acid as the major products (Bohmer et al. 1998).

Xenobiotics in air pollution

A large number of populations in almost all countries are exposed to high concentrations of air pollutants. It causes serious health issues including cardiovascular and respiratory diseases, neurodegenerative disorders and cancer (Pope and Dockery 2006). Interestingly, Prasetyo et al. (2016) observed the *T. villosa* laccase were directly oxidize many air pollutants compounds such as hydroquinone, methylhydroquinone, trimethylhydroquinone, resorcinol, *p*-cresol, o-cresol, m-cresol, 4-methylcatechol, catechol, and polyaromatic hydrocarbons (2-naphthylamine, 2-methylnaphthalene, and benzo[a]pyrene). After laccase oxidation, some of the air pollutants were further acted as mediators for oxidation/coupling with none-laccase air pollutants. For example, 2-naphthylamine, hydroquinone and methylhydroquinone (laccase substrates) reacted with 3-aminobiphenyl, 4-aminobiphenyl, benzo[a]pyrene and benz[a]anthracene, 3-(1-nitrosopyrrolidin-2-yl)pyridine (NNN), 4-(methyl-nitrosamino-1-(3-pyridyl)-1-butanone (NNK), crotonaldehyde, nitric oxide, and acrylonitrile resulting in new coupling products. Based on these findings, Prasetyo et al. (2016) suggested that the laccase based filter might be an efficient bioremediation system for the *in situ* degradation of toxicants in the food industries.

Conclusions

During the past decade, a great number of researches on laccase catalyzed xenobiotic compounds biodegradation have been evaluated. Fungal laccases are capable of transforming a large group of hazardous xenobiotic compounds. The laccase catalyzed degradation process may produce non-toxic by-products. This eco-friendly approach is very helpful for a substantial restoration of the polluted eco-system and may result in evident benefits and safety for living things. Research advances are continuously upgrading in the terms of laccase stability enhancement with the help of nanomaterials and metabolic pathways involved in the degradation of xenobiotic compounds. However, still more research is needed to ensure the complete detoxification of xenobiotic compounds in order to scale up for *in vivo* applications.

Acknowledgments

P. Sathishkumar is grateful to South China Normal University for providing the research grant to carry out this study successfully.

References

Adnan, L.A., P. Sathishkumar, T. Hadibarata and A.R. Mohd Yusoff. 2015. Metabolites characterisation of laccase mediated Reactive Black 5 biodegradation by fast growing ascomycete fungus *Trichoderma atroviride* F03. Int. Biodeter. Biodegr. 104: 274–282.

Adnan, L.A., P. Sathishkumar, A.R. Mohd Yusoff, T. Hadibarata and F. Ameen. 2017. Rapid bioremediation of Alizarin red S and Quinizarine green SS dyes using *Trichoderma lixii* F21 mediated by biosorption and enzymatic processes. Bioprocess Biosyst. Eng. 40: 85–97.

Ahmed, S., M.G. Rasul, R. Brown and M.A. Hashib. 2011. Influence of parameters on the heterogeneous photocatalytic degradation of pesticides and phenolic contaminants in wastewater: a short review. J. Environ. Manag. 92: 311–330.

Asgher, M., S. Noreena and M. Bilal. 2017. Enhancing catalytic functionality of *Trametes versicolor* IBL-04 laccase by immobilization on chitosan microspheres. Chem. Eng. Res. Des. 119: 1–11.

Bohmer, S., K. Messner and E. Srebotnik. 1998. Oxidation of Phenanthrene by a fungal laccase in the presence of 1-hydroxybenzotriazole and unsaturated lipids. Biochem. Biophys. Res. Commun. 244: 233–238.

Bokare, V., K. Murugesan, Y.M. Kim, J.R. Jeon, E.J. Kim and Y.S. Chang. 2010. Degradation of triclosan by an integrated nano-bio redox process. Bioresour. Technol. 101: 6354–6360.

Bourbonnais, R. and M.G. Paice. 1990. Oxidation of non-phenolic substrates: an expanded role for laccase in lignin biodegradation. FEBS Lett. 267: 99–102.

Brijwani, K., A. Rigdon and P.V. Vadlani. 2010. Fungal laccases: production, function, and applications in food processing. Enz. Res. 1–10.

Cabana, H., J.L.H. Jiwan, R. Rozenberg, V. Elisashvili, M. Penninckx, S.N. Agathos et al. 2007. Elimination of endocrine disrupting chemicals nonylphenol and bisphenol A and personal care product ingredient triclosan using enzyme preparation from the white rot fungus *Coriolopsis polyzona*. Chemosphere 67: 770–778.

Cabana, H., A. Ahamed and R. Leduc. 2011. Conjugation of laccase from the white rot fungus *Trametes versicolor* to chitosan and its utilization for the elimination of triclosan. Bioresour. Technol. 102: 1656–1662.

Canbolat, M.F., H.B. Savas and F. Gultekin. 2017. Enzymatic behavior of laccase following interaction with g-CD and immobilization into PCL nanofibers. Anal. Biochem. 528: 13–18.

Collins, P.J., M. Kotterman, J.A. Field and A. Dobson. 1996. Oxidation of anthracene and Benzo [a]pyrene by laccases from *Trametes versicolor*. Appl. Environ. Microbiol. 62: 4563–4567.

Daassi, D., A. Prieto, H. Zouari-Mechichi, M. Jesús Martínez, M. Nasri and T. Mechichi. 2016. Degradation of bisphenol A by different fungal laccases and identification of its degradation products. Int. Biodeter. Biodegr. 110: 181–188.

Dai, Y., Y. Song, S. Wang and Y. Yuan. 2015. Treatment of halogenated phenolic compounds by sequential tri-metal reduction and laccase-catalytic oxidation. Water Res. 71: 64–73.

Das, A., J. Singh and Y.K.N. Yogalakshmi. 2017. Laccase immobilized magnetic iron nanoparticles: Fabrication and its performance evaluation in chlorpyrifos degradation. Int. Biodeter. Biodegr. 117: 183–189.

Davies, G.J. and V. Ducros. 2006. Laccase. pp. 1359–1368. *In*: Messerschmidt, A., R. Huber, K. Wieghardt and T. Poulos [eds.]. Handbook of Metalloproteins. Hoboken: Wiley.

Deblonde, T., C. Cossu and P. Harteman. 2011. Emerging pollutants in wastewater: a review of the literature. Int. J. Hyg. Environ. Health 214: 442–448.

Dong, J.L., Y.W. Zhang, R.H. Zhang, W.Z. Huang and Y.Z. Zhang. 2005. Influence of culture conditions on laccase production and isozyme patterns in the white-rot fungus *Trametes gallica*. J. Basic Microbiol. 45: 190–198.

Du, P., H. Zhao, C. Liu, Q. Huang and H. Cao. 2016. Transformation and products of captopril with humic constituents during laccase-catalyzed oxidation: Role of reactive intermediates. Water Res. 106: 488–495.

Eldridge, H.C., A. Milliken, C. Farmerd, A. Hamptone, N. Wendland, L. Coward et al. 2017. Efficient remediation of 17α-ethinylestradiol by *Lentinula edodes* (shiitake) laccase. Biocatal. Agric. Biotechnol. 10: 64–68.

Elisashvili, V. and E. Kachlishvili. 2009. Physiological regulation of laccase and manganese peroxidase production by white-rot Basidiomycetes. J. Biotechnol. 144: 37–42.

Fortes, C.C.S., A.L. Daniel-da-Silva, A.M.R.B. Xavier and A.P.M. Tavare. 2017. Optimization of enzyme immobilization on functionalized magnetic nanoparticles for laccase biocatalytic reactions. Chem. Eng. Process. 117: 1–8.

Fukuda, T., H. Uchida, Y. Takashima, T. Uwajima, T. Kawabata and M. Suzuki. 2001. Degradation of bisphenol A by purified laccase from *Trametes villosa*. Biochem. Biophys. Res. Commun. 284: 704–706.

Fukuda, T., H. Uchida, M. Suzuki, H. Miyamoto, H. Morinaga, H. Nawata et al. 2004. Transformation products of bisphenol A by a recombinant *Trametes villosa* laccase and their estrogenic activity. J. Chem. Technol. Biotechnol. 79: 1212–1218.

Gonzalez-Coronel, L.A., M. Cobas, M.J. Rostro-Alanis, R. Parra-Saldívar, C. Hernandez-Lun, M. Pazos et al. 2016. Immobilization of laccase of *Pycnoporus sanguineus* CS43. New Biotechnol. http://dx.doi.org/10.1016/j.nbt.2016.12.003.

Gullotto, A., S. Branciamore, I. Duchi, M.F.P. Caño, D. Randazzo, S. Tilli et al. 2008. Combined action of a bacterial monooxygenase and a fungal laccase for the biodegradation of mono- and poly-aromatic hydrocarbons. Bioresour. Technol. 99: 8353–8359.

Hata, T., H. Shintate, S. Kawai, H. Okamura and T. Nishida. 2010. Elimination of carbamazepine by repeated treatment with laccase in the presence of 1-hydroxybenzotriazole. J. Hazard. Mater. 181: 1175–1178.

Hu, X., P. Wang and H. Hwang. 2009. Oxidation of anthracene by immobilized laccase from *Trametes versicolor*. Bioresour. Technol. 100: 4963–4968.

Huang, M.T., Y.C. Lu, S. Zhang, F. Luo and H. Yang. 2016. Rice (*Oryza sativa*) laccases involved in modification and detoxification of herbicides Atrazine and Isoproturon residues in plants. J. Agric. Food Chem. 64: 6397–406.

Jasni, M.J.F., P. Sathishkumar, S. Sornambikai, A.R. Mohd Yusoff, F. Ameen, N.A. Buang et al. 2017. Fabrication, characterization and application of laccase-nylon 6,6/Fe^{3+} composite nanofibrous membrane for 3,3′-dimethoxybenzidine detoxification. Bioprocess Biosyst. Eng. 40: 191–200.

Ji, C., J. Hou, K. Wang, Y. Zhang and V. Chen. 2016. Biocatalytic degradation of carbamazepine with immobilized laccase mediator membrane hybrid reactor. J. Membrane Sci. 502: 11–20.

Jin, X., X. Yu, G. Zhu, Z. Zheng, F. Feng and Z. Zhang. 2016. Conditions optimizing and application of laccase-mediator system (LMS) for the laccase-catalyzed pesticide degradation. Sci. Rep. 6: 35787.

Kalia, S., K. Thakur, A. Kumar and A. Celli. 2014. Laccase-assisted surface functionalization of lignocellulosics. J. Mol. Catal. B-Enzym. 102: 48–58.

Karaki, N., A. Aljawish, L. Muniglia, S. Bouguet-Bonnet, S. Leclerc, C. Paris et al. 2017. Functionalization of pectin with laccase-mediated oxidation products of ferulic acid. Enzyme Microb. Technol. 104: 1–8.

Khomutov, S.M., A.A. Shutov, A.M. Chernikh and N.M. Myasoedova. 2016. Laccase-mediated oxyfunctionalization of 3β-hydroxy-Δ^5-steroids. J. Mol. Catal. B-Enzym. 123: 47–52.

Kitahara, Y., S. Takahashi, M. Tsukagoshi and T. Fujii. 2010. Formation of bisphenol A by thermal degradation of poly(bisphenol A carbonate). Chemosphere 80: 1281–1284.

Laufer, Z., R.P. Beckett, F.V. Minibayeva, S. Luthje and M. Bottger. 2009. Diversity of laccases from lichens in suborder Peltigerineae. Bryologist 112: 418–26.

Lin, J., Q. Wen, S. Chen, X. Le, X. Zhou and L. Huang. 2017. Synthesis of amine-functionalized Fe_3O_4@C nanoparticles for laccase immobilization. Int. J. Biol. Macromol. 96: 377–383.

Lloret, L., G. Eibes, G. Feijoo, M.T. Moreira and J.M. Lema. 2012. Degradation of estrogens by laccase from *Myceliophthora thermophila* in fed-batch and enzymatic membrane reactors. J. Hazard. Mater. 213-214: 175–183.

Majcherczyk, A., C. Johannes and A. Hüttermann. 1998. Oxidation of polycyclic aromatic hydrocarbons (PAH) by laccase of *Trametes versicolor*. Enzyme Microb. Technol. 22: 335–341.

Majeau, J.A., S.K. Brar and R.D. Tyagi. 2010. Laccases for removal of recalcitrant and emerging pollutants. Bioresour. Technol. 101: 2331–2350.

Marysková, M., I. Ardao, C.A. García-González, L. Martinová, J. Rotková and A. Sevcu. 2016. Polyamide 6/chitosan nanofibers as support for the immobilization of *Trametes versicolor* laccase for the elimination of endocrine disrupting chemicals. Enzyme Microb. Technol. 89: 31–38.

Mate, D.M. and M. Alcalde. 2015. Laccase engineering: From rational design to directed evolution. Biotechnol. Adv. 33: 25–40.

Mayer, A.M. and R.C. Staples. 2002. Laccase: new functions for an old enzyme. Phytochemistry 60: 551–565.

Melo, C.F., M. Dezotti and M.R.C. Marques. 2016. A comparison between the oxidation with laccase and horseradish peroxidase for triclosan conversion. Environ. Technol. 37: 335–343.

Méndez, E., M.A. González-Fuentes, G. Rebollar-Perez, A. Méndez-Albores and E. Torres. 2017. Emerging pollutant treatments in wastewater: Cases of antibiotics and hormones. J. Environ. Sci. Health A Tox. Hazard. Subst. Environ. Eng. 52: 235–253.

Mikiashvili, N., S.P. Wasser, E. Nevo and V. Elisashvili. 2006. Effects of carbon and nitrogen sources on *Pleurotus ostreatus* ligninolytic enzyme activity. World J. Microbiol. Biotechnol. 22: 999–1002.

Minussi, R.C., G.M. Pastore and N. Duran. 2007. Laccase induction in fungi and laccase/N-OH mediator systems applied in paper mill effluent. Bioresour. Technol. 98: 158–164.

Morozova, O.V., G.P. Shumakovich, M.A. Gorbacheva, S.V. Shleev and A.I. Yaropolov. 2007. "Blue" laccases. Biochemistry 72: 1136–1150.

Mot, A.C. and R. Silaghi-Dumitrescu. 2012. Laccases: complex architectures for one-electron oxidations. Biochemistry 77: 1395–1407.

Murugesan, K., Y.Y. Chang, Y.M. Kim, J.R. Jeon, E.J. Kim and Y.S. Chang. 2010. Enhanced transformation of triclosan by laccase in the presence of redox mediators. Water Res. 44: 298–308.

Nguyen, L.N., F.I. Hai, W.E. Price, F.D.L. Leusch, F. Roddick, E.J. McAdam et al. 2014. Continuous biotransformation of bisphenol A and diclofenac by laccase in an enzymatic membrane reactor. Int. Biodeter. Biodegr. 95: 25–32.

Nicotra, S., A. Intra, G. Ottolina, S. Riva and B. Danieli. 2004. Laccase-mediated oxidation of the steroid hormone 17β-estradiol in organic solvents. Tetrahed. Asym. 15: 2927–2931.

Palmieri, G., P. Giardina, C. Bianco, B. Fontanella and G. Sannia. 2000. Copper induction of laccase isoenzymes in the ligninolytic fungus *Pleurotus ostreatus*. Appl. Environ. Microbiol. 66: 920–924.

Palvannan, T. and P. Sathishkumar. 2010. Production of laccase from *Pleurotus florida* NCIM 1243 using Plackett-Burman design and Response surface methodology. J. Basic Microb. 50: 325–335.

Pang, R., M. Li and C. Zhang. 2015. Degradation of phenolic compounds by laccase immobilized on carbon nanomaterials: Diffusional limitation investigation. Talanta 131: 38–45.

Poonkuzhali, K., P. Sathishkumar, R. Boopathy and T. Palvannan. 2011. Aqueous state laccase thermostabilization using carbohydrate polymers: Effect on toxicity assessment of azo dye. Carbohydr. Polym. 85: 341–348.

Pope, C.A. 3rd and D.W. Dockery. 2006. Health effects of fine particulate air pollution: lines that connect. J. Air Waste Manage. Assoc. 56: 709–742.

Prasetyo, E.N., S. Semlitsch, G.S. Nyanhongo, Y. Lemmouchi and G.M. Guebit. 2016. Laccase oxidation and removal of toxicants released during combustion processes. Chemosphere 144: 652–660.

Rahmani, K., M.A. Faramarzi, A.H. Mahvi, M. Gholami, A. Esrafili, H. Forootanfar et al. 2015. Elimination and detoxification of sulfathiazole and sulfamethoxazole assisted by laccase immobilized on porous silica beads. Int. Biodeter. Biodegr. 97: 107–114.

Reddy, C.A. and Z. Mathew. 2001. Bioremediation potential of white rot fungi: fungi in bioremediation. G.M. Gadd [ed.]. Cambridge University Press, Cambridge, UK.

Rodríguez-Delgado, M., C. Orona-Navar, R. García-Morales, C. Hernandez-Luna, R. Parra, J. Mahlknecht et al. 2016. Biotransformation kinetics of pharmaceutical and industrial micropollutants in groundwaters by a laccase cocktail from *Pycnoporus sanguineus* CS43 fungi. Int. Biodeter. Biodegr. 108: 34–41.

Santhanam, N., J.M. Vivanco, S.R. Decker and K.F. Reardon. 2011. Expression of industrially relevant laccases: prokaryotic style. Trends Biotechnol. 29: 480–489.

Sathishkumar, P., K. Murugesan and T. Palvannan. 2010. Production of laccase from *Pleurotus florida* using agro-wastes and efficient decolorization of Reactive blue 198. J. Basic Microb. 50: 360–367.

Sathishkumar, P., M. Arulkumar and T. Palvannan. 2012a. Utilization of agro-industrial waste *Jatropha curcas* pods as an activated carbon for the adsorption of reactive dye Remazol Brilliant Blue R (RBBR). J. Clean. Prod. 22: 67–75.

Sathishkumar, P., C. Chae, A.R. Unnithan, T. Palvannan, H.Y. Kim, K.J. Lee et al. 2012b. Laccase-poly(lactic-co-glycolic acid) (PLGA) nanofiber: highly stable, reusable, and efficacious for the transformation of diclofenac. Enzyme Microb. Tech. 51: 113–118.

Sathishkumar, P., T. Palvannan, K. Murugesan and S. Kamala-Kannan. 2013a. Detoxification of malachite green by *Pleurotus florida* laccase produced under solid-state fermentation using agricultural residues. Environ. Technol. 34: 139–147.

Sathishkumar, P., K. Balan, T. Palvannan, S. Kamala-Kannan, B.T. Oh and S. Rodríguez-Couto. 2013b. Efficiency of *Pleurotus florida* laccase on decolorization and detoxification of the reactive dye Remazol Brilliant Blue R (RBBR) under optimized conditions. CLEAN-Soil Air Water 41: 665–672.

Sathishkumar, P., A. Mythili, T. Hadibarata, T. Palvannan, R. Jayakumar, M.S. Kanthimathi et al. 2014a. Laccase mediated diclofenac transformation and cytotoxicity assessment on mouse fibroblast 3T3-L1 preadipocytes. RSC Adv. 4: 11689–11697.

Sathishkumar, P., S. Kamala-Kannan, M. Cho, J.S. Kim, T. Hadibarata, M.R. Salim et al. 2014b. Laccase immobilization on cellulose nanofiber: The catalytic efficiency and recyclic application for simulated dye effluent treatment. J. Mol. Catal. B-Enzym. 100: 111–120.

Sathishkumar, P., M. Arulkumar, V. Ashokkumar, A.R. Mohd Yusoff, K. Murugesan, T. Palvannan et al. 2015. Modified phyto-waste *Terminalia catappa* fruit shells: a reusable adsorbent for the removal of micropollutant diclofenac. RSC Adv. 5: 30950–30962.

Schwarz, J., M.O. Aust and S. Thiele-Bruhn. 2010. Metabolites from fungal laccase-catalysed transformation of sulfonamides. Chemosphere 81: 1469–1476.

Shi, H., J. Peng, J. Li, L. Mao, Z. Wang and S. Gao. 2016. Laccase-catalyzed removal of the antimicrobials chlorophene and dichlorophen from water: Reaction kinetics, pathway and toxicity evaluation. J. Hazard. Mater. 317: 81–89.

Skoronski, E., D.H. Souza, C. Ely, F. Broilo, M. Fernandes, A.F. Júnior et al. 2017. Immobilization of laccase from *Aspergillus oryzae* on graphene nanosheets. Int. J. Biol. Macromol. 99: 121–127.

Sun, K., Q. Huang and S. Li. 2017. Transformation and toxicity evaluation of tetracycline in humic acid solution by laccase coupled with 1-hydroxybenzotriazole. J. Hazard. Mater. 331: 182–188.

Sutar, R.S. and V.K. Rathod. 2015. Ultrasound assisted laccase catalyzed degradation of ciprofloxacin hydrochloride. J. Ind. Eng. Chem. 31: 276–282.

Tahmasbi, H., M.R. Khoshayand, M. Bozorgi-Koushalshahi, M. Heidary, M. Ghazi-Khansari and M.A. Faramarzi. 2016. Biocatalytic conversion and detoxification of imipramine by the laccase-mediated system. Int. Biodeter. Biodegr. 108: 1–8.

Torres-Duarte, C., M.T. Viana and R. Vazquez-Duhalt. 2012. Laccase-mediated transformations of endocrine disrupting chemicals abolish binding affinities to estrogen receptors and their estrogenic activity in zebrafish. Appl. Biochem. Biotechnol. 168: 864–876.

Tsutsumi, Y., T. Haneda and T. Nishida. 2001. Removal of estrogenic activities of bisphenol A and nonylphenol by oxidative enzymes from lignin-degrading basidiomycetes. Chemosphere 42: 271–276.

Uchida, H., T. Fukuda, H. Miyamoto, T. Kawabata, M. Suzuki and T. Uwajima. 2001. Polymerization of bisphenol A by purified laccase from *Trametes villosa*. Biochem. Biophys. Res. Commun. 287: 355–358.

Vallecillos, L., Y. Sadef, F. Borrull, E. Pocurull and K. Bester. 2017. Degradation of synthetic fragrances by laccase-mediated system. J. Hazard. Mater. 334: 233–243.

Verrastro, M., N. Cicco, F. Crispo, A. Morone, M. Dinescu, M. Dumitru et al. 2016. Amperometric biosensor based on laccase immobilized onto a screen-printed electrode by matrix assisted pulsed laser evaporation. Talanta 154: 438–445.

Weng, S.S., K.L. Ku and H.T. Lai. 2012. The implication of mediators for enhancement of laccase oxidation of sulfonamide antibiotics. Bioresour. Technol. 113: 259–264.

Xu, F. 1996. Oxidation of phenols, anilines, and benzenethiols by fungal laccases: correlation between activity and redox potentials as well as halide inhibition. Biochemistry 35: 7608–7614.

Xuan-Zhen, L., C. Qian, W. Yu-Cheng, F. You-Zhi, L. Wei-Wei and L. Xian-Gui. 2014. Influencing factors and product toxicity of anthracene oxidation by fungal laccase. Pedosphere 24: 359–366.

Yang, J., Y. Lin, X. Yang, T.B. Ng, X. Ye and J. Lin. 2017. Degradation of tetracycline by immobilized laccase and the proposed transformation pathway. J. Hazard. Mater. 322: 525–531.

Yang, Y., Q. Wei, J. Zhang, Y. Xi, H. Yuan, C. Chen et al. 2015. Degradation of MXC by host/guest-type immobilized laccase on magnetic tubular mesoporous silica. Biochem. Eng. J. 97: 111–118.

Yoshida, H. 1883. LXIII—Chemistry of lacquer (Urushi). Part I. Communication from the Chemical Society of Tokio. J. Chem. Soc. Trans. 43: 472–486.

Yousefi-Ahmadipour, A., M. Bozorgi-Koshalshahi, M. Mogharabi, M. Amini, M. Ghazi-Khansari and M.A. Faramarzi. 2016. Laccase-catalyzed treatment of ketoconazole, identification of biotransformed metabolites, determination of kinetic parameters, and evaluation of micro-toxicity. J. Mol. Catal. B-Enzym. 133: 77–84.

Zeng, S., X. Qin and L. Xia. 2017. Degradation of the herbicide isoproturon by laccase-mediator systems. Biochem. Eng. J. 119: 92–100.

9

Genes Involved in Microbial Bioremediation

K.N. ArulJothi

INTRODUCTION

Growing industrialization and modernization leads to an excess of toxic byproducts and waste products which accumulate and pose a big challenge and hazard to the environment. The tremendous growth in various industrial sectors like metal refinery, mining, tanning, dye, textile, pesticide production, pharmaceutical, electronics, and hardware in recent days are overabundant and the wastes and byproducts let out by them are enormous. These man-made artificial substances released in the environment are called xenobiotic wastes. For example, in textile and dyeing industries, the residual dyes and the metals which were used as a mordant are disposed of as waste products. Most of the pesticides formulated in recent days are organophosphates (OP) based. These organophosphates are released into soil directly in the form of pesticides in agricultural lands. They are degraded by being exposed to sunlight and are less threatening to underground water since they are immobile through the soil. But the heavy rainfall or floods may carry these pesticides from the surface of the soil to aboveground water bodies, where they form a threat to humans and aquatic organisms.

The bioremediation process has become an indispensable method to overcome the pollution generated due to the industrialization and urbanization across the world. Bioremediation is defined as the conversion of toxic substances into non-toxic or less harmful substances with the aid of biological organisms. If plants are used for the

Department of Genetic Engineering, SRM University, Chennai, India.
Emails: aruljothibiotech@gmail.com, aruljothi.n@ktr.srmuniv.ac.in

process it is said to be phytoremediation; when algae and fungi are used it is called as phycoremediation and mycoremediation, respectively. This chapter mainly discusses genes and their regulation in the process of xenobiotic bioremediation.

There are specific genes in microbes that play a functional role in bioremediation of xenobiotic compounds in the environment. The bacteria, in general, tends to develop new phenotypic characters by the operation of regulatory elements or by forming new plasmids with new set of genes. These genes may be acquired by the bacteria through the vertical or horizontal gene transfer route. These genes are specialized to perform bioremediation based on the pollutant present in their environment. Based on the type of environmental stress, the remediation genes and the mechanism of remediation varies; sometimes the genes are controlled positively or negatively by structural elements in form of an operon or transposon. The remediation of metals by bacteria is mediated through bioaccumulation, biotransformation, biomineralization, and biosorption (Dixit et al. 2015) by biocatalysis and enzyme-mediated remediation for petroleum products (Peixoto et al. 2011) and organophosphate pesticides.

Heavy Metal Pollution

Heavy metals are the metals which have a molecular weight more than 53, atomic number more than 20, and atomic weight more than 6. They occur mostly in soil and rocks but due to the pollution, the concentrations of these metal gets increased which may pose a threat to the biotic forms. High amounts of heavy metals are introduced into the environment through metal smelting, metalliferous mining, activities of metallurgical industries, waste disposals, corrosion of metals in use, and agriculture and petroleum exploration among others. The discharge of effluents containing heavy metals mounts pressures on the ecosystem and consequently causes health hazards to plants, animals, aquatic life, and humans. Some of the metals are essential for living organisms for active redox mechanisms, osmotic regulation, and other enzymatic process, yet they are toxic at higher concentrations (Joshi and Modi 2013).

Cadmium is extremely toxic for all living organisms even when present at a low concentration. It has been reported that Cd21 damages the cells by a broad spectrum of effects on cell metabolism for instances by binding to essential respiratory enzymes (Nies 1995), inducing oxidative stress or inhibiting DNA repair. Cd^{2+} can easily enter bacterial cells by the transport systems for essential divalent cations such as Mn^{2+} or Zn^{2+}, so almost all prokaryotes and eukaryotes have developed mechanisms to prevent excessive accumulation of Cd^{2+} in the cells. The relatively low intracellular Cd^{2+} concentration is maintained through the regulation of sequestration or efflux. Microbial resistance to Cd^{2+} is usually based on energy-dependent efflux mechanisms (Nies 1995). Three major groups are responsible for this cadmium efflux: CBA-type chemiosmotic antiporters, P-type ATPases, and cation diffusion facilitators (CDF). The CBA-type antiporters are protein complexes that are able to span the complete cell wall of a gram-negative bacterium, while CDF proteins and P-type ATPases are located in the cytoplasmic membrane and are single-subunit systems.

Chromium is the seventh most abundant metal on earth. Cr(VI) and Cr(III) are more stable and abundant forms of chromium. Chromium(III) is less toxic to organisms because its solubility at the physiological pH is less, whereas Cr(VI) is highly soluble and available as a contaminant in soil and water bodies. Cr(VI) causes mutagenic and carcinogenic effects in biological organisms. Cr(VI) does not interact directly with DNA, hence its genotoxicity is attributed to its intracellular reduction to Cr(III) via reactive intermediates. The resulting types of DNA damage that are produced can be grouped into two categories: (1) oxidative DNA damage and (2) Cr(III)-DNA interactions (Joutey et al. n.d.). The Chromium hexavlent is reduced to trivalent chromium by enzymatic or non-enzymatic reactions in a few bacterial species. A variety of chromate-resistant bacterial isolates has been reported, and the mechanisms of resistance to this ion may be encoded either by plasmids or by chromosomal genes. Usually, the genes located in plasmids encode membrane transporters, which directly mediate efflux of chromate ions from the cell's cytoplasm (Ramı and Jesu 2008).

Mercury is one of the major contaminants of water bodies and the sources of mercury pollution include the industries producing chlor-alkali, disinfectants, paints, pharmaceuticals and paper. Apart from the human contamination, natural processes such as soil erosion, geothermal activities, hydrological cycle, volcanic eruption and wild fires contribute to the global mercury load. The mercuric resistance mechanism is mediated through two different ways (1) synthesis of Thiol compounds which could bind to mercury and reduce the toxicity (Huang et al. 2010), (2) the presence of barriers that stop the entry of mercury into the cells. There is broad range of bacteria to possess the resistance against mercury and reducing it to non-toxic forms. The important resistance mechanism for the mercury is controlled by the mer operon, which may be located in genomic DNA, transposons or extra chromosomal DNA or on integrons. There are 2 different types of mer determinants: the one which confers the resistance of inorganic mercury salts only (narrow spectrum) and the other type is resistant to organomercurials (methyl- and phenylmercury) and inorganic mercury salts (broad spectrum). The mer operon contains operator, promoter and regulator genes, the operon also has functional genes such as merP, merT, merD, merA, merF, merC and merB in broad spectrum strains. (Dash and Das 2012).

Lead (Pb) is considered as one of the chief pollutants among the metalloids. Industrial activities, such as production of batteries and pigments and metal smelting, as well as manufacture of products such as lead arsenate insecticides or lead water pipes are the main sources of Pb in the environment. Natural processes including soil erosion, volcanic emission, and mobilization of Pb from minerals contribute only to a minor degree to Pb pollution of the environment. Pb(II) enters the cell by the machinery used by Mn(II) and Zn(II); however, lead even at low concentration becomes toxic to the host cell. Pb(II) toxicity occurs as a result of changes in the conformation of nucleic acids and proteins, inhibition of enzyme activity, disruption of membrane functions, and oxidative phosphorylation, as well as alterations of the osmotic balance. Pb(II) also shows a stronger affinity for thiol and oxygen groups than essential metals such as calcium and zinc (Jarosławiecka and Piotrowska-Seget 2014). *Bacillus cereus, Arthrobacter* sp., and *Corynebacterium* sp.; the Gram-negative bacteria *Pseudomonas marginalis, Pseudomonas vesicularis*, and *Enterobacter* sp.; and fungi *Saccharomyces cerevisiae* and *Penicillium* sp. Psf-2 have been reported for exhibiting lead resistance.

Pesticide pollution

Worldwide pesticide use has increased dramatically due to the changing farming practices. This leads to the contamination of agricultural and non-agricultural soil. Pesticides are the agents or substances that control pests, especially in agricultural fields. The term includes herbicide, insecticide, nematicide, molluscicide, rodenticide, fungicide, bactericide, and so on. Generally, pesticides are chemical or biological agent that kills and controls growth of pests. Pesticides are chemically classified into three groups, organochlorines, organophosphates, and carbamates. Organochlorine compounds disturb the sodium/potassium channels of the nervous system and forces the nerves to transmit continuously. Organochlorides are not used these days since their bioaccumulation is a persistent problem. Organophosphates and carbamates replace organochlorides and they work by targeting the acetylcholine esterase enzyme which leads to the continuous signalling of neurons, thus causing paralysis. Carbamates are less toxic to vertebrates compared to organophosphorus pesticides. Organophosphorus compounds, namely, glyphosate, chlorpyrifos, parathion, methyl parathion, diazinon, coumaphos, monocrotophos, fenamiphos, and phorate have been extensively used as pesticides. The organophosphate pesticide constitutes a group of widely used very heterogeneous compound monochrotophos, phosphoric acid, monocron, and monostar that share a phosphoric acid derivative chemical structure, exposure to which contaminates a wide range of water and terrestrial ecosystems. The essence of the pesticide remains basically the poisonous impacts of their mixed constant substances on the target organisms while its impact on nontarget organisms and on the environment is the degradation of soil constituents due to retention (Zhang et al. 2011).

Aquatic animals and invertebrates are also affected by the pesticides which get accumulated through the food chain. This can lead to severe allergic reactions and irritations. These pesticides can be a potential carcinogen in some cases. Hypersensitivity, asthma, and skin problems are other ailments caused by the pesticide poisoned food intake. It can also lead to cancer, fertility-related problems in humans, and disruption of hormone production (Pandit et al. 2011). It has also been reported that the toxins in pesticides interact with sodium channels by producing a hyperexcitable state on the central nervous system (Iwanicka et al. 2010).

The genes responsible for pesticide resistant genes spread from one bacteria to another by the means of horizontal transfer mode. The other mechanisms of bacterial gene transfer are conjugation, transformation, or transduction. The genes conferring pesticide resistance gets integrated into the host chromosome by the mechanism of recombination or remains as an extrachromosomal plasmid. The genes that are transferred may have induced mutations that is responsible for encountering the pesticide problem (Kazama 1998). In most cases, the resistance is conferred by the bacterial genes that are associated with plasmids or transposable elements which are often found to be conjugation positive. Among Gram-negative and Gram-positive bacteria, conjugative transposons are predictable as important mediators of genetic exchange on large R-plasmids. The conjugative factors and transposable elements increase the incidence of spreading pesticide resistance to a variety of bacteria in the soil. The pesticide resistance genes that are capable of transferring genes horizontally are mostly present in gene clusters. The resistant genes are transferred by specific DNA structures called integrons, which are the elements which capture the genes by site-specific recombination with the element located on either bacterial chromosomes or plasmid DNA. The integrons flanking the pesticide resistance genes are not repeats; it contains a site-specific integration machinery within the same species or different species which favour transposition. Integrons promote the capture of one or more gene cassettes within the same attachments, thereby forming clusters of antibiotic and pesticide resistance genes (Roy 1995; Salisbury et al. 1972). Resistance gene cassettes have been found for most classes of pesticides and the gene products are also involved in various resistance mechanisms. Integron movements allow the transfer of the gene cluster-associated resistance gene from one DNA replicon to another. Horizontal transfer of resistance genes can be achieved when an integron is incorporated in a broad post range of plasmids. The resistant encoding gene establish resistance gene clusters, which represents a potential source for horizontal gene transfer between bacteria. The horizontal gene transfer within bacterial species such as pesticide-resistant genes or transposable elements (Spratt 1992). The fact is that conjugative bacteria have been evolving for over fifty million years, and they can exchange gene transfer, which has strong implications within the evolution of resistance genes.

When the pesticide resistant strains are exposed to the human microbial flora, though food or other entries, the human microbial flora may also get chance to receive the pesticide resistant genes through conjugation or by another horizontal gene transfer mode. This results in the development of resistant strains in the human enteric site. The pesticide and antibiotic resistance share similar mechanisms to some extent. This results in acquiring a novel resistance in the microbial flora which turns out to be a multi-drug resistant bacteria due to adaptive resistance. The exposure of human microbial flora with other microorganisms from various ecosystems will enhance the chances of genetic variations and the emergence of novel resistance mechanisms which will be introduced into the human system through microbial flora.

There are many genetic loci identified in large variety of microorganisms, which are associated with pesticide resistance. The pesticide resistance mechanism can be a result of mutation in chromosomal genes or by the acquisition of a new gene from plasmids or by the combination of both, hence, there is a wide range of genetic determinants for pesticide resistance. This raises several questions about the evolution and ecology of pesticide-resistant genes. The evolution and diversity of pesticide-resistant genes suggest that at least some of these genes have a long evolutionary diversification (Beeman 1982). The mutation events occur randomly as replication errors of incorrect repair of a DNA damaged activity within the cells. Pesticide resistance occurs by single nucleotide mutations that are at the same time growth permissive and able to produce a resistance phenotype in genetic materials. A variety of pesticide resistance enzymes can be involved with several different targets, access, and protection pathways in the bacterial cells. The significant numbers of the biochemical mechanism of pesticide resistance are based on mutational events. The unpredictable mutation of the sequences of genes encoding the target of multiresistant properties among pesticides was exposed to soil microbes. The variation in expression of pesticides that degraded the efflux systems may also be modified by mutation. Some of the resistance associated with uptake and efflux systems is caused by mutations in regulatory genes or their promoter regions. Also, the mutations leading to increased expression of the efflux systems confer multi-resistance (pesticides and antibiotics).

Dye pollution

There are many types of reactive dyes with different reactive groups. Azo and anthraquinone dyes are the two most important classes used in industries. Azo dyes have structural properties that are not easily degradable under natural conditions and are not removed from water by conventional wastewater treatment systems due to their complex structures and xenobiotic nature. Various azo dyes and their degradation intermediates contribute to the mutagenic activity of ground and surface waters that are polluted by textile effluents. The existence of azo dyes and their by-products in aqueous ecosystems lead to aesthetically unacceptable colouration of waters, along with obstruction of light penetration and diminution of dissolved oxygen, consequently leading to the death and putrefaction of aquatic animals. Several combinations of treatment methods have been developed in order to effectively process textile wastewater. Among them, the decolourization technique is an efficient one. The treatment of dye wastewater involves physical/chemical methods such as coagulation, precipitation, and adsorption (Selvakumar 2011) by activated charcoal, oxidation by ozone, ionizing radiation, and ultrafiltration which forms costly ways of remediation. Microbial decolourization and degradation is an environment-friendly and cost-competitive alternative to chemical decomposition processes. The methods like physical methods, such as adsorption, biological methods (biodegradation) and chemical methods (chlorination, ozonation) are the most frequently used for dye degradation or dye removal from the water bodies. Methods such as combined coagulation, electrochemical oxidation, and active sludge have recently been investigated and proved to be adequate. Other methods such as flocculation, reverse osmosis, and adsorption on activated carbon have also been tested. The drawbacks of these methods are mainly the creation of a more concentrated pollutant-containing phase (Lachheb et al. 2002).

There are many enzymes isolated from bacteria and fungi which are capable of dye degradation. Microorganisms or enzymes have been discovered to degrade both azo and triphenylmethane dyes except for fungi-harboured laccase or lignin peroxidase. A new family of microbial peroxidases, known as dye-decolorizing peroxidases (DyPs), was shown to successfully degrade not only high redox anthraquinone-based but also azo dyes, β-carotene, and aromatic sulphides.

Petrochemical pollution

Petroleum is naturally available in an enormous quantity in the earth's crust. Petroleum pollution can be due to accidental spills by the accidents of cargos or containers in sea bodies. Petroleum is composed of hydrocarbon fractions of paraffin and naphthalenes which carry aliphatic and aromatic branches. Microorganisms are the chief biocatalysts for bioremediation of petroleum products. The bioremediation of petroleum products are mainly done by enzymatic reactions. The degradation occurs in microbes by sequential metabolism of the compounds. The genes involved in petroleum degradation are found in chromosomal or plasmid DNA. The degradation of aliphatic and aromatic hydrocarbon compounds occur under aerobic and anaerobic conditions. The introduction of oxygen atoms into hydrocarbon by oxygenase enzyme occurs under aerobic condition. The aerobic process in the degradation are usually faster due to the availability of oxygen as an electron acceptor. The final product of oxidation cycle is acetyl CoA which will enter into the Krebs cycle. This reaction will be repeated on further degradation of hydrocarbons. The aromatic compounds are degraded by aerobic condition, which forms catechol as the first product, after which it will be degraded and the resulting compounds are introduced into the Krebs cycle. Once it is formed, it will be degraded and the resulting compounds are introduced into the Krebs cycle. These compounds are further converted into carbon-dioxide. Alkane hydroxylases are alkane-degrading enzymes which are distributed in various microorganisms. There are three categories of alkane-degrading enzyme systems (i) C1 to C4 degrading, (ii) C5–C16 degrading, and (iii) C17+ degrading enzymes. There is a well-characterized mechanism of alkane degradation described in *Pseudomonas putida* Gpo1. In this species, the degradation is mediated by membrane monooxygenases, rubredoxin, and rubredoxin convertase that converts alkane to alcohol. The catechol dioxygenase is an iron-containing enzyme that is involved in the degradation of aromatic hydrocarbons under aerobic conditions. This class of enzymes are highly responsible for the microbial mediated degradation of aromatic hydrocarbons.

Phytoremediation

Phytoremediation is a process of degradation of toxic substances such as metals, dyes, and other xenobiotic substances with the help of plants. Phytoremediation is an efficient strategy that can be used to remove metal pollutants from the contaminated site. This approach is eco-friendly and also cost effective. There are many methods that ere used in remediation of metals but those come with high cost and adverse by products. The chemical methods generate large volumes of sludge and are very expensive with technically difficulties compared to other methods, moreover these methods can also degrade valuable components of soil. Conventionally, the remediation of heavy-metal contaminated soils involves either onsite management or excavation and subsequent disposal to a landfill site. This method of disposal solely shifts the contamination problem elsewhere along with the hazards associated with the transportation of contaminated soil and the migration of contaminants from the landfill into an adjacent environment.

Phytoremediation uses the privilege of selective uptake abilities of the plant root systems along with the transport, bioaccumulation and biodegradation feature of the plant (Tangahu 2011). The Plants can either accumulate or exclude the contaminants. The accumulators concentrate the contaminant in an inert form in their tissues whereas the excluders prevent the uptake of contaminant into the plant biomass. The excluders stop the translocation of metals into other tissues from the root

Baker (1981). The uptake by the plants can be enhanced by adding suitable chelating agents in and around the plant for faster and expedite remediation.

Phycoremediation

Phycoremediation is the process of the clean up of contaminants like heavy metals, pesticides, and dyes using micro and macroalgae. Since, this method of bioremediation is very effective only in cleaning up the water bodies and surrounding areas, this forms a limitation of phycoremediation that it cannot be used in other dry areas. This method of remediation overcomes many practical problems such as pH correction, BOD, and COD removal and sludge removal. This methodology is efficient and saves 90% operation cost since no other chemicals are used for remediation; thus, this method is very cheap and environmental friendly. Since few algae are also engaged in biofuel production, the phycoremediation done with such algae can have a dual purpose and has economic value. *Chlorella, Scenedesmus, Phormidium, Botryococcus, Chlamydomonas,* and *Spirulina* microalgae have been widely used in wastewater management effectively (Fathi et al. 2013). Microalgae offer a low-cost and effective approach to remove excess nutrients and other contaminants in tertiary wastewater treatment, while producing potentially valuable biomass, because of a high capacity for inorganic nutrient uptake.

Mycoremediation

Mycoremediation is the use of fungi to clean the contaminated site. Fungi have been proved to play a role in degradation of phenolic, chlorinated, hydrocarbon compounds, pesticides, and dyes. White-rot fungi have been experimented for nearly half a century in the aspect of biodegradation. It was proved to exhibit the ligninolytic enzymatic activity which degrades the wood. Advancement in genetic engineering and molecular biology has improved enzymes and fungi in the process of bioremediation. Fungi has been modified with desirable qualities using genetic engineering and recombinant DNA technology that suits efficient bioremediation. Sometimes fungal genes involved in bioremediation have been incorporated in bacterial species to promote efficient remediation. *Candida utilis* have been widely used for bioremediation since it can use a wide variety of carbon and nitrogen sources and tolerance of low pH. *Saccharomyces fragilis, Rhodotorula, Myceliophthora thermophila, Candida tropicalis, Thermomyces lanuginosus,* and *Aspergillus oryzae* were some of the fungi and yeast employed in wastewater management (Singh 2000).

Heavy Metal Bioremediation

The bacteria develop resistance against heavy metals by converting them into non-toxic forms by bioaccumulation or biotransformation mechanisms (Kamika and Momba 2013). There are a number of metal resistant genes developed in various bacterial species for heavy metal remediation purpose. The metal resistant genes and its regulatory elements are discussed in this section.

Cadmium resistance and its regulation

Cadmium resistance in bacteria was well characterized in *Pseudomonas putida* 06909 (Lee et al. 2001) and two important genes cadA (cadmium—transporting ATPase) and *cadR* (cadmium resistant) were discovered in its genomic DNA. The cadA promoter respond to Cadmium(II), Zinc(II), and lead(II) metal ions. Whereas the *cadR* gene is inducible only by Cd(II) ions. In addition, these genes also confer zinc resistance properties to the bacteria. The *cadA* codes for cadmium transporting ATPase protein, a cation-transporting protein which transports cadmium(II) outside the cell with the help of ATP hydrolysis (Fig. 1A).

Another class of cadmium resistance genes namely *cadB* and *cadD* were identified in *Staphylococcus aureus*; the *cadB* gene product makes the cadmium to bind to the membrane thus protecting the cells from cadmium stress (Chu and Tapan 1989) (Fig. 1B) while *cadD* gene results in a cadmium-binding transmembrane protein being formed on the cell membrane. The *cadD* gene is a component of an operon which has *cadD* and *cadX*. Whereas the *cadX* serves as a positive regulator gene for cadD gene. Generally, the cadD gene shows low level of resistance towards cadmium(II) compared to the resistance provided by cadA and B genes. The cadX gene positively regulates the cadD gene and increases the resistance by upto 10 folds. Hydropathy Studies suggest that the cadD forms an integral membrane protein with five transmembrane domains (Crupper et al. 1999). This transmembrane protein facilitates the export of cationic compounds like cadmium inside the cell.

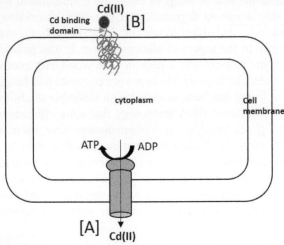

Fig. 1: Various mechanisms of cadmium resistance; [A] Cadmium transporting ATPase protein; [B] Cadmium binding transmembrane protein.

Chromium resistance and its regulation

Microbial chromium(VI) reduction was first reported in *Pseudomonas dechromaticans*. Later several genus such as *Aerococcus, Micrococcus* and *Aeromonas, Thermus scotoductus, Achromobacter, Streptomyces, Amycolatopsis,* and *Flexivirga alba* have also been widely exploited for their chromium reduction properties. The chromium resistance is mediated by a transmembrane efflux mechanism and cytoplasmic conversion of Cr(VI) to Cr(III) by chromium reductase enzymes. Various mechanisms have been described for chromium resistance by the bacterial systems; they are: (i) reduction of Cr(VI) by extracellular mechanism, (ii) reduction of chromium on the cell surface preceded by adsorption of Cr(VI) to the cell membrane, and (iii) intracellular chromium reduction followed by efflux of Cr(III) outside the cell (Joutey et al. 2015). There are six different genes involved in chromium resistance, they are *chrR, chrA, chrB, chrC, chrD, chrE,* and *chrF*. The mobile element (TnOtChr) carrying chromium resistant genes (*chrA, chrB, chrC,* and *chrF*) which form a chr operon was identified in *Ochrobactrum tritici* 5bvl1. The *chrA* gene encodes for the ChrA transporter protein that efflux the chromate ions outside the cell as a mechanism of chromium resistance (Fig. 2A). The *chrC* encodes for a putative superoxide dismutase enzyme and *chrF* codes for a protein with uncertain functions (Branco and Morais 2013). The *chrB* plays regulatory role in the *chr* operon, it is induced by the presence of chromium ions in the environment (Branco and Morais 2013). Previous reports suggest that the ChrB protein acts as a negative regulator of *chr* operon by repressing the *chr* gene expression (Figs. 3A–C). The chromium reductase enzyme is coded by the *chrR* gene. This enzyme is involved in the reduction of Cr(VI) to Cr(III) by utilizing the NADPH molecule. The chromium (VI) ions are transported inside by means of an ion gradient and converted into trivalent ions and are released outside through the ChrA efflux pump (Figs. 2B and A). In some cases, the extracellular reduction of chromium is also carried out by chromium reductase; in this mechanism the hexavalent ions are adsorbed on the bacterial cell surface facilitating the reduction by the enzyme (Fig. 2C). Chromium trivalent is less harmful compared to hexavalent; hence, chromate reduction is an effective mechanism of remediation.

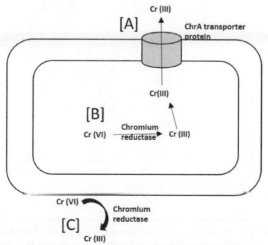

Fig. 2: Various mechanisms of chromium resistance: [A] Chromium(III) efflux by ChrA transporter protein; [B] Chromium reductase enzyme reducing Cr(VI) to Cr(III); [C] Extracellular reduction of Cr(VI) to Cr(III).

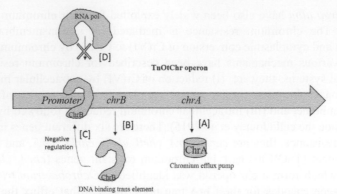

Fig. 3: TnOtChr operon and its regulation mechanism in chromium resistance: [A] *chrA* gene encoding ChrA transporter protein; [B] *chrB* gene encoding for ChrB DNA binding trans element; [C] negative regulation of *chr* operon by ChrB trans element.

Mercury-resistant genes

Mercury reductase is an oxidoreductase enzyme (MerA) coded by *merA* gene, which reduces Hg^{2+} from the toxic form to the Hg^0 non-toxic form (Fig. 4A). The mercury resistant genes were identified from a *mer* transposon which also had other genes like *merB*, *merT*, *merR*, and *merD* (Dash and Das 2012). MerD and MerR proteins regulate the expression of the mRNA gene. MerP is a periplasmic transporter protein which uptakes Hg from the surrounding into the cell through the periplasmic membrane (Fig. 4B) and facilitates the mercury(II) reductase with

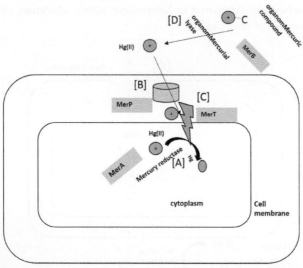

Fig. 4: Mechanisms of Mercury(II) in bacteria; [A] The mercury reductase (MerA) converts Hg^{2+} to Hg^0; [B] transport of Hg^{2+} inside the periplasmic space by MerP; [C] transport of Hg^{2+} inside the cytoplasm by MerT; [D] organomercurial lyase breaks Hg-C bonds from organomercury compounds.

Hg^{2+} for reduction. MerT is a transmembrane protein which binds to free-floating mercury, where the extracellular reductase acts on the bound Hg^{2+} (Fig. 4C). MerB breaks the Hg-C bonds in organomercury compounds after which the Hg transported to mercury reductase for further reduction; it is also called as organomercurial lyase (Fig. 4D). MerR is a trans element which binds to the *merO* operator region of *mer* loci, which positively and negatively regulates the functional expression. The presence of Hg^{2+} ions at optimal concentration promote the binding of MerR trans factor to the mer operator region and activates the transcription, whereas the Hg^{2+} at lower concentrations, represses the activation of *mer* gene.

Lead resistant genes and its regulation

The lead resistance is seen in gram-positive and gram-negative bacteria. Intracellular lead accumulation was first reported in *B. megaterium* and later this property was discovered in *S. aureus* and *C. freundii* whereas the extracellular reduction was reported in *P. marginalis*. Pb(II) was also effluxed by CadA ATPase and ZntA ATPase proteins. In *R. metallidurans* CH34, the genetic determinant *pbr* responsible for lead resistance was isolated and characterized for the first time (Borremans et al. 2001). When this pbr region was analyzed, it had seven ORF regions in an operon-like structure which encodes for seven different peptide products with varied sizes. Out of seven ORFs, there are five distinct genes *pbrT* (ORF652), *pbrA* (ORF798), *pbrB* (ORF112), *pbrC* (ORF206), and *pbrD* (ORF241) which are involved in lead resistance. The gene *pbrT* codes for lead uptake protein, *pbrA* codes for lead P-type lead ATPase efflux protein, *pbrB* codes for an integral protein with undetermined role, *pbrC* codes for prolipoprotein signal peptidase, and *pbrD* codes for lead-binding protein which is essential for sequestration of lead. There was a promoter sequence and a regulatory gene *pbrR* present in this operon. This regulatory gene is similar to the *merR* gene involved in mercury resistance. The *pbr* operon is induced by the presence of Pb(II) in the environment and regulated by the PbrR regulatory protein to transcribe the structural genes in the operon.

Other metal resistance gene

The genes *ZntA* codes for a P-type ATPase transporter protein, which transports zinc inside the cell. The genes nccA and nikA are responsible for nickel resistance. The copper resistance is conferred by copA and copB genes (Margaryan et al. 2013). For the molybdenum remediation, *E. coli* has a gene cluster which has many structural genes, namely *moaA, moeA, moaC,* and *moaE* which are involved in synthesizing the protein molybdopterin cofactor (MoCo) and another gene cluster in *Arthrobacter* with structural genes *modA, modB,* and *modC* (which codes for the molybdate transport protein) was identified and characterized (Otto et al. 1997). The czc system, which is responsible for the resistance of cobalt(II), Zinc(II) and cadmium(II) metal ions, was first identified in *Alcaligenes eutrophus* (Nies 1992).

Xenobiotic Bioremediation of Other Compounds

There are various other compounds like petroleum spills, industrial dye effluents, and pesticide used in agricultural fields which pose a great threat to the environment and human beings. These xenobiotic compounds are effectively degraded by bacterial-derived enzymes and proteins. The following sections enumerate the genes involved in the degradation of pesticides, petroleum derivatives, and industrial dye effluents.

Pesticide degradation

Organophosphate (OP) pesticides are more commonly used against pests including termites, mosquitoes, and roundworms. Pesticides like parathion, methyl parathion, paraoxon, chlorpyrifos, diazinon, malathion, and dichlorvos are types of OP pesticides. The increased use of OP leaves residues in agricultural fields and water bodies. This type of pesticides poses a great threat to the aquatic and terrestrial organisms including higher animals which receive these pesticides through the food chain, that is, through the process of bioaccumulation. In pesticide degradation, three classes of enzymes are involved; hydrolases, esterases, and oxidases. The pesticides are generally degraded in three phases: the first phase is facilitated by oxidation/reduction or hydrolysis of pesticide to a water-soluble less toxic metabolite; in phase two, the less toxic metabolite is conjugated with a sugar molecule; and the third phase converts the conjugate into non-toxic form.

There are a wide range of enzymes present in different bacterial species that encounter the diverse chemical pesticides and catalyze their degradation. The oxidoreductase (from *Pseudomonas* sp. LBr) degrades glyphosate, monooxygenases degrade (*Mycobacterium* sp.) DDT and malathion, haloalkane dehalogenase (from *Sphingobium* sp.) act on hexachlorocyclohexane, and OP hydrolases work on organophosphorus derived pesticides. There are number of bacteria, which synthesize the Organophosphorus hydrolase (OPH) enzyme coded by the *opd* gene, the enzyme degrades the OP pesticides effectively (Neti and Zakkula 2012). The phosphotriesterases are a type of OPH enzyme which was first isolated and characterized from *P. diminuta* MG (Ortiz-hernández et al. 2013). The bacterial strains possessing this gene uses OP as a carbon source. A similar hydrolase gene *opaA* was reported in *B. glumae* PG1. *Pseudomonas monteilii* C11 consists of *hocA* gene which codes for OP hydrolase enzyme. *Imh* is a gene codes for OP hydrolase in *Arthrobacter* sp., Methyl parathion hydrolase (MPH) is coded by *mpd* gene present in *Ochrobactrum* sp. Yw28 capable of degrading methyl parathion, which is an another OP class pesticide. The gene opdA which codes for putative monooxygenase is isolated from *Sphingomonas* sp. PWE1 strain that degrade octylphenol pesticide (Porter and Hay 2007).

Petrochemical degradation and its regulation

Pollution caused by petrochemical products pose a huge threat to the aquatic and terrestrial organisms. There are number of strategies employed to manage the petrochemical pollution, bioremediation emerges as an effective mode, among various methods, for degrading petrochemical pollutants. There are various microorganisms which possess catabolic genes that can degrade or hydrolyse the petrochemical products and its derivatives. Unbranched n-alkanes are major hydrocarbon constituents of petroleum. Multimeric alkane monooxygenase is the enzyme which carries out the first step of the degradation of alkanes. The alkanes are first converted to primary alcohol which are reduced to aldehydes. These aldehydes are further converted to fatty acids. The first reaction is carried out by an integral membrane non-heme di-iron monooxygenase enzyme that hydroxylates the substrate to alcohol. This multimeric protein has particulate membrane-bound hydroxylase (pAH), a rubredoxin (AlkG), and a rubredoxin reductase (AlkT). *Pseudomonas putida* GPo1 expresses a well-characterized AlkB, a model of pAH enzyme that converts n-butane, propane, and other alkanes. The *alkB*-like genes are mostly seen in the genera of *Proteobacteria, Burkholderia, Acinetobacter, Pseudomonas, Alcanivorax, Oleiphilus, Mycobacterium, Rhodococcus, Nocardia, Prauserella*, and *Actinobacteria*. The alkane-degrading genes are present in *alkBFGHJKL* and *alkST* operons located in the OCT plasmid from *P. putida* GPo1. The *alkBFGHJKL* encodes for the enzymes to convert the n alkane to acetyl CoA and *alkST* codes for AlkT and a transcriptional activator AlkS. In *Acinetobacter baylyi* ADP1, the alkane hydroxylation is mediated by monooxygenase AlkM (*alkM*) which is similar to AlkB like alkane monooxygenases. This gene is activated by a regulatory gene *alkR* in *A. baylyi* ADP1.

In *P. putida* mt-2, a gene *xyl* was identified which degrades toluene (n-xylene and p-xylene). The operon *xylUWCMABN* which codes for the enzymes that converts toluene to benzoate followed by deoxygenation which leads to catechol that enters the Kreb's cycle. The second step is controlled by the enzymes of operon *xylXYZLTEGFJQKIH* gene cluster. The positive regulators XylR and XylS coordinate the expression of these two operons. In *Acinetobacter, Kocuria,* and *Pseudomonas* strains, a novel gene *xylE*, which encodes for catechol 2–3 dioxygenase, has been reported recently (Fuentes et al. 2014). The *nah* gene codes for naphthalene degrading enzyme and is regulated by the *nahAaAbAcAdBFCED* operon present in *P. putida* G7. This enzyme converts naphthalene into salicylate.

Dye degradation

Dyes can be degraded by variety of enzymes based on the nature of dye. The enzymes used in dye degradation are peroxidases, azoreductases, mono or dioxygenases, and laccases. A family of Heme containing peroxidases from microbial sources known as dye decolorizing peroxidases (DyPs) were capable of degradation of azo-based, anthraquinone based, phenolic-based, aromatic sulfur based, and beta-carotene based dyes. The DyPs were first identified in fungal sources and later, from bacterial sources. *Bacillus subtilis* and *Pseudomonas putida* have been a good source of DyPs for dye degradation and the genes responsible for dye degradation ***bsDyP*** and ***ppDyP*** were cloned and characterized completely in *E. coli* BL21 (Santos et al. 2014). This study proved that the Dye decolouring peroxidases can be effective in dye degradation around pH 4–5 and 10 to 30°C. Both the enzymes resulted in the bleaching of dyes; however, ppDyp enzymes showed increased activity as compared to bsDyP.

Laccase is an oxidoreductase enzyme present in bacteria, fungi, and plants which can catalyse the oxidation of aromatic substrates. The substrate spectrum for the laccase enzyme is very wide which makes it a suitable candidate for bioremediation purposes. *Lenzite gibbosa*, a white rot fungi, is the chief source of the laccase enzyme which is encoded by *laccase* gene. In a study made on recombinant laccase, the alizarin dye was degraded 98.8% in 30 minutes and the dyes like azo, anthraquinone, triphenylmethane, and heterocyclic were also degraded effectively by the purified recombinant laccase enzyme (Zheng et al. 2014). In another study, *CotA-laccase* gene was identified and characterized for the degradation of azo and anthraquinone dyes. There is another gene identified in *Pseudomonas putida* MET94, that is, the *PPAzoR* gene which can also degrade azo dyes effectively (Santos et al. 2014).

Summary

This chapter discuss the outlines of bioremediation and explains the key contents of bio, phyto, myco, and phycoremediation. The different types of pollution caused by heavy metals, textile dyes, petrochemicals, and pesticides have been discussed in this chapter. This chapter mainly explains the genes and their roles involved in various xenobiotic remediations. There are various genes present in bacteria for metal resistance, dye degradation, pesticide degradation, and petrochemical degradation purposes which are mostly present in an extrachromosomal plasmids or in transposons. The metal ions are effluxed mostly by an ATPase transporter membrane protein and by means of enzyme-mediated oxidation-reduction reaction inside and outside the cell. The metal resistant structural genes are positively regulated by the presence of the respective metal ions in higher concentrations. Most of the metal resistant genes are regulated through inducible operons and regulatory trans factors. The pesticides are degraded by catabolic enzymes like peroxidases, phosphodiesterases, and hydrolases. The petrochemicals are degraded by monooxygenases, rubredoxin, and rubredoxin reductases. The textile dyes are degraded using various enzymes like laccases and peroxidases. The genes can be effectively used in a combination for a better xenobiotic remediation purpose.

Acknowledgements

I acknowledge Dr. M.K. Harishankar and Ms. Suruthi Abirami for their suggestions and help in writing this chapter.

References

Baker, A.J.M. 1981. Accumulators and excluders -strategies in the response of plants to heavy metals. J PLANT NUTR 3(1–4): 643–654. doi:10.1080/01904168109362867.

Beeman, R. 1982. Recent advances in mode of action of insecticides. Annu. Rev. Entomol. 27: 253–281.

Bieby Voijant Tangahu, Siti Rozaimah Sheikh Abdullah, Hassan Basri, Mushrifah Idris, Nurina Anuar and Muhammad Mukhlisin. A Review on Heavy Metals (As, Pb, and Hg) Uptake by Plants through Phytoremediation, IJChER, vol. 2011, Article ID 939161, 31 pages, 2011.

Borremans, B., J.L. Hobman, A. Provoost, N.L. Brown and D. van Der Lelie. 2001. Cloning and functional analysis of the pbr lead resistance determinant of *Ralstonia metallidurans* CH34. J. Bacteriol. 183: 5651–5658.

Branco, R. and P.V. Morais. 2013. Identification and characterization of the transcriptional regulator ChrB in the chromate resistance determinant of *Ochrobactrum tritici* 5bvl1. PLOS ONE 8(11): e77987.

Chu, L. and K. Tapan. 1989. Cadmium resistance from *Staphylococcus aureus* plasmid pI258. Proc. Natl. Acad. Sci. 86: 3544–3548.

Crupper, S.S., V. Worrell and G.C. Stewart. 1999. Cloning and expression of cadD, a new cadmium resistance gene of *Staphylococcus aureus*. J. Bacteriol. 181: 4071–4075.

Dash, H.R. and S. Das. 2012. International biodeterioration & biodegradation bioremediation of mercury and the importance of bacterial mer genes. Int. Biodeter. Biodegr. 75: 207–213.

Dixit, R., Wasiullah, D. Malaviya, K. Pandiyan, U.B. Singh, A. Sahu, R. Shukla, B.P. Singh, J.P. Rai, P.K. Sharma, H. Lade and D. Paul. 2015. Bioremediation of heavy metals from soil and aquatic environment: an overview of principles and criteria of fundamental processes. Sustainability-Basel 7: 2189–2212.

Fathi, A.A., M.M. Azooz and M.A. Al-fredan. 2013. Phycoremediation and the potential of sustainable algal biofuel production using wastewater. Am. J. Appl. Sci. 10: 189–194.

Fuentes, Valentina Méndez, Patricia Aguila and Michael Seeger. 2014. Bioremediation of petroleum hydrocarbons: catabolic genes, microbial communities, and applications. Appl. Microbiol. Biotechnol. 98: 4781–4794.

Huang, C.C., M.F. Chien and Lin K.H. 2010. Bacterial Mercury Resistance of TnMERI1Únd Its' Application in Bioremediation. Interdisciplinary Studies on Environmental Chemistry. vol. 3, Terrapub, Tokyo, pp. 23–29.

Iwanicka, N.B. and A. Borzęcki. 2010. Influence of fenpropathrin on memory and movement in mice after transient incomplete cerebral ischemia. Journal Toxicol. Environ. Health. 73: 1166–1172.

Jarosławiecka, A. and Z. Piotrowska-Seget. 2014. Lead resistance in micro-organisms. Microbiology (United Kingdom) 160(Part 1): 12–25.

Joshi, B.H. and K.G. Modi. 2013. Screening and characterization of heavy metal resistant bacteria for its prospects in bioremediation of contaminated soil. J. Environ. Res. Develop. 7: 1531–1538.

Joutey, N.T., H. Sayel, W. Bahafid and N. El Ghachtouli. 2015. Mechanisms of hexavalent chromium resistance and removal by microorganisms. Rev. Environ. Contam. Toxicol. 233: 45–70.

Kamika, I. and M.N.B. Momba. 2013. Assessing the resistance and bioremediation ability of selected bacterial and protozoan species to heavy metals in metal-rich industrial wastewater. BMC Microbiology 13: 28.

Kazama, H., H. Hamashima and M. Sasatsu. 1998. Distribution of the antiseptic-resistance gene qacE Delta 1 in Gram-positive bacteria. FEMS Microbiol. Lett. 165: 295–299.

Lachheb, H., P. Eric, H. Ammar, K. Mohamed, E. Elimame, G. Chantal et al. 2002. Photocatalytic degradation of various types of dyes (Alizarin S, Crocein Orange G, Methyl Red, Congo Red, Methylene Blue) in water by UV-irradiated titania. Appl. Catal. B: Environmental 39: 75–90.

Lee, S., E. Glickman and D.A. Cooksey. 2001. Chromosomal locus for cadmium resistance in Pseudomonas putida consisting of a cadmium-transporting ATPase and a MerR family response regulator. Appl. Environ. Microb. 67: 1437–1444.

Margaryan, A.A., H.H. Panosyan, N.K. Birkeland and A.H. Trchounian. 2013. Heavy metal accumulation and the expression of the CopA and nikA genes in *Bacillus subtilis* AG4 isolated from the SOTK gold mine in Armenia. Biolog. Journal of Armenia 3: 51–57.

Menéndez, C.1., A. Otto, G. Igloi, P. Nick, R. Brandsch, B. Schubach et al. 1997. Molybdate-uptake genes and molybdopterin-biosynthesis genes on a bacterial plasmid characterization of MoeA as a filament-forming protein with adenosinetriphosphatase activity. Eur. J. Biochem. 250: 524–531.

Neti, N. and V. Zakkula. 2012. Cloning and expression of organophosphate pesticide, chlorpyrifos degrading opd gene of *Kocuria* sp. Current Biotica. 6: 282–293.

Nies, D.H. 1992. CzcR and CzcD, gene products affecting regulation of resistance to cobalt, zinc, and cadmium (czc system) in *Alcaligenes eutrophus*. J. Bacteriol. 174: 8102–8110.

Nies, D.H. and S. Silver. 1995. Ion efflux systems involved in bacterial metal resistances. J. Ind. Microbiol. 14: 186–199.

Ortiz-hernández, M.L., E. Sánchez-Salinas, E. Dantán-González and M.L. Castrejón-Godínez. 2013. Pesticide Biodegradation: Mechanisms, Genetics and Strategies to Enhance the Process. INTECH Publishers.

Pandit, S.V. and A.Y. Mundhe. 2013. Monocrotophos induced behavioural stress, biochemical and histological alterations in Lamellidens Marginalis (Lamarck). The International Quality J. Life Sci. 8: 1053–1056.

Peixoto, R.S., A.B. Vermelho and A.S. Rosado. 2011. Petroleum-degrading enzymes: bioremediation and new prospects. Enzyme Res. 2011: 1–7.

Porter, A.W. and A.G. Hay. 2007. Identification of opdA, a gene involved in biodegradation of the endocrine disrupter Octylphenol. Appl. Environ. Microbiol. 73: 7373–7379.

Ramı, M.I., C. Díaz-Pérez, E. Vargas, H. Riveros-Rosas, J. Campos-García and C. Cervantes. 2008. Mechanisms of bacterial resistance to chromium compounds. Biometals 21: 321–332.

Roy, P.H. 1995. Integrons: novel mobile genetic elements mediating antibiotic resistance on *enterobacteria* and *Pseudomonas*. Alliance Prudent Use Antibiotics Newsletter 13: 1–6.

Salisbury, V., R.W. Hedges and N. Datta. 1972. Two modes of curing transmissible bacterial plasmids. J. Gen. Microbiol. 70: 443–445.

Santos, A., S. Mendes and V. Brissos. 2014. New dye-decolorizing peroxidases from *Bacillus subtilis* and *Pseudomonas putida* MET94: towards biotechnological applications. Appl. Microbiol. Biotechnol. 98: 2053–2065.

Selvakumar, R., N. Arul Jothi, V. Jayavignesh, K. Karthikaiselvi, G.I. Antony, P.R. Sharmila et al. 2011. As(V) removal using carbonized yeast cells containing silver nanoparticles. Water Res. 45: 583–592.

Singh, H. 2006. Mycoremediation: Fungal Bioremediation. John Wiley and Sons, New Jersey.

Spratt, B.G. 1994. Resistance to antibiotics mediated by target alterations. Science 264: 388–393.

Zheng, M., Yujie Chi, Hongwei Yi and Shuli Shao. 2014. Decolorization of Alizarin Red and other synthetic dyes by a recombinant laccase from *Pichia pastoris*. Biotechnol. Lett. 36: 39–45.

10

Bioremediation of Heavy Hydrocarbons and Polycyclic Aromatic Hydrocarbons

Environmental Implications and Technical Constraints

Edmo M. Rodrigues * and *Marcos R. Tótola*

INTRODUCTION

Pollutants are widespread in the environment due to human activities. Crude oil is the main source of global energy, meeting approximately 50% of the world's energy demand at present times. Given the amount of oil currently produced worldwide (approximately 96 million barrels per day, according to the U.S. Energy Information Administration) and the complexity of the operations of exploration, transportation, and distribution, it is not surprising that hydrocarbons are one of the main environmental contaminants.

Departamento de Microbiologia, Laboratório de Biotecnologia e Biodiversidade para o Meio Ambiente, Universidade Federal de Viçosa, Viçosa, Minas Gerais Brasil CEP 36570-900.
Email: totolaufv@gmail.com
* Corresponding author: edmomontes@yahoo.com.br

Although by definition not all hydrocarbons are classified as xenobiotics, polycyclic aromatic hydrocarbons (PAH) and high molecular weight hydrocarbons are included in this class of pollutants. Among the PAHs, 16 molecules are classified as priority pollutants by US EPA (Bojes and Pope 2007). According to the U.S. EPA (2010), PAHs occur primarily in the environment as complex mixtures generated from the incomplete combustion of fossil fuels or as result of wildfires, and rarely occur in the environment as a single contaminant molecule. Therefore, if bioremediation of these compounds is to be considered for environmental cleanup, it must be planned so that living organisms, especially microorganisms, through their metabolic pathways, are capable not only of biodegradation of the target contaminants, but also to tolerate the presence of many different PAHs that may be present.

The hydrocarbons known as BTEX (benzene, toluene, ethyl-benzene, and xylenes) are highly harmful volatile low molecular weight aromatic molecules, both for microorganisms and for organisms of higher complexity (Jo et al. 2008). The water solubility of these small hydrocarbons is relatively high when compared with other hydrocarbons, which makes them highly bioavailable and potentially toxic (Aguilera et al. 2009). When in contact with microbial cells, soluble hydrocarbons can cause disruption of the cell's membrane (Chiou et al. 1998), reducing the biomass of active microbial populations that may exert crucial functions in the biodegradation of less soluble and more recalcitrant hydrocarbons also present in the impacted environment.

The problems related to oil contamination are easily noticeable. However, even after the removal of the visible part of the oil, chronic toxic effects may remain for decades (Haritash and Kaushik 2009). Soil contamination generally occurs in regions close to extraction fields, refineries, and fuel stations, while contamination of aquatic environments, more specifically of marine environments, may be related to offshore oil extraction or to oil transportation to continental receiving bases. When contamination occurs in coastal areas, it may bring a serious environmental and socio-economic problem, since such regions tend to have high primary productivity and are home to a great diversity of animals, plants, and microorganisms, with many of these beings exploited commercially or as subsistence items.

Environmental consequences of oil spills depend on several factors, such as the type of environment (soil or aquatic environments), local temperatures, precipitation, oil dispersion patterns, among others. A key factor influencing the consequences of oil spills is the number of carbon atoms in the hydrocarbon molecules, which is directly proportional to the density of the oil. According to the American Petroleum Institute, depending on its density, oil is categorized in API levels. The API gravity is an inverse measure of a petroleum liquid's density relative to the water. The API gravity is referred to as being in "degrees". Its measure is expressed as:

- Light oil: > 31.1° API gravity (density less than 870 kg/m³);
- Medium Oil: 22.3° < API ≤ 31.1° (density between 870 and 920 kg/m³);
- Heavy Oil: 10.0° ≤ API ≤ 22.3° (density between 920 and 1,000 kg/m³);
- Ultra-Heavy Oil: API < 10.0° (density greater than 1,000 kg/m³).

Heavy and ultra-heavy oils are enriched in long carbon-chain hydrocarbons of low bioavailability, what impairs their assimilation by hydrocarbonoclastic microorganisms. This causes these molecules to be more recalcitrant and, thus, more persistent in the environment. Conversely, light oils have a higher proportion of small to medium carbon-chain hydrocarbons, which are more readily biodegraded (Salanitro et al. 1997). These differences determine, in large part, the rate of biodegradation of crude oil in and the fate of its components in the environment.

Environmental Remediation of Hydrocarbons

Available technologies for environmental remediation of hydrocarbons include physical, chemical, and biological ones. Physical and chemical treatments usually lead to the dispersal of a part of the pollutants, which limits their effectiveness. These technologies also present disadvantages such as high costs and complex logistics. Bioremediation, on the other hand, under favorable conditions, present lower costs and generally results in an almost complete removal of pollutants, whose final concentration meets legal standards (Bao et al. 2012).

Chemical technologies for oil remediation

Chemicals are sometimes used to assist in the mitigation of impacts arising from the environmental contamination by hydrocarbons. However, these chemicals may present high toxicity, which characterizes an ecological risk (Murado et al. 2011). For example, synthetic surfactants are often used to increase the solubility of hydrocarbons, thereby making them more bioavailable and more susceptible to biodegradation. However, these surfactants and other products can interfere with embryogenesis, growth of larvae, and survival of many species of organisms, both in aquatic and in soil environments. Species with short life cycles are more endangered, because these chemicals can cause the death of an entire generation (Fucik et al. 1994; Beiras et al. 1999).

Besides the potential direct damage of chemicals used to mitigate oil pollution on local biota, surfactants tend to disperse the oil when the contamination occurs in aquatic environments, increasing the concentration of soluble—and thus bioavailable—hydrocarbons in the water. In soils, this solubilizing effect is also potentially harmful, as it can promote dispersal of contaminants through a larger soil volume and increase hydrocarbon concentration in the soil solution (Fucik et al. 1994). The dispersion of hydrocarbons by surfactants, when performed without rigorous control and prior knowledge of the contaminated environment, can result in even more harmful effects than the contamination itself. When dispersion of hydrocarbons is not connected with effective contaminant removal, the chance of these contaminants getting in contact with a greater numbers of living organisms increases. Contact can occur with the cell membrane, diffusion by the skin and the mucous membranes, bioaccumulation in living tissue (Ramachandran et al. 2004) or even covering of the entire animal body and plants by the spilled oil. In some cases, deemulsifiers and solidifying are also used to aid in oil-spilling cleanup, depending on specific characteristics of the contaminated site and the type of oil. Again, these products can also be toxic to microorganisms, with the consequent reduction of the effectiveness of natural attenuation by the local microbiota (Patrick et al. 2012).

Physical technologies for oil remediation

Containment barriers are a physical remediation alternative that can be used both in aquatic and soil environments. The goal of physical remediation can be to contain surface dispersion of contaminants and direct the contamination plume to areas where the excess of oil can be collected by skimmers and absorbent granules, preventing pollution from reaching other unaffected areas (Wolfe et al. 1999; Zhu et al. 2001). This strategy depends on the environmental conditions at the time of the intervention, as well as the properties of the oil. Although not harmful to the environment, oil removal by containment barriers is not efficient for hydrocarbons that are soluble in the water, adhere to soil aggregates, or adsorb to particulate matter and migrate to the benthic zone. In the sea, another alternative can be to burn the oil on the surface; however, the soluble fraction that gets retained in the water column has a greater potential for acute toxicity to some of the biota (Gundersen et al. 1996). In the soil, it is possible to use the vitrification process, which consists of using very high temperatures (1.600–2.000°C) to melt and fuse contaminants and soil into a glass-like solid. The heat of soil is usually delivered via molybdenum electrodes, using too graphitic or glass material mixed in to initiate the melting process (Vidonish et al. 2016).

Oil washing is often used to remove oil adhering to rocks, soil particles, and sediments. Hot or cold pressurized water is used for *in situ* oil washing (Patrick et al. 2012). For *ex situ* washing, the soil or sediment is removed from the environment and transported to specific locations where surfactants or biosurfactants can be employed to facilitate the release of oil from the soil aggregates and particles (Pavel and Gavrilescu 2008).

Biological technologies for oil remediation: Bioremediation

Microorganisms with the capacity to degrade hydrocarbons (hydrocarbonoclastic) are ubiquitous, but usually occur in small proportions of the microbial communities found in non-contaminated environments. Several species of bacteria, archaea, and fungi can degrade petroleum hydrocarbons. While metabolizing hydrocarbons, the microorganisms convert them into energy, cellular biomass, and biological wastes (Rahman et al. 2002). The bioremediation of hydrocarbons by microorganisms aims at the mineralization or at least the conversion of hydrocarbons into less toxic and/or less mobile forms in the environment. Enzymes responsible for incorporating oxygen atoms into hydrocarbons, collectively named oxygenases (mono and dioxygenases), are generally essential to the catabolism of these compounds (Karigar and Rao 2011). Therefore, the presence of oxygen facilitates the bioremediation of hydrocarbon-contaminated environments (Huesemann and Truex 1996), not only because it is the preferential electron acceptor, but also because its incorporation into hydrocarbon molecules is a critical step to prepare the molecule for further catabolism.

Among the most recalcitrant hydrocarbons, we highlight resins, asphaltenes, and PAHs, which can be metabolized by a small number of microbial species (Rodrigues and Tótola 2015). Although PAHs occur at a low concentration in crude oil, they are highly toxic molecules and difficult to remove. After oxygen is incorporated into PAHs by the action of oxygenases, microbial biodegradation starts. However, the biodegradation of these compounds may also occur, albeit more slowly, in anaerobic environments (Tian et al. 2002; Yu et al. 2005; Atlas and Hazen 2011).

Obtaining Hydrocarbonoclastic Microorganisms

Microorganisms may use both naturally occurring and synthetic chemicals in their metabolism. Hydrocarbons are composed mainly of atoms of carbon and hydrogen, and a small fraction additionally contains nitrogen, sulfur, and oxygen. Carbon oxidation provides most of the metabolic energy needed for microbial growth through cellular respiration. The remaining carbon is preserved as new biomass for the growing populations.

A procedure commonly used to evaluate if a given microbial population can use hydrocarbons for growth and to isolate hydrocarbonoclastic microorganisms consists in the cultivation of microbial species of interest or a consortium of microorganisms in a mineral broth containing the target hydrocarbon as the sole source of carbon and energy (Rodrigues et al. 2015a). The method is based on the selective advantage of microorganisms that can use that compound for growth, resulting in a dominance of those over the non-hydrocarbonoclastic populations. This enrichment step will facilitate the isolation of hydrocarbonoclastic microorganisms upon inoculation of this mixed culture in mineral solid medium containing or not containing the same carbon source. Sometimes, this procedure does not result in microbial growth, even if different compositions of the culture medium are used in combination with a range of variables such as temperature, pH, salinity, and oxygen concentration. When this happens, one can assume that microorganisms with the capacity to use the target hydrocarbon are not present, or the concentration of the carbon source—or other essential elements—is below or above the ideal for cell development.

Members of several bacterial and fungal genera have been isolated by the enrichment technique in mineral media containing hydrocarbons (Gargouri et al. 2015; Rodrigues et al. 2015a). The development of different species on the same solid media containing a single hydrocarbon as carbon source is indicative that toxic intermediates are not released during its catabolism by the isolated populations. These populations are, therefore, promising for use in bioremediation, especially when considering the safety of the process regarding the generation of toxic intermediates during the biodegradation of hydrocarbons.

Metabolic Pathways Involved in the Biodegradation of PAHs

PAHs pose an economic and socio-environmental risk as they are toxic, carcinogenic, mutagenic, and teratogenic molecules resistant to biodegradation (Fernández-Luqueño et al. 2008; Sayara et al. 2011). From an environmental perspective, the main PAHs found as contaminants are those that contain between two and five aromatic rings in the molecule. PAHs constitute only a small fraction of crude oil and its derivatives, and can also result from the incomplete combustion of organic matter. It is thus not surprising that PAHs are frequently detected in aquatic environments, air, and soil (Abdel-Shafy and Mansour 2016). The recalcitrance of this class of hydrocarbons can be attributed in part to the electron density around the ring structures, making them resistant to hydrolytic enzymes (Ukiwe et al. 2013). They are also thermodynamically stable and practically non-soluble in water, which diminishes the bioavailability and makes them difficult to be accessed by microorganisms (Dabrowska et al. 2008; Shukla et al. 2014). As a consequence of their recalcitrance, PAHs tend to accumulate both in aquatic and terrestrial environments, leading to chronic toxicity of the local biota.

The biodegradation of PAHs occurs through cleavage of one, several, or all the benzene rings that make up the molecule. Enzymes such as PAH dioxygenase and catechol oxygenase exert key roles during the biodegradation of PAHs (Tian et al. 2002). In order for the oxidation of the substrates to occur in the cell interior by cytoplasmic dioxygenases, the molecules of PAHs must first be transported through the cytoplasmic membrane, which generally occurs through passive diffusion (Wattiau 2002). The initial oxidation of the molecule by a multicomponent enzyme system generally results in the formation of *cis*-dihydrodiol, after two atoms of oxygen are incorporated into the PAH molecule. These intermediates can be processed through the ortho-cleavage or meta-cleavage in addition to going through dehydrogenase actions. The action of dehydrogenase results in the formation of intermediaries such as salicylate, gentisate, protocatecuato, and catechol, which will be then converted to intermediates of the tricarboxylic acid cycle (Kanaly and Harayama 2000; Samanta et al. 2002).

Considering the biodegradation of PAHs as a means to obtain metabolic energy, the entire molecules, or just parts thereof, is mineralized. Cerniglia (1984) concluded that, in some organisms, especially some fungi, PAHs are used only partially as substrate, implicating in that intermediate compounds are excreted due to incomplete oxidation and hydroxylation. In bacteria, these oxidation and hydroxylation reactions seem to be more directly connected to carbon assimilation into cell components, due to their greater plasticity in obtaining energy and carbon sources.

Besides using PAHs as their single carbon and energy source, some microorganisms can metabolize these molecules by co-metabolism. In such circumstances, PAHs are not a useful source of either carbon or energy. In this process, the PAH is subject to a non-specific enzymatic reaction, competing with the primary substrate of a similar structure for the active site of the enzyme (Johnsen et al. 2005). Recently, Khan et al. (2015) isolated soil bacteria capable of co-metabolic transformation of PAHs, including acenaphthylene, anthracene, and phenanthrene, using naphthalene as their primary substrate.

Prospecting recalcitrant hydrocarbons-degrading bacteria is of great practical and scientific interest, because they may be candidates for use in bioaugmentation of contaminated sites. Rodrigues et al. (2015a) isolated different bacterial species from a microbial community developed on the surface of weathered oil left as a "trap" in the coastal water of the tropical Trindade Island, located in an area far from industrial centers. Among the isolates recovered upon enrichment, two actinobacteria stood out in growth assays using individual hydrocarbons as sole sources of carbon and energy, namely *Rhodococcus rhodochrous* TRN7 and *Nocardia farcinica* TRH1. These isolates were able to grow using all aliphatic hydrocarbons and PAHs added as carbon and energy sources to the mineral growth medium. Other bacteria such as *Pseudomonas* spp., *Burkholderia cepacia*, *Sphingomonas* spp., *Flavobacterium* spp., *Mycobacterium* spp., and *Tistrella* spp., have also been reported as being able to degrade PAHs (Zhang et al. 2006).

Naphthalene—The catabolism of naphthalene by bacteria begins with a multicomponent enzyme system, naphthalene-dioxygenase. This enzyme is responsible for the dihydroxylation of the molecule, producing cis-1,2 dihydroxynaphthalene (Fig. 1). The action of cis-dihydrodiol dehydrogenase converts the resulting compound into 1,2-dihydroxynaphthalene, and then the cleavage of the aromatic ring occurs, resulting in the formation of salicylate (Seo et al. 2009; Tomás-Gallardo et al. 2013). This can be metabolized to form catechol or gentisate, depending on the pathway used by the microorganism. The pathway of naphthalene degradation is well known for *Pseudomonas* spp. and *Ralstonia* spp. In *Pseudomonas*, the degradation of

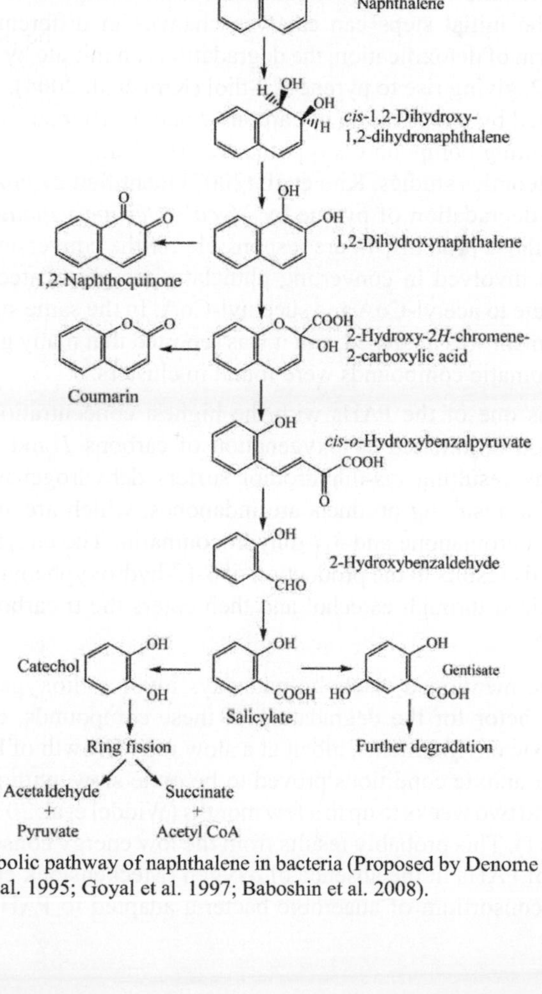

Fig. 1: Aerobic catabolic pathway of naphthalene in bacteria (Proposed by Denome et al. 1993; Kiyohara et al. 1994; Auger et al. 1995; Goyal et al. 1997; Baboshin et al. 2008).

naphthalene occurs through the catechol pathway (Eaton and Chapman 1992), while *Ralstonia* sp. U2 uses the gentisate pathway (Jeon et al. 2006).

Phenanthrene and Anthracene—The biodegradation of phenanthrene and anthracene, both having three aromatic rings, also initiates by a multicomponent dioxygenase system that incorporates two oxygen atoms into the rings to produce cis-dihydrodiol. In some species of *Pseudomonas*, two different pathways of phenanthrene degradation are described. In the first one, the initial dioxygenation occurs at carbons *3* and *4*, resulting in the formation of catechol or salicylate, phthalate, and protocatecoate. The second pathway starts with dioxygenation of carbons *1* and *2* and results in the formation of cis-*1,2*-dihydrodiol (Evans et al. 1965; Moody et al. 2001; Prabhu and Phale 2003). Moody et al. (2001) reported that *Mycobacterium* sp. PYR-1 has several dioxygenases, what can in part explain the ability of this isolate in promoting the degradation of a variety of PAHs, such as anthracene, phenanthrene, and pyrene. In this isolate, similarities but also differences were observed in the ortho- and meta-fission pathways of the aromatic rings in relation to other bacterial strains, pointing to the existence of alternative metabolic pathways for the degradation of PAHs.

Pyrene—The degradation of pyrene is somehow more complex, since the enzymes responsible for the initial steps can catalyze changes in different regions of the molecule. As a form of detoxification, the degradation can initiate by the oxygenation of carbons *1* and *2*, giving rise to pyrene-*1,2*-diol (Kim et al. 2004). The degradation can also be initiated by oxygenation of carbons *4* and *5*. After several steps of rings cleavage, the resulting compound is *o*-phthalate (Heitkamp et al. 1988; Kim et al. 2005). Using proteomics studies, Kim et al. (2007) identified 27 enzymes necessary for the complete degradation of pyrene by *Mycobacterium vanbaalenii* PYR-1, in 25 steps. Among those proteins, 14 are responsible for the conversion of pyrene into phthalate, six are involved in converting phthalate into protocatecoate, and seven from protocatecoate to acetyl-CoA and succinyl-CoA. In the same study, the genome of this microorganism was analyzed, and it was reported that many genes involved in degradation of aromatic compounds were found in clusters.

Fluorene—This is one of the PAHs with the highest concentration in fossil fuels and its degradation is initiated by oxygenation of carbons *1* and *2* or *3* and *4* of the molecule. The resulting *cis*-dihydrodiol suffers dehydrogenation and finally, meta-cleavage. The resulting products are indanones, which are substrates for the production of *3*-isocromanone and *3,4*-dihydrocoumarin. The enzymatic hydrolysis of these compounds results in the production of *3*-(*2*-hydroxyphenyl) propionic acid, which is metabolized through catechol and then enters the tricarboxylic acid cycle (Peng et al. 2008).

All the above-mentioned catabolic pathways involve dioxygenases. Although oxygen is a key factor for the degradation of these compounds, degradation may also occur in anoxic environments, albeit at a slow rate. Growth of bacteria on these compounds under anoxic conditions proved to be quite slow, with generation times ranging from one to two weeks to up to a few months (Widdel et al. 2010; Meckenstock and Mouttaki 2011). This probably results from the low energy conservation coupled to the oxidation of PAHs in the absence of oxygen (Meckenstock et al. 2016). After inoculation of a consortium of anaerobic bacteria adapted to PAHs in microcosm,

Chang et al. (2002) observed the disappearance of 80 to 100% of phenanthrene, acenaphthlene, anthracene, fluorine, and pyrene after 90 days of incubation. Biodegradation of PAHs under anaerobiosis has also been observed in contaminated aquifers and marine sediments (Tabak et al. 2003; Chang et al. 2008; Bahr et al. 2015).

Kümmel et al. (2015) evaluated the capacity of bacteria to degrade PAH using sulfate as the electron acceptor. The results obtained in that work indicate that one possible reaction occurring under this condition is $C_{10}H_8 + 6\ SO_4^{2-} + 2\ H^+ + 6\ H_2O \rightarrow 10\ HCO_3^- + 6\ H_2S$. The dominant groups in the microbial community were members of families *Anaerolineaceae, Spirochaetaceae,* and *Desulfobacteraceae.* Degradation of phenol by representatives of *Anaerolineaceae* under anaerobiosis had already been reported by Rosenkranz et al. (2013), indicating that this family may contain members with the ability to degrade aromatic compounds under anaerobiosis. Although a great incubation period (500 days) was necessary in the former study, naphthalene degradation under anaerobic condition is an indication that, even in the absence of oxygen, natural attenuation may occurs in the environment, albeit slowly.

Kleemann and Meckenstock (2011) reported the growth of the bacterial strain N49 using ferric iron as the electron acceptor and naphthalene as the sole source of carbon and energy. The balance of biochemical reactions pointed to the complete mineralization of naphthalene. The concentration of the main metabolite of naphthalene degradation pathway, 2-naphthoic acid, increased along the incubation period. The study pointed to the fact that the degradation of PAHs can also be coupled to the reduction of ferric iron.

The degradation of PAHs has also been reported under methanogenic condition. Zhang et al. (2012) evaluated the degradation of anthracene and the concomitant production of methane in microcosms inoculated with a contaminated aquifer sample. After 120 days, 88% of the anthracene had been removed. Using ^{13}C-anthracene and stable-isotope probing (SIP) technique, *Methylibium, Legionella*, and Rhizobiales were pointed out as the main groups responsible for anthracene biodegradation in the methanogenic microcosms.

The biodegradation of petroleum hydrocarbons, both under aerobic and anaerobic conditions, makes feasible the bioremediation in many different environments, including deep soils and aquifers. Groundwater represents the largest reserve of freshwater in the world. Due to the growing deterioration of surface waters, modern societies may rely on groundwater to supply their actual needs for clean water; part of the water available is worthless for human consumption today. The possibility of bioremediate underground environments contaminated with PAHs and other hydrocarbons is crucial to ensure the supply from these sources of water, which tend to become increasingly important for mankind.

Consequences of Incomplete Biodegradation of PAHs

PAHs are lipophilic and chemically inert. However, upon contact with living organisms, they can be metabolized and generate toxic and reactive forms. Some metabolic intermediates formed during PAH metabolism may be more harmful to the biota than the molecule of the contaminant (Abdel-Shafy and Mansour 2016). The presence of PAHs in the humans and other animals is highly harmful, insofar as activation reactions of these molecules can leads to reactive intermediates. The high reactivity of some of these intermediates can cause a number of problems, ranging from narcotic to mutagenic effects (La Rocca et al. 1996; Ritter et al. 2002; Chen and White 2004; Shimada and Fujii-Kuriyama 2004; Mehdinia et al. 2015; Abdel-Shafy and Mansour 2016).

Some PAHs do not show genotoxic effects. However, after being metabolized, they can give rise to diol epoxides, a group of highly reactive molecules, which contain cyclic ether with three carbon atoms that form a ring. The conformation of these molecules confers then a high density and makes epoxides more reactive than other ethers. Due to these characteristics, diol epoxides can bind covalently to DNA and induce genotoxic damage, which may induce carcinogenesis in humans and other animals (Schwerdtle et al. 2010; Rengarajan et al. 2015).

Luckert et al. (2013) investigated the ability of PAHs to induce the activity of the enzyme cytochrome P450 3A4 (CYP3A4). It was found that benzo[c]phenanthrene (BcP) and dibenzo[a, l]pyrene (DBalP), as well as their respective dihydrodiols and diol epoxides, can induce CYP3A4 activity. The reactions of these compounds with cytochrome P450 result in reactive molecules that can induce mutations and, as a consequence, carcinogenesis.

Catechols produced during the catabolism of PAHs can result in both environmental and human health problems. Catechols can form stable complexes with many di- and trivalent metal ions, allowing them to remain in the environment without suffering further degradation. When enzymatically oxidized or in the presence of oxygen and heavy metals, electrons from catechol are transferred to molecular oxygen, resulting in the formation of a superoxide ion. In the presence of heavy metals, such as iron and copper, superoxide is reduced to hydrogen peroxide and hydroxyl radicals. The reactive oxygen species (ROS) formed can be highly damaging to cells and organisms if their quenching does not occur quickly (Schweigert et al. 2000). Catechols can damage cell membranes by means of redox activity and damage DNA and proteins. These compounds do not cause direct oxidative damage to DNA. However, when combined with heavy metals and in the presence of molecular oxygen, they may induce a DNA double-strand break. This occurs after a sequence of redox reactions (Fig. 2), which results in an intense release of hydroxyl radicals in the vicinity of DNA (Schweigert et al. 2000). Other effects of the presence of catechol in animal tissues have also been reported. The generation of superoxide and quinones during auto-oxidation of catechol has been reported as a cytotoxic factor for glial cells (Pereira et al. 2004), which can result in the induction of cell apoptosis (Oliveira et al. 2010). Additionally, ROS generation is a common cause of problems linked to degenerative diseases affecting the central nervous system (Oliveira et al. 2012).

The partial degradation of PAHs and of the mono-aromatic hydrocarbon benzene, can result in the formation of hydroquinone (benzene-*1,4*-diol). Hydroquinone is a type of phenol, having two hydroxyl groups bonded to a benzene ring in a para position. When oxidized, it results in the formation of reactive quinones (e.g., *p*-benzoquinone) (Bahrs et al. 2013). These reactive quinones induce oxidative stress and form non-specific bonds in DNA and proteins (Bolton et al. 2000). Hydroquinone is known for its highly toxic potential for aquatic organisms, as

catechol	$+$ Cu^{2+}		$\Rightarrow$ catechol$^+$	$+$ Cu^{1+}					(1)
O_2	$+$ Cu^{1+}		$\Rightarrow$ O_2^-	$+$ Cu^{2+}					(2)
O_2^-	$+$ Cu^{1+}	$+$ $2 H_2O$	$\Rightarrow$ H_2O_2	$+$ Cu^{2+}	$+$ $2 OH^-$				(3)
O_2^-	$+$ H_2O_2		$\Rightarrow$ OH^-	$+$ OH	$+$ $^1O_2$				(4)
H_2O_2	$+$ Cu^{1+}		$\Rightarrow$ $Cu^{1+}OOH$	$+$ H^+	$\Rightarrow$ free OH	$+$ Cu^{2+}		$+$ OH^-	(5)
H_2O_2	$+$ $DNA\text{-}Cu^{1+}$		$\Rightarrow$ $DNA\text{-}Cu^{1+}OOH$	$+$ H^+	$\Rightarrow$ OH	$+$ $DNA\text{-}Cu^{2+}OH$			(6)

Fig. 2: Hydroxyl radical generation by catechol and DNA degradation in the presence of oxygen and copper (adapted from Schweigert et al. 2000).

Pimephales promelas, Brachydanio rerio, Daphnia magna, Desmodesmus armatus, Synechocystis sp., *Nostoc* sp., and *Microcystis aeruginosa* (Bahrs et al. 2013). In mammalian cells, hydroquinone can induce leukemia of mononuclear cells, tumors of renal tubular cells, and cancer of liver in rodents (Kari 1992). The synergic action of hydroquinone and catechol causes *in vivo* oxidative damage to DNA (Marrazzini et al. 1994).

Recalcitrance and Biodegradation of Hydrocarbons

The recalcitrance of hydrocarbons is related to the size and complexity of the molecule. Low molecular weight aromatic hydrocarbons, such as benzene, toluene, and xylenes, cause acute toxicity, but can be easily degraded by environmental microorganisms (Cury 2002).

Molecules having two or more condensed aromatic rings or branches are more difficult to be degraded, and a small number of microbial species is capable of metabolizing them (Hernández-López et al. 2015). When compared to mono-aromatic hydrocarbons, they present less toxicity. However, they are more persistent in the environment, thus causing chronic effects (Aske et al. 2002; Ghazali et al. 2004). They may also form nanostructures between 2–20 nm that results from strong interactions between them and other constituents of crude oil. These nanostructures show strong adhesion to surfaces (Mostowfi et al. 2009).

PAHs containing six or more aromatic rings are recalcitrant to microbial attack (Das and Chandran 2011). Although there is no consensus in the literature, some authors consider PAHs containing four or more aromatic rings as high molecular weight PAHs, while others consider only PAHs with more than six aromatic rings as belonging to this group. Recently, Marín-Spiotta et al. (2013) suggested that the term "recalcitrant" must refers specially to the low solubility of some organics, instead of molecular complexity.

Asphaltenes are large hydrocarbon molecules present in crude oil. In light oils, asphaltenes represent approximately only 1% of the components, while in heavy oils this value can exceed 20% (Hernández-López et al. 2016). Asphaltenes are represented mainly by molecules containing from 4 to 10 rings, either aromatic or non-aromatic, linked to branched alkyl chains (Groenzin and Mullins 2000), including aromatic molecules of high molecular weight (Mullins 2011), whose complex structures and high hydrophobicity makes them resistant to biodegradation. They are soluble in aromatic solvents, such as toluene and benzene, and insoluble in *n*-alkanes, such as *n*-heptane or n-pentane (Strausz et al. 2006; Gray et al. 2011).

The fungus *Neosartorya fischeri* was the first microorganism described as able to mineralize asphaltenes. After 11 weeks, the fungus was capable of metabolizing 15.5% of the asphaltene present in the growth medium as the sole carbon source (Uribe-Alvarez et al. 2011). Yanto and Tachibana (2013, 2014) reported that the halotolerant fungus *Pesalotiopsis* sp. could also degrade asphaltenes. Recently, Hernández-López et al. (2016) reported that genes responsible for the synthesis of monooxygenases of the cytochrome P450 system are induced when *Neosartorya fischeri* is grown in the presence of asphaltenes. Induction of these genes has also been described in response to different PAHs, as well as their role in the biotransformation of these compounds, including PAHs with six aromatic rings (Syed et al. 2010, 2011, 2013). Using confocal microscopy, Hernández-López et al. (2016) reported the presence of high molecular weight PAHs inside fungal hyphae of *Neosartorya fischeri*. The catabolism of these compounds was confirmed by high performance liquid chromatography (HPLC).

The metabolism of high molecular weight PAHs seems to occur inside the fungal hyphae (Hernández-López et al. 2016). Fayeulle et al. (2014) detected benzo(a) pyrene inside the cells of *Fusarium solani*, suggesting that lipid bodies are sites of intracellular storage of PAHs, which enter the cell via energy-dependent mechanisms (Fayeulle et al. 2014).

Representatives of *Pseudomonas, Citrobacter, Enterobacter, Staphylococcus, Bacillus,* and *Lysinibacillus* are usually mentioned in the literature by their ability to degrade hydrocarbons (Atlas 1981; Cerniglia 1984; Denome et al. 1993; Johnsen et al. 2005; Chang et al. 2008; Das and Chandran 2011; Bao et al. 2012; Baruah et al. 2016), including PAHs with up to seven aromatic rings. Baruah et al. (2016) isolated 14 hydrocarbonoclastic bacterial strains of soil contaminated with crude oil in India. The isolates *Enterobacter* sp. RC4 and *Bacillus* sp. RC6 removed 80% and 64% of an asphaltene-rich crude oil from soil microcosms, respectively. *Enterobacter* sp. RC4 removed 78% and 73% of resin and asphaltenes, respectively. The presence of the cytochrome P450 gene in this isolate was suggested to be a key factor for its high biodegradation activity.

The growth of bacteria in habitats where hydrocarbons are the main source of carbon and energy can be limited by the low availability of the carbon substrate, due to the generally low solubility of these molecules (Sekelsky and Shreve 1999). Surfactants produced by a variety of microorganisms (biosurfactants) may increase the bioavailability of hydrophobic compounds, as they increase their solubility in water (Marcoux 2000). Mnif et al. (2011) isolated bacteria from oil fields with the capacity to degrade hydrocarbons and produce biosurfactants. A hydrocarbonoclastic *Bacillus cereus* isolated from an oil-contaminated soil was able to produce biosurfactants using molasses as carbon source. The isolate degraded 40% of asphaltene during incubation in a mineral medium (5 g/L of asphaltene) for 60 days (Asadollahi et al. 2016). The application of biosurfactants in an environment contaminated with hydrocarbons can also stimulate the biodegradation of hydrocarbons. Rodrigues et al. (2015b) observed a greater catabolism of hydrocarbons by the indigenous microbial community of seawater from Trindade Island purposely contaminated with crude oil upon the application of rhamnolipids. The authors concluded that the higher biodegradation was due to the higher bioavailability of the contaminants resulting from the biosurfactant application. The biodegradation of hydrocarbons can be accelerated when a group of specialized microorganisms, acting synergistically, is already present or is introduced intentionally in contaminated sites (Ghazali et al. 2004). Introduction of non-indigenous populations must, however, be viewed with particular caution, because they may not adapt to the new environment (Tyagi et al. 2011).

Environmental Peculiarities and Bioremediation Strategies for Hydrocarbon-Contaminated Sites

Risk assessment must always precede any other step when a contamination occurs. It will provide information on modes of action of contaminants, the potential effects on biological communities and priority actions to be taken in order to minimize the negative effects on the local biota. A critical factor is to anticipate most of the problems that the contamination may cause, as well as the delineation of the extent of contamination (Roy and McGill 2000). Although in a first survey, the extension of the affected location can be delineated, the mobility of contamination plumes will depend on intrinsic conditions of the target environment. A comprehensive evaluation of the extent of contamination will support decision making about the potential for remediation and environmental damage (Das and Chandran 2011).

Bioremediation technologies are intended to accelerate the biodegradation of contaminants and the recovery of the contaminated site to a condition found before contamination (Röling et al. 2002; Stroud et al. 2007; Minai-Tehrani et al. 2015). To achieve these goals, environmental constraints that limit the catabolic activities of the microbial populations responsible for the biodegradation of the contaminants must be eliminated or reduced, as far as possible (Stroud et al. 2007). This biostimulation approach generally presents high efficiency when one is dealing with contamination by hydrocarbons (Gallego et al. 2001; Ruberto et al. 2009; Abed et al. 2015). If biodegradation does not improve after eliminating physical and chemical constraints, we may conclude that the contaminated site dos not host microorganisms with the metabolic pathways necessary for the catabolism of the contaminants. In this case, inoculation with selected hydrocarbonoclastic or biosurfactant-producing microorganisms (bioaugmentation) may be beneficial (Hassanshahian et al. 2014; Mao et al. 2015; Rodrigues et al. 2015b; Chang et al. 2015). As an additional bioremediation strategy, we can rely on natural attenuation, a methodology of passive remediation based on monitoring the *in situ* contaminant degradation. This technique is efficient when contamination occurs in sites where the microbial community hosts some populations able to efficiently degrade the contaminants and environmental conditions are adequate to keep these populations metabolically active, guaranteeing that the biodegradation of contaminants will occur in a period considered reasonable by the managers of the decontamination program (Kao et al. 2006).

Accounting to the fact that the most susceptible environments to contamination by petroleum hydrocarbons are soils and marine waters, we will describe the peculiarities of the bioremediation strategies in each of these environments.

Bioremediation of hydrocarbons in soils

Soil contamination by hydrocarbons is a serious environmental problem (Sutton et al. 2013; Aller et al. 2014), caused mainly by leakage during the extraction, transportation, and refining of the oil, or during the storage of oil derivatives in fuel stations (Lin and Mendelssohn 2012; Mosaed et al. 2015). Soil remediation has been intensely studied, and efficient strategies for the mitigation of the environmental impacts and efficient elimination of hydrocarbons are now well established (Gogoi et al. 2003; Lin and Mendelssohn 2012; Gordon et al. 2015; Mosaed et al. 2015).

When dealing with soil contamination, key factors which affect the fact and biodegradation of contaminants by soil microorganisms, such as particle size, soil porosity, hydraulic conductivity, moisture, pH, organic matter, and inorganic nutrients content must be characterized. This will be necessary to decide about what bioremediation strategies will be more adequate. Oxygen is a key factor for the action of oxygenase enzymes on the hydrocarbon molecules, and one should also consider its role as preferential electron acceptor in the respiratory chain of microorganisms responsible for biodegradation (Karigar and Rao 2011). In a compacted soil, for example, a simple soil mixing to improve aeration may be efficient to accelerate biodegradation. Moisture is also determinant of microbial activity, including the biodegradation of organic contaminants. If water content is too low, it will limit microbial activity and growth; if in excess, it will reduce diffusion of oxygen. Generally, optimal aerobic microbial activity occurs when soil moisture is between 50 and 80% of its water retention capacity (WRC), which, as suggested by Baker (1994), corresponds to the amount of water that remains in the soil after draining by gravitational force. In bioremediation experiments conducted by Viñas et al. (2005), it was concluded that both, autochthonous microbial populations and soil affect significantly the biodegradation of contaminants (Table 1).

The biodegradation of petroleum hydrocarbons tends to be less efficient in soils with low organic matter content (Sullivan and Krieger 2001). The effect seems to be attributed to the fact that microorganisms are more active in organic matter-rich soils, since these molecules serve as substrate to sustain microbial primary metabolism (Fontaine et al. 2003; Johnsen et al. 2005). In addition, organic matter adsorb hydrocarbon molecules, reducing their concentration in the aqueous and gaseous phases of the soil. This effect may protect microbial populations from toxic effects of some of the contaminants, and consequently reduce the inhibition of the microbial populations responsible for their biodegradation (Labud et al. 2007). On the other hand, the adsorption of hydrocarbons by soil organic matter can reduce their bioavailability, and this will in turn affect biodegradation, especially if local microorganisms are not capable of producing biosurfactants or adhere to the molecules of the contaminants (Labud et al. 2007; Marín-Spiotta et al. 2013; Li et al. 2016). If bioavailability is a cause for low biodegradation, the use of surfactants, including biosurfactants, can

Table 1: Hydrocarbon biodegradation after incubation for 200 days.*

	Initial concentration (mg/Kg dry)	% biodegradation after 200 days		
		Control (5,8% of WRC)	Autoclaved soil	Treatment (40% of WRC)
TPH	8196 ± 285	1	8	79
Fluorene	182 ± 2	25	36	100
Phenanthrene	496 ± 24	0	7	96
Anthracene	114 ± 15	26	17	84
Fluoranthene	693 ± 48	0	0	92
Pyrene	387 ± 30	0	1	87
Chrysene	144 ± 10	0	0	62
Total PAHs	2724 ± 134	3	8	87

* Adapted from Viñas et al. (2005).

stimulate the biodegradation. Surfactants act increasing the solubility and desorption of hydrocarbons and, consequently, their bioavailability (Chang et al. 2015; Li et al. 2016). The use of synthetic surfactants can be an issue, because they are considered environmental contaminants *per se* and may inhibit microbial growth and activity, thus affecting the catabolism of the target hydrocarbons (Bramwell and Laha 2000).

The size and types of soil mineral particles may also interfere with the biodegradation of organic contaminants, since this affects soil porosity and oxygen transfer into the soil matrix. One must also consider that carbon availability increases abruptly after contamination by hydrocarbons, and the biodegradation of this excess of organic substrate may be limited by the availability not only of oxygen, but also of mineral nutrients, especially nitrogen and phosphorus (Das and Chandran 2010). When these nutrients are limiting, the addition of fertilizers to bring the ratio C:N:P to values close to 100:10:1 tends to stimulate the biodegradation of the contaminants (Hoeppel and Hinchee 1994; Bouchez et al. 1995; Cookson 1995; Leys et al. 2005).

Hydrocarbonoclastic microorganisms are widely distributed in soil habitats (Pandey et al. 2016). Soil bacteria are capable of adapting and metabolizing a wide number of different hydrocarbon molecules to obtain carbon and energy (Viñas et al. 2005; Wang et al. 2008). Morais et al. (2016) evaluated changes in the microbial community structure in pristine soils intentionally contaminated with oil. The abundance of some *taxa*, including Rubrobacteraceae, Streptomycetaceae, *Nocardia, Pimelobacter, Acinetobacter*, and *Perlucidibaca*, among others, increased significantly, leading researchers to suggest that these members of the microbial community are taking advantage of the hydrocarbons in their metabolism. Other reports of the metabolism of PAHs by autochthonous microorganisms of pristine soils are found in the literature (e.g., Tervahauta et al. 2009; Yrjälä et al. 2010), suggesting that hydrocarbon biodegradation is a widespread attribute of soil microorganisms.

Although biodegradation of hydrocarbons in soils generally occurs without the need for inoculation of specialized populations, bioaugmentation has been reported to increase the efficiency and speed of bioremediation (Tyagi et al. 2011). The use of bacterial consortia may be even more effective than inoculation with individual populations. Gargouri et al. (2014) obtained a bacterial consortium from a continuous stirred tank reactor treating refinery wastewater and used it in the bioremediation of a soil contaminated with petroleum hydrocarbons. After 30 days of treatment, concentration of all aliphatic hydrocarbons having between 12 and 30 carbons was reduced significantly. Population density of heterotrophic aerobic bacteria also increased and TPHs reduction reached 96.05%. The treated soil was considered as non-phytotoxic. Leal et al. (2017) showed that the municipal solid waste compost, from successive additions of gasoline, is a suitable substrate for the development of microbial inoculants for the bioremediation of hydrocarbon contaminates soils. The use of the inoculants in the contaminated-soil increased the CO_2 emission from the soil as well as the heterotrophic bacterial count and promoted the development of fungi and Gram-negative bacteria in the soil.

Nevertheless, a successful example of bioaugmentation with non-indigenous populations is reported. Leal (2009) used a bacterial consortium composed of *Pseudomonas aeruginosa, Bacillus subtilis, Ochrobaetrum anthropic,* and two strains of *Acinetobacter baumannii* isolated from hydrocarbon-contaminated sites in a diesel biodegradation study in soil microcosms. Genes coding for enzymes involved in the biodegradation of hydrocarbons and biosurfactant production were some of the attributes of the members of the consortium. In the culture medium, the degradation of alkanes reached more than 90% in only seven days. In diesel-contaminated soil, 90% of initial Total Petroleum Hydrocarbons (TPH) (20 ml Kg^{-1}) were degraded by the consortium, and all bacterial members were still present ($> 2.0 \times 10^7$ $CFU.g^{-1}$) after 60 days.

A recently developed bioremediation strategy relies on the use of electrodes inserted directly into the contaminated soil. Electro-bioremediation is a hybrid technology for the remediation of organic contaminants encompassing both electrokinesis and bioremediation approaches, in order to increase the mobilization of pollutants and microbial cells capable of carrying out the catabolism of the target compounds (Wick et al. 2007). The electrodes establish weak electric fields (between 0.2 and 2 V/cm) in the soil matrix (Hassan et al. 2016). The electric field can result on development of pH and voltage gradients, variation of electric current and voltage gradient, formation of zones with different current density, and an increase in temperature of the soil, can play a significant role in the outcome of an electrokinetic bioremediation processes (Hassan et al. 2016). This technique provides a uniform distribution of nutrients added to the soil to stimulate biodegradation (Xu et al. 2010) and increase oxygen availability in the region of the anode due to water electrolysis (Ramirez et al. 2014). Some limitations of this technique include the high energy required to establish and maintain the electric field, the potential to increase competition and other negative interactions between the microbial populations (autochthonous or introduced), and the need for chemicals to keep the pH in an acceptable range, which tends to be variable between the cathode and the anode (Wick et al. 2007). The application of the electric current can also affect the transport of solutes across cell membranes due to the reorientation of membrane lipids (Groves et al. 1997) and disrupts membrane potential. Therefore, although promising, the electro-bioremediation must still be more studied so that large-scale assays can be promoted in the future.

Remediation of soils contaminated by hydrocarbons can become even more difficult when there are heavy metals contaminating the same site, since these inhibit microbial activity and consequently, the biodegradation of organic pollutants (Sandrin and Maier 2003; Thavamani et al. 2012a, 2012b). After obtaining isolates from soil contaminated with hydrocarbons and heavy metals in enrichment media containing phenanthrene and pyrene, Thavamani et al. (2012c) conducted PAHs degradation tests using a microbial consortium composed of representatives of genera *Pandoraea Alcaligenes, Pseudomonas*, and *Paenibacillus*. Even in the presence of cadmium, the consortium was capable of effectively degrading PAHs, but the degradation efficiency was reduced when they used metal concentrations greater than 5 mg L^{-1}.

The presence of plants can aid hydrocarbon biodegradation since they can stimulate the development of hydrocarbonoclastic microorganisms in the rhizospheric region (Kirk et al. 2005; Kim et al. 2006; Zhuang et al. 2007; Gaskin et al. 2008). Pagé et al. (2015) reported that specific bacterial genes related to xenobiotic degradation have the expression stimulated by plant species *Salix purpurea* when hydrocarbons are present in the soil. The plant exudates released in the radicular region select the microbiota, so that the microorganisms protect the plant against the toxic contaminants and contribute to their growth (Segura et al. 2009; Hayat et al. 2010). The prevalence of catabolic genes of hydrocarbons has been reported when some species of plants are found in the contaminated environment (Siciliano et al. 2001; Siciliano et al. 2003; Palmroth et al. 2005; Palmroth et al. 2007; Marecik et al. 2015). On the basis of the results of these and other studies, it is possible to suggest that the rhizorremediation is a promising alternative to the removal of hydrocarbons from contaminated soil.

Bioremediation of hydrocarbons in marine environments

Marine environments are subject to contamination by many compounds related to human activity. The crude oil and its derivatives are some of the most important organic pollutants in the marine environment because they cause a series of economic, social, and environmental problems (Peterson et al. 2003). The marine contaminations by hydrocarbons can be derived both from activities in land, which cause the dragging of material to the coastal areas, as well as activities of maritime transportation and in the very offshore industry of oil production, where there are risks from the time of the extraction until the arrival of the oil in the onshore refineries. Accidents involving oil spill are even more harmful when they contaminate coastal regions, which are among the most productive environments on Earth, causing accelerated degradation and loss of biodiversity of these ecosystems (Lotze et al. 2006).

In 1989, there was a major accident in history involving oil spills at sea by the ship *Exxon Valdez* (Peterson et al. 2003). At the time, 42 million gallons of crude oil into the sea were discharged, which spread throughout 1,990 km of coastal area, with the greatest impact on the *Prince William* cove. The contamination was responsible for the negative influence on the development of algae, due to the toxicity exerted by the hydrocarbons and, as a result, of lower primary productivity, the whole food chain was affected. There was rapid reduction in the ichthyofauna, crabs, starfish, and invertebrates populations in that area. Some groups of organisms resistant to the effects of the contamination became more abundant, such as, for example, oligochaetes and polychaetes, probably as a result of the reduction of their predators; also an increase in the abundance of hydrocarbon degrading bacteria was also evidenced (Peterson 2001). Even after six years from the disaster, plumes and sediments contaminated with oil persisted, so that punctual toxic effects could still occur (Carls et al. 2001). After 13 years from the accident, oil residues were found mainly in the upper intertidal regions, associated with fine sediments and sand (Taylor and Reimer 2008). In 2010, the accident involving the platform of the company Deepwater Horizon in the Gulf of Mexico affected a great coastal area of the U.S.A. and exhibited the fragility of coastal ecosystems. Several measures were taken to remove the oil from the affected areas, but it is estimated that 25% of the oil still remained in the environment (Tate et al. 2011).

Bioremediation techniques are more effective and of lesser cost in the removal of the contaminants from aquatic environments when compared with the physico-chemical techniques (Khan et al. 2004; Bao et al. 2012). However, in order to increase the efficiency of the actions, the techniques must be used as complementary (Harayama et al. 1999; Aburto-Medina et al. 2015). The removal of hydrocarbon molecules, solubilized or dispersed in the water column in marine environments for subsequent treatment in specific places is impracticable. Therefore, *in situ* techniques are used at sea, understanding both the bioaugmentation and the biostimulation as to the natural attenuation (Yu et al. 2005). When the oil reaches the marine environment, changes occur in its physico-chemical composition as a result of the performance of several environmental factors, which directly affect on the permanence of the hydrocarbons of that environment, such as evaporation, dissolution, dispersion, photochemical oxidation, emulsification, adsorption in particles, and sedimentation (Miller et al. 1988; Atlas 1981; Garret et al. 1998; Dutta and Harayama 2000; Medina-Bellver et al. 2005; Haritash and Kaushik 2009; Karakoç et al. 2015).

Although ubiquitous, microorganisms capable of degrading hydrocarbons constitute only about 0.1% of the total microbial community in non-contaminated environments. After a contamination event, however, the proportion of hydrocarbonoclastic microorganisms in the microbial community can reach up to 100% (Atlas 1981). Rodrigues et al. (2015b) showed that, in ocean water, representatives of the *Bacteria* domain prevail in the colonization of the spilled oil, followed by fungi and then *Archaea*.

Recently, Rodrigues et al. (2015a) used enrichment medium to obtain isolated from the coastal region of an oceanic tropical island considered as pristine for hydrocarbons contamination. Later, Rodrigues et al. (2015b) used one of these microorganisms, bacterium *Rhodococcus rhodochrous* TRN7, in microcosms simulating bioaugmentation in the presence of oil previously heated to remove the more volatile hydrocarbons. Microbial activity in microcosms containing cells of *R. rhodochrous* TRN7 increased significantly during the incubation period, suggesting that such isolate can use the fractions of hydrocarbons made available. Similarly, only the biostimulation simulated in this environment was responsible for the increase in bacterial activity, in order to suggest that biostimulation is an effective technique in bioremediation of oligotrophic environments considered pristine.

Pyrene and indene (1,2,3-cd) pyrene are the most abundant PAHs in moist areas of the estuary Liaohe, in China. *Pseudomonas putida* PYR1 and *Acinetobacter baumannii* INP1 immobilized in ash dust spheres or as free cells had been used as strategy of bioaugmentation in microcosms with both the contaminants (Huang et al. 2016). After 30 days of incubation, the efficiency of the biodegradation of the PAHs was greater when the cells were immobilized in comparison with the free cells, probably related to the fact that the pyrene initially adsorbed in the ash spheres, promoting greater contact surface between the contaminant and the bacterial cells. Technologies capable of increasing the efficiency of biodegradation of hydrocarbons in oceanic and coastal environments have been proposed as well as biostimulation agents; for example, when one uses floating flakes of clay fertilized with inorganic nutrients (Warr et al. 2013) in open oligotrophic waters. The use of oleophlic fertilizers, such as Inipol EAP 22, was proposed as a way to increase the efficiency of biodegradation in oceans. However, after using it in regions affected by the accident of the ship *Exxon Valdez*, some reports questioned the efficacy of the product (Nikolopoulou and Kalogerakis 2008) because its formula contains 2-butoxyethanol, a potentially toxic compound. In addition, in the case of Inipol EAP 22, when the product comes into contact with the water, emulsion breakage occurs, releasing all the urea in the aqueous phase, which is lost quickly (Ron and Rosenberg 2014). Nikolopoulou et al. (2007) proposed, as an alternative, the use of uric acid and soy lecithin (compounds of low solubility in water) as sources of nitrogen and phosphorus, respectively, to stimulate the biodegradation of hydrocarbons in aquatic oligotrophic environments.

Influence of Environmental Factors on the Biodegradation of Hydrocarbons

High molecular weight hydrocarbons and PAHs do not persist in the environment only by intrinsic characteristics of the molecule, but also by environmental factors that may not make its biodegradation possible (Taylor and Reimer 2008). If the presence of hydrocarbonoclastic microorganisms is required to eliminate them, as well as the existence of environmental conditions so that the catabolism of these compounds occurs, interventions in the contaminated environment should be carried out. The contamination by other compounds or by toxic metals can hinder microbial development. However, prospecting and screening of microorganisms that are effective in metabolizing hydrocarbons in the presence of these possible inhibitory agents can make bioremediation even under such conditions, apparently inhibitory.

The particularity of each environment often requires different remediation techniques to be adopted to attain an efficient removal of contaminants from a given site. Therefore, knowing the biotic and abiotic factors of the contaminated environment is essential to initiate the planning of which actions should be taken. Hydrocarbonoclastic microorganisms isolated from soils may not adapt to aquatic environments and vice versa. Therefore, bioaugmentation using indigenous microorganisms can be more efficient, because it is known that these isolates can readapt themselves with ease to their original environment (Aburto-Medina et al.

2015). One should also take into consideration factors such as predation by protozoa, competition by nutrients, local temperature, incidence of solar radiation, among others. For example, biostimulation by compounds sources of nitrogen and phosphorus may not result in any effect on removal of hydrocarbons when these micronutrients are not the limiting factor of microbial activity in the contaminated environment. On the other hand, the natural attenuation can be the most effective way and the cheapest one, if the environment has hydrocarbonoclastic microorganisms and still presents significant values of nitrogen and phosphorus, in addition to micronutrients (El Fantroussi and Agathos 2005; Yu et al. 2005).

The persistence of a contaminant in the environment can be reduced if actions are taken to provide greater bioavailability and, consequently, greater biodegradation speed. However, the cost of these actions, which may require time, becomes the main impediment of continuous action in contaminated environments. In this case, a careful assessment of the cost/benefit ratio and a proper risk analysis are essential to coherent decision-making, taking the maximum advantage of the metabolic potential of microorganisms with a view to preserving the environments and the health of men, animals, and other members of the biosphere.

References

Abdel-Shafy, H.I. and M.S.M. Mansour. 2016. A review on polycyclic aromatic hydrocarbons: Source, environmental impact, effect on human health and remediation. Egyptian J. Petroleum 25: 107–123.

Abed, R.M.M., S. Al-Kharusi and M. Al-Hinai. 2015. Effect of biostimulation, temperature and salinity on respiration activities and bacterial community composition in an oil polluted desert soil. Int. Biodeterior. Biodegr. 98: 43–52.

Aburto-Medina, A., E.M. Adetutu, S. Aleer, J. Weber, S.S. Patil, P.J. Sheppard et al. 2015. Comparison of indigenous and exogenous microbial populations during slurry phase biodegradation of long-term hydrocarbon-contaminated soil. Biodegradation 23: 813–822.

Aguilera, R.F., R.G. Eggert, C.C.G. Lagos and J.E. Tilton. 2009. Depletion and the future availability of petroleum resources. Energy J. 30: 141–174.

Aller, S., E.M. Adetutu, J. Weber, A.S. Ball and A.L. Juhasz. 2014. Potential impact of soil microbial heterogeneity on the persistence of hydrocarbons in contaminated subsurface soils. J. Environ. Manage. 136: 27–36.

Asadollahi, L., H. Salehizadeh and N. Yan. 2016. Investigation of biosurfactant activity and asphaltene biodegradation by *Bacillus cereus*. J. Polym. Environ. 24: 119–128.

Aske, N., H. Kallevik and J. Sjöblom. 2002. Water-in-crude oil emulsion stability studied by critical electric field measurements. Correlation to physico-chemical parameters and near-infrared spectroscopy. J. Pet. Sci. Eng. 36: 1–17.

Atlas, R.M. 1981. Microbial degradation of petroleum hydrocarbons: an environmental perspective. Microbiol. Rev. 45: 180–209.

Atlas, R.M. and T.C. Hazen. 2011. Oil biodegradation and bioremediation: a tale of the two worst spills in U.S. history. Environ. Sci. Technol. 45: 6709–6715.

Auger, R.L., A.M. Jacobson and M.M. Domach. 1995. Effect of nonionic surfactant addition on bacterial metabolism of naphthalene: Assessment of toxicity and overflow metabolism potential. J. Hazard Mater. 43: 263–272.

Baboshin, M., V. Akimov, B. Baskunov, T.L. Born, S.U. Khan and L. Golovleva. 2008. Conversion of polycyclic aromatic hydrocarbons by *Sphingomonas* sp. VKM B-2434. Biodegradation 19: 567–576.

Bahr, A., A. Fischer, C. Vogt and P. Bombach. 2015. Evidence of polycyclic aromatic hydrocarbon biodegradation in a contaminated aquifer by combined application of *in situ* and laboratory microcosms using (13)C-labelled target compounds. Water Res. DOI: 10.1016/j. watres.2014.10.045.

Bahrs, H., A. Putschew and C.E. Steinberg. 2013. Toxicity of hydroquinone to different freshwater phototrophs is influenced by time of exposure and pH. Environ. Sci. Pollut. R. 20: 146–154.

Baker, K.H. 1994. Biodegradation of surface and subsurface soils. pp. 203–259. *In*: Baker, K.H. and D.S. Herson [eds.]. Bioremediation. McGraw-Hill, Ins. New York.

Bao, M.T., L.N. Wang, P.Y. Sun, L.X. Cao, J. Zou and Y.M. Li. 2012. Biodegradation of crude oil using an efficient microbial consortium in a simulated marine environment. Mar. Pollut. Bull. 64: 1177–1185.

Baruah, R., D.J. Kalita, B.K. Saikia, A. Gautam, A.K. Singh and H.P.D. Boruah. 2016. Native hydrocarbonoclastic bacteria and hydrocarbon mineralization processes. Int. Biodeter. Biodegrad. 112: 18–30.

Beiras, E.H.R., M.N.L. Seaman and A. France. 1999. The assessment of marine pollution— bioassays with bivalve embryos and larvae. Adv. Mar. Biol. 37: 1–178.

Bojes, H.K. and P.G. Pope. 2007. Characterization of EPAs 16 priority pollutant polycyclic aromatic hydrocarbons (PAHs) in tank bottom solids and associated contaminated soils at oil exploration and production sites in Texas. Regul. Toxicol. Pharmacol. 47: 288–295.

Bolton, J.L., M.A. Trush, T.M. Penning, G. Dryhurst and T.J. Monks. 2000. Role of quinones in toxicology. Chem. Res. Toxicol. 13: 135–160.

Bouchez, M., D. Blanchet and J.P. Vandecasteele. 1995. Degradation of polycyclic aromatic hydrocarbons by purê strains and by defined strain associations: inhibition phenomena and cometabolism. Appl. Microbiol. Biot. 43: 156–164.

Bramwell, D.P. and S. Laha. 2000. Effects of surfactant addition on the biomineralization and microbial toxicity of phenanthrene. Biodegrad. 11: 263–277.

Carls, M.G., M.M. Babcock, P.M. Harris, G.V. Irvine, J.A. Cusick and S.D. Rice. 2001. Persistence of oiling in mussel beds after the Exxon Valdez oil spill. Mar. Environ. Res. 51: 167–190.

Cerniglia, C.E. 1984. Microbial metabolism of polycyclic aromatic hydrocarbons. Adv. Appl. Microbiol. 30: 31–71.

Chang, B.V., L.C. Shiung and S.Y. Yuan. 2002. Anaerobic biodegradation of polycyclic aromatic hydrocarbon in soil. Chemosphere 48: 717–724.

Chang, B.V., I.T. Chang and S.Y. Yuan. 2008. Anaerobic degradation of phenanthrene and pyrene in mangrove sediment. Bull. Environ. Contam. Toxicol. 80: 145–149.

Chang, J.S., D.K. Cha, M. Radosevich and Y. Jin. 2015. Effects of biosurfactants-producing bacteria on biodegradation and transport of phenanthrene in subsurface soil. J. Environ. Sci. Heal. A. 50: 611–616.

Chen, G. and P.A. White. 2004. The mutagenic hazards of aquatic sediments: a review. Mutat. Res. 567: 151–225.

Chiou, C.T., S.E. Mcgroddy, D.E. Kile, U.S.G. Survey and D. Federal. 1998. Partition characteristics of polycyclic aromatic hydrocarbons on soils and sediments. Environ. Sci. Technol. 32: 264–269.

Cookson, J.R.J. 1995. Bioremediation Engineering: Design and Application. McGraw-Hill Inc., New York.

Cury, J.D.E.C. 2002. Atividade microbiana e diversidades metabólica e genética em solo de mangue contaminado com petróleo. Master's Dissertation. Universidade de São Paulo. São Paulo, Brazil.

Dabrowska, D., A. Kot-Wasik and J. Namiesnik. 2008. Stability studies of selected polycyclic aromatic hydrocarbons in different organic solvents and identification of their transformation products. Polish. J. Environ. Stud. 17: 17–24.

Das, N. and P. Chandran. 2011. Microbial degradation of petroleum hydrocarbon contaminants: an overview. Biotechnol. Res. Int. 2011: 1–13.

Denome, S.A., D.C. Stanley, E.S. Olson and K.D. Young. 1993. Metabolism of dibenzothiophene and naphthalene in *Pseudomonas* strains: Complete DNA sequence of an upper naphthalene catabolic pathway. J. Bacteriol. 175: 6890–6901.

Dutta, T.K. and S. Harayama. 2000. Fate of crude oil by the combination of photooxidation and 543 biodegradation. Environ. Sci. Technol. 34: 1500–1505.

Eaton, R.W. and P.J. Chapman. 1992. Bacterial metabolism of naphthalene: construction and use of recombinant bacteria to study ring cleavage of 1,2-dihydroxynaphthalene and subsequent reactions. J. Bacteriol. 174: 7542–7554.

El Fantroussi, S. and S.N. Agathos. 2005. Is bioaugmentation a feasible strategy for pollutant removal and site remediation? Curr. Opin. Microbiol. 8: 268–275.

Evans, W.C., H.N. Fernley and E. Griffiths. 1965. Oxidative metabolism of phenanthrene and anthracene by soil pseudomonads. Biochem. J. 95: 819–831.

Fayeulle, A., E. Veignie, C. Slomianny, E. Dewailly, J.C. Munc and C. Rafin. 2014. Energy-dependent uptake of benzo[a]pyrene and its cytoskeleton-dependent intracellular transport by the telluric fungus *Fusarium solani*. Environ. Sci. Pollut. Res. 21: 3515–3523.

Fernández-Luqueño, F., R. Marsch, D. Espinosa-Victoria, F. Thalasso, M.E. Hidalgo Lara, A. Munive et al. 2008. Remediation of PAHs in a saline-alkaline soil amended with wastewater sludge and the effect on dynamics of C and N. Sci. Total. Environ. 402: 18–28.

Fontaine, S., A. Mariotti and L. Abbadie. 2003. The priming effect of organic matter: a question of microbial competition? Soil Biol. Biochem. 35: 837–843.

Fucik, K.W., K.A. Carr and B.J. Balcom. 1994. Dispersed Oil Toxicity Tests with Biological Species Indigenous to the Gulf of Mexico. U.S. Department of the Interior, New Orleans.

Gallego, J.L.R., J. Loredo, J.F. Llamas, F. Vázquez and J. Sánchez. 2001. Bioremediation of diesel-contaminated soils: Evaluation of potential *in situ* techniques by study of bacterial degradation. Biodegradation 12: 325–335.

Gargouri, B., F. Karray, N. Mhiri, F. Aloui and S. Sayadi. 2014. Bioremediation of petroleum hydrocarbons-contaminated soil by bacterial consortium isolated from an industrial wastewater treatment plant. J. Chem. Technol. Biotechnol. 89: 978–987.

Gargouri, B., N. Mhiri, F. Karray, F. Aloui and S. Sayadi. 2015. Isolation and characterization of hydrocarbon-degrading yeast strains from petroleum contaminated industrial wastewater. Biomed. Res. Int. 929424.

Garrett, R.M., I.J. Pickering, C.E. Haith and R.C. Prince. 1998. Photooxidation of crude oils. Environ. Sci. Technol. 32: 3719–3723.

Gaskin, S., K. Soole and R. Bentham. 2008. Screening of Australian native grasses for rhizoremediation of aliphatic hydrocarbon-contaminated soil. Int. J. Phytoremed. 10: 378–389.

Ghazali, F.M., R.N.Z.A. Rahman, A.B. Salleh and M. Basri. 2004. Biodegradation of hydrocarbons in soil by microbial consortium. Int. Biodeterior. Biodegr. 54: 61–67.

Gogoi, B.K., N.N. Dutta, P. Goswami and T.R. Krishna Mohan. 2003. A case study of bioremediation of petroleum-hydrocarbon contaminated soil at a crude oil spill site. Adv. Environ. Res. 7: 767–782.

Gordon, A., H.D. Zakpaa and MakMensah. 2015. Biodegradation potentials of bacterial isolates from petroleum storage facilities within the Kumasi Metropolitan area. Afr. J. Microbiol. Res. 9: 433–447.

Goyal, A.K. and G.J. Zylstra. 1997. Genetics of naphthalene and phenanthrene degradation by *Comamonas testosteroni*. J. Ind. Microbiol. Biotechnol. 19: 401–407.

Gray, R., R.R. Tykwinski, J.M. Stryker and X. Tan. 2011. Supramolecular assembly model for aggregation of petroleum asphaltenes. Energy Fuel 25: 3125–3134.

Groenzin, H. and O.C. Mullins. 2000. Molecular size and structure of asphaltenes from various sources. Energy Fuels 14: 677–684.

Groves, J.T., S.G. Boxer and H.M. McConnel. 1997. Electric field-induced reorganization of two-component supported bilayer membranes. P. Natl. Aca. Sci. USA 94: 13390–13395.

Gundersen, D.T., S.W. Kristanto, L.R. Curtis, S.N. Al-Yakoob, M.M. Metwally and D. Al-Ajmi. 1996. Subacute toxicity of the water-soluble fractions of kuwait crude oil and partially combusted crude oil on *Menidia beryllina* and *Palaemonetes pugio*. Arch. Environ. Con. Tox. 31: 1–8.

Harayama, S., H. Kishira, Y. Kasai and K. Shutsubo. 1999. Petroleum biodegradation in marine environments. J. Mol. Microbiol. Biotechnol. 1: 63–70.

Haritash, A.K. and C.P. Kaushik. 2009. Biodegradation aspects of polycyclic aromatic hydrocarbons (PAHs): a review. J. Hazard. Mater. 169: 1–15.

Hassan, I., E. Mohamedelhassan, E.K. Yanful and Z.C. Yuan. 2016. A review article: electrokinetic bioremediation current knowledge and new prospects. Adv. Microb. 6: 57–72.

Hassanshahian, M., G. Emtiazi, G. Caruso and S. Cappello. 2014. Bioremediation (bioaugmentation/biostimulation) trials of oil polluted seawater: A mesocosm simulation study. Mar. Environ. Res. 95: 28–38.

Hayat, R., S. Ali, U. Amara, R. Khalid and I. Ahmed. 2010. Soil beneficial bacteria and their role in plant growth promotion: a review. Annu. Microbiol. 60: 579–598.

Heitkamp, M.A., J.P. Freeman, D.W. Miller and C.E. Cerniglia. 1988. Pyrene degradation by a *Mycobacterium* sp.: identification of ring oxidation and ring fission products. Appl. Environ. Microbiol. 54: 2556–2565.

Hernández-López, E.L., M. Ayala and R. Vazquez-Duhalt. 2015. Microbial and enzymatic biotransformations of asphaltenes. Pet. Sci. Technol. 33: 1019–1027.

Hernández-López, E.L., L. Perezgasga, A. Huerta-Saquero, R. Mouriño-Pérezand and R. Vazquez-Duhalt. 2016. Biotransformation of petroleum asphaltenes and high molecular weight polycyclic aromatic hydrocarbons by *Neosartorya fischeri*. Environ. Sci. Pollut. Res. Int. 32: 10773–10784.

Hoeppel, R.E. and R.E. Hinchee. 1994. Enhanced Biodegradation for Onsite Remediation of Contaminated Soils and Groundwater. Marcel Dekker Inc., New York.

Huang, R.Y., Q. Liu, H. Yu, X. Jin, Y.G. Zhao, Y.H. Zhou et al. 2016. Enhanced biodegradation of pyrene and indeno(1,2,3-cd)pyrene using bactéria immobilized in cinder beads in estuarine wetlands. Mar. Pollut. Bull. 102: 128–133.

Huesemann, M.H. and M.J. Truex. 1996. The role of oxygen diffusion in passive bioremediation of petroleum contaminated soils. J. Hazard Mater. 51: 93–113.

Jeon, C.O., M. Park, H.S. Ro, W. Park and E.L. Madsen. 2006. The naphthalene catabolic (nag) genes of *Polaromonas naphthalenivorans* CJ2: evolutionary implications for two gene clusters and novel regulatory control. Appl. Environ. Microbiol. 72: 1086–1095.

Jo, M.S., E.R. Rene, S.H. Kim and H.S. Park. 2008. An analysis of synergistic and antagonistic behavior during BTEX removal in batch system using response surface methodology. J. Hazard. Mater. 152: 1276–1284.

Johnsen, A.R., L.Y. Wick and H. Harms. 2005. Principles of microbial PAH-biodegradation in soil. Environ. Pollut. 133: 71–84.

Kanaly, R.A. and S. Harayama. 2000. Biodegradation of high-molecular-weight polycyclic aromatic hydrocarbons by bactéria. Appl. Environ. Microbiol. 182: 2059–2067.

Kao, C.M., W.Y. Huang, L.J. Chang, T.Y. Chen, H.Y. Chien and F. Hou. 2006. Application of monitored natural attenuation to remediate a petroleum-hydrocarbon spill site. Water Sci. Technol. 53: 321–328.

Karakoç, F.T., H. Atabay, L. Tolun and E. Kuzyaka. 2015. Fast scanning of illegal oil discharges for forensic identification: a case study of Turkish coasts. Environ. Monit. Assess. 187: 211.

Kari, F.W., J. Bucher, S.L. Eustis, J.K. Haseman and J.E. Huff. 1992. Toxicity and carcinogenicity of hydroquinone in F344/N rats and B6C3F1 mice. Food Chem. Toxicol. 30: 737–747.

Karigar, C.S. and S.S. Rao. 2011. Role of microbial enzymes in the bioremediation of pollutants: A review. Enzyme Res. doi: http://dx.doi.org/10.4061/2011/805187.

Khan, F.I., T. Husain and R. Hejazi. 2004. An overview and analysis of site remediation technologies. J. Environ. Manag. 71: 95–122.

Khan, S.A., N. Bibi and S.K. Shrewani. 2015. Isolation, screening and co-metabolism of polycyclic aromatic hydrocarbons by soil bacteria. American-Eurasian J. Agric. Environ. Sci. 15: 800–812.

Kim, J., S.H. Kang, K.A. Min, K.S. Cho and I.S. Lee. 2006. Rhizosphere microbial activity during phytoremediation of diesel-contaminated soil. J. Environ. Sci. Health Part A. 41: 2503–2516.

Kim, Y.H., J.D. Moody, J.P. Freeman, K.H. Engesser and C.E. Cerniglia. 2004. Evidence for the existence of PAH-quinone reductase and catechol-*O*-methyltransferase in *Mycobacterium vanbaalenii* PYR-1. J. Ind. Microbiol. Biotechnol. 31: 507–516.

Kim, Y.H., J.P. Freeman, J.D. Moody, K.H. Engesser and C.E. Cerniglia. 2005. Effects of pH on the degradation of phenanthrene and pyrene by *Mycobacterium vanbaalenii* PYR-1. Appl. Microbiol. Biotechnol. 67: 275–285.

Kim, S.J., O. Kweon, R.C. Jones, J.P. Freeman, R.D. Edmondson and C.E. Cerniglia. 2007. Complete and integrated pyrene degradation pathway in *Mycobacterium vanbaalenii* PYR-1 based on systems biology. J. Bacteriol. 189: 464–472.

Kirk, J.L., J.N. Klironomos, H. Lee and J.T. Trevors. 2005. The effects of perennial ryegrass and alfafa on microbial abundance and diversity in petroleum contaminated soil. Environ. Pollut. 133: 455–465.

Kiyohara, H., S. Torigoe, N. Kaida, T. Asaki, T. Iida, H. Hayashi et al. 1994. Cloning and characterization of a chromosomal gene cluster, *pah*, that encodes the upper pathway for phenanthrene and naphthalene utilization by *Pseudomonas putida* OUS82. J. Bacteriol. 176: 2439–2443.

Kleemann, R. and R.U. Meckenstock. 2011. Anaerobic naphthalene degradation by Gram-positive, ironreducing bacteria. FEMS Microbiol. Ecol. 78: 488–496.

Kümmel, S., F.A. Herbst, A. Bahr, M. Duarte, D.H. Pieper, N. Jehmlich et al. 2015. Anaerobic naphthalene degradation by sulfate-reducing Desulfobacteraceae from various anoxic aquifers. FEMS Microbiol. Ecol. 91: fiv006.

La Rocca, C., L. Conti, R. Crebelli, B. Crochi, N. Iacovella, F. Rodriguez et al. 1996. PAH content and mutagenicity of marine sediments from the Venice lagoon. Ecotoxicol. Environ. Saf. 33: 2362–45.

Labud, V., C. Garcia and T. Hernandez. 2007. Effect of hydrocarbon pollution on the microbial properties of a sandy and a clay soil. Chemosphere 66: 1863–1871.

Leal, A.J., E.M. Rodrigues, P.L. Leal, A.D.L. Júlio, R.C.R. Fernandes, A.C. Borges et al. 2017. Changes in the microbial community during bioremediation of gasoline-contaminated soil. Braz. J. Microbiol. 48: 342–351.

Leal, P.L. 2009. Atividade e dinâmica populacional de um consórcio bacteriano durante biodegradação de óleo diesel no solo. Ph.D. Thesis, Universidade Federal de Viçosa, Viçosa, BR.

Leys, N.M., L. Bastiaens, W. Verstraete and D. Springael. 2005. Influence of the carbon/nitrogen/phosphorus ratio on polycyclic aromatic hydrocarbon degradation by Mycobacterium and Sphingomonas in soil. Appl. Microbiol. Biotechnol. 66: 726–736.

Li, X., L. Zhao and M. Adam. 2016. Biodegradation of marine crude oil pollution using a salt-tolerant bacterial consortium isolated from Bohai Bay, China. Mar. Pollut. Bull. 105: 43–50.

Lin, Q. and I.A. Mendelssohn. 2012. Impacts and recovery of the deepwater horizon oil spill on vegetation structure and function of coastal salt marshes in the Northern Gulf of Mexico. Environ. Sci. Technol. 46: 3737–3743.

Lotze, H.K., H.S. Lenihan, B.J. Bourque, R.H. Bradbury, R.G. Cooke, M.C. Kay et al. 2006. Depletion, degradation, and recovery potential of estuaries and coastal seas. Science 312: 1806–1809.

Luckert, C., A. Ehlers, T. Buhrke, A. Seidel, A. Lampen and S. Hessel. 2013. Polycyclic aromatic hydrocarbons stimulate human CYP3A4 promoter activity via PXR. Toxicol. Lett. 222: 180–188.

Mao, X., R. Jiang, W. Xiao and J. Yu. 2015. Use of surfactants for the remediation of contaminated soils: A review. J. Hazard. Mater. 285: 419–435.

Marcoux, J. 2000. Optimization of high-molecular-weight polycyclic aromatic hydrocarbons' degradation in a two-liquid-phase bioreactor. J. Appl. Microbiol. 88: 655–662.

Marecik, R., L. Chrzanowski, A. Piotrowska-Cyplik, W. Juzwa and Biegańska. 2015. Rhizosphere as a tool to introduce a soil-isolated hydrocarbon-degrading bacteriaç consortium into a wetland environment. Int. Biodeter. Biodegr. 97: 135–142.

Marín-Spiotta, E., K.E. Gruley, J. Crawford, E.E. Atkinson, J.R. Miesel, S. Greene et al. 2013. Paradigm shift in soil organic matter research affect interpretations of aquatic carbon cycling: transcending disciplinary and ecosystem boundaries. Biogeochemistry 117: 279–297.

Marrazzini, A., L. Chelotti, I. Barrai, N. Loprien and R. Barale. 1994. *In vivo* genotoxic interactions among three phenolic benzene metabolites. Mutat. Res. 341: 29–46.

Meckenstock, R.U. and H. Mouttaki. 2011. Anaerobic degradation of non-substituted aromatic hydrocarbons. Curr. Opin. Biotechnol. 22: 406–414.

Meckenstock, R.U., M. Boll, H. Mouttaki, J.S. Koelschbach, P. Cunha-Tarouco, P. Weyrauch et al. 2016. Anaerobic degradation of benzene and polycyclic aromatic hydrocarbons. J. Mol. Microbiol. Biotechnol. 26: 92–118.

Medina-Bellver, J.I., P. Marin, A. Delgado, A. Rodriguez-Sanchez, E. Reyes, J.L. Ramos et al. 2005. Evidence for *in situ* crude oil biodegradation after the Prestige oil spill. Environ. Microbiol. 7: 773–779.

Mehdinia, A., V. Aghadadashi and N.S. Fumani. 2015. Origin, distribution and toxicological potential of polycyclic aromatic hydrocarbons in surface sediments from the bushehr coast, the persian gulf. Mar. Pollut. Bull. 90: 334–338.

Miller, R.M., G.M. Singer, J.D. Rosen and R. Bartha. 1988. Photolysis primes biodegradation of benzo[a]pyrene. Appl. Environ. Microbiol. 54: 1724–1730.

Minai-Tehrani, D., P. Rohanifar and S. Azami. 2015. Assessment of bioremediation of aliphatic, aromatic, resin, and asphaltene fractions of oil-sludge-contaminated soil. Int. J. Environ. Sci. Technol. 12: 1253–1260.

Mnif, S., M. Chamkha, M. Labat and S. Sayadi. 2011. Simultaneous hydrocarbon biodegradation and biosurfactant production by oilfield-selected bacteria. J. Appl. Microbiol. 111: 525–536.

Moody, J.D., J.P. Freeman, D.R. Doerge and C.E. Cerniglia. 2001. Degradation of phenanthrene and anthracene by cell suspensions of *Mycobacterium* sp. strain PYR-1. App. Env. Microbiol. 67: 1476–1483.

Morais, D., V. Pylro, I.M. Clark, P.R. Hirsch and M.R. Tótola. 2016. Responses of microbial community from tropical pristine coastal soil to crude oil contamination. PeerJ doi: 10.7717/peerj.1733.

Mosaed, H.P., S. Sobhanardakani, H. Merrikhpour, A. Farmany, M. Cheraghi and C. Ashorlo. 2015. The effect of urban fuel stations on soil contamination with petroleum hydrocarbons. Iran J. Toxicol. 9: 1378–1384.

Mostowfi, F., K. Indo, O.C. Mullins and R. McFarlane. 2009. Asphaltene nanoaggregates studied by centrifugation. Energy Fuels 23: 1194–1200.

Mullins, O.C. 2011. The asphaltenes. Annu. Rev. Anal. Chem. 4: 393–418.

Murado, M.A., J.A. Vázquez, D. Rial and R. Beiras. 2011. Dose-response modelling with two agents: application to the bioassay of oil and shoreline cleaning agents. J. Hazard. Mater. 185: 807–817.

Nikolopoulou, M., N. Pasadakis and N. Kalogerakis. 2007. Enhanced bioremediation of crude oil utilizing lipophilic fertilizers. Desalination 211: 286–295.

Nikolopoulou, M. and N. Kalogerakis. 2008. Enhanced bioremediation of crude oil utilizing lipophilic fertilizers combined with biosurfactants and molasses. Mar. Pollut. Bull. 56: 1855–1861.

Oliveira, D.M., B.P.S. Pitanga, M.S. Grangeiro, R.M.F. Lima, M.F.D. Costa, S.L. Costa et al. 2010. Catechol cytotoxicity *in vitro*: induction of glioblastoma cell death by apoptosis. Hum. Exp. Toxicol. 29: 199–212.

Oliveira, D.M., R.M. Ferreira Lima and R.S. El-Bachá. 2012. Brain rust: recent discoveries on the role of oxidative stress in neurodegenerative diseases. Nutr. Neurosci. 15: 94–102.

Pagé, A.P., É. Yergeau and C.W. Greer. 2015. *Salix purpurea* stimulates the expression of specific bacterial xenobiotic degradation genes in a soil contaminated with hydrocarbons. PLos One 10: e0132062.

Palmroth, M.R.T., U. Munster, J. Pitchel and J.A. Puhakka. 2005. Metabolic responses of microbiota to diesel fuel addition in vegetated soil. Biodegradation 16: 91–101.

Palmroth, M.R.T., P.E.P. Koskinen, A.H. Kaksonen, U. Munster, J. Pitchel and J.A. Puhakka. 2007. Metabolic and phylogenetic analysis of microbial communities during phytoremediation of soil contaminates with weathered hydrocarbons and heavy metals. Biodegradation 18: 769–782.

Pandey, P., H. Pathak and S. Dave. 2016. Microbial ecology of hydrocarbon degradation in the soil: a review. Res. J. Environ. Toxicol. 10: 1–15.

Patrick, A., D.L. Craig, E. Sena, L. Magalhães and M. Canielas. 2012. Técnicas de limpeza de vazamentos de petroleo em alto mar. Cad. Grad. Ciências Exatas e Tecnológicas 1: 75–86.

Pavel, L.V. and M. Gavrilescu. 2008. Overview of *ex situ* decontamination techniques for soil cleanup. Env. Engin. Manag. J. 7: 815–834.

Peng, R.H., A.S. Xiong, Y. Xue, X.Y. Fu, F. Gao, W. Zhao et al. 2008. Microbial biodegradation of polyaromatic hydrocarbons. FEMS Microbiol. Rev. 32: 927–955.

Pereira, M.R.G., E.S. de Oliveira, F.A.G.A de Villar, M.S. Grangeiro, J. Fonseca, A.R. Silva et al. 2004. Cytotoxicity of catechol towards human glioblastoma cells via superoxide and reactive quinones generation. J. Bras. Patol. Clín. 40: 281–286.

Peterson, C.H. 2001. The "Exxon Valdez" oil spill in Alaska: acute, indirect and chronic effects on the ecosystem. Adv. Mar. Biol. 39: 1–103.

Peterson, C.H., S.D. Rice, J.W. Short, D. Esler, J.L. Bodkin, B.E. Ballachey et al. 2003. Long-term ecosystem response to the Exxon Valdez oil spill. Science 302: 2082–2086.

Prabhu, Y. and P.S. Phale. 2003. Biodegradation of phenanthrene by *Pseudomonas* sp. strain PP2: novel metabolic pathway, role of biosurfactant and cell surface hydrophobicity in hydrocarbon assimilation. Appl. Microbiol. Biotechnol. 61: 342–351.

Rahman, K.S.M., J. Thahira-Rahman, P. Lakshmanaperumalsamy and I.M. Banat. 2002. Towards efficient crude oil degradation by a mixed bacterial consortium. Bioresour. Technol. 85: 257–261.

Ramachandran, S.D., P.V. Hodson, C.W. Khan and K. Lee. 2004. Oil dispersant increases PAH uptake by fish exposed to crude oil. Ecotoxicol. Environ. Saf. 59: 300–308.

Ramirez, E.M., J.V. Camacho, M.A.R. Rodrigo and P.C. Canizares. 2014. Feasibility of electrokinetic oxygen supply for soil bioremediation purposes. Chemosphere 117: 382–387.

Rengarajan, T., P. Rajendran, N. Nandakumar, B. Lokeshkumar, P. Rajendran and I. Nishigaki. 2015. Exposure to polycyclic aromatic hydrocarbons with special focus on cancer. Asian Pac. J. Trop. Biomed. 5: 182–189.

Ritter, K.S. and L. Paul Sibley. 2002. Sources, pathways, and relative risks of contaminants in surface water and groundwater: a perspective prepared for the Walkerton inquiry. J. Toxicol. Environ. Health A. 65: 1–142.

Rodrigues, E.M. and M.R. Tótola. 2015. Petroleum: From basic features to hydrocarbons bioremediation in oceans. OALib 2: e2136.

Rodrigues, E.M., K.H.M. Kalks and M.R. Tótola. 2015a. Prospect, isolation, and characterization of microorganisms for potential use in cases of oil bioremediation along the coast of Trindade Island. Braz. J. Environ. Manag. 156: 15–22.

Rodrigues, E.M., K.H.M. Kalks, P.L. Fernandes and M.R. Tótola. 2015b. Bioremediation strategies of hydrocarbons and microbial diversity in the Trindade Island shoreline—Brazil. Mar. Pollut. Bull. 101: 517–525.

Röling, W.F.M., M.G. Milner, D.M. Jones, K. Lee, F. Daniel, R.J.P. Swannell et al. 2002. Robust hydrocarbon degradation and dynamics of bacterial communities during nutrient-enhanced oil spill bioremediation. App. Environ. Microbiol. 68: 5537–5548.

Ron, E.Z. and E. Rosenberg. 2014. Enhanced bioremediation of oil spills in the sea. Curr. Opin. Biotechnol. 27: 191–194.

Rosenkranz, F., L. Cabrol, M. Carballa, A. Donoso-Bravo, L. Cruz, G. Ruiz-Filippi et al. 2013. Relationship between phenol degradation efficiency and microbial community structure in an anaerobic SBR. Water Res. 47: 6739–6749.

Roy, J.L. and W.B. McGill. 2000. Investigation into mechanisms leading to the development, spread and persistence of soil water repellency following contamination by crude oil. Can. J. Soil Sci. 80: 595–606.

Ruberto, L., R. Dias, B.A. Lo, S.C. Vazquez, E.A. Hernandez and W.P. Mac Cormack. 2009. Influence of nutrients addition and bioaugmentation on the hydrocarbon biodegradation of a chronically contaminated Antarctic soil. J. Appl. Microbiol. 106: 1101–1110.

Salanitro, J.P., P.B. Dorn, M.H. Huesemann, K.O. Moore, I.A. Rhodes, L.M.R. Jackson et al. 1997. Crude oil hydrocarbon bioremediation and soil ecotoxicity assessment. Environ. Sci. Technol. 31: 1769–1776.

Samanta, S.K., O.V. Singh and R.K. Jain. 2002. Polycyclic aromatic hydrocarbons: Environmental pollution and bioremediation. Trends Biotechnol. 20: 243–248.

Sandrin, T.R. and R.M. Maier. 2003. Impact of metals on the biodegradation of organic pollutants. Environ. Health. Perspect. 111: 1093–1101.

Sayara, T., E. Borràs, G. Caminal, M. Sarrà and A. Sánchez. 2011. Bioremediation of PAHs-contaminated soil through composting: influence of bioaugmentation and biostimulation on contaminant biodegradation. Int. Biodeterior. Biodegrad. 65: 859–865.

Schweigert, N., J.L. Acero, U. von Gunten, S. Canonica, A.J.B. Zehnder and R.I.L. Eggen. 2000. DNA degradation by the mixture of copper and catechol is caused by DNA-copperoxo complexes, probably DNA-Cu (I) OOH. Environ. Mol. Mutagen. 36: 5–12.

Schwerdtle, T., F. Ebert, C. Thuy, C. Richter, L.H. Mullenders and A. Hartwig. 2010. Genotoxicity of soluble and particulate cadmium compounds: impact on oxidative DNA damage and nucleotide excision repair. Chem. Res. Toxicol. 23: 432–442.

Segura, A., S. Rodríguez-Conde, C. Ramos and J.L. Ramos. 2009. Bacterial responses and interactions with plants during rhizoremediation. Microb. Biotecnol. 2: 452–464.

Sekelsky, A.M. and G.S. Shreve. 1999. Kinetic model of biosurfactant-enhanced hexadecane biodegradation by *Pseudomonas aeruginosa*. Biotechnol. Bioeng. 63: 401–409.

Seo, J.S., Y.S. Keum and Q.X. Li. 2009. Bacterial degradation of aromatic compounds. Int. J. Environ. Res. Public Health. 6: 278–309.

Shimada, T. and Y. Fujii-Kuriyama. 2004. Metabolic activation of polycyclic aromatic hydrocarbons to carcinogens by cytochromes P450 1A1 and 1B1. Cancer Sci. 95: 1–6.

Shukla, S.K., N. Mangwani, T.S. Rao and S. Das. 2014. Biofilm-mediated bioremediation of polycyclic aromatic hydrocarbons. *In*: Microbial Biodegradation and Bioremediation. 1st edition.

Siciliano, S.D., N. Fortin, A. Mihoc, G. Wisse, S. Labelle, D. Beaumier et al. 2001. Selection of specific endophytic bacterial genotypes by plants in response to soil contamination. Appl. Environ. Microbiol. 67: 2469–2475.

Siciliano, S.D., J.J. Germida, M.K. Banks and C.W. Greer. 2003. Changes in microbial community composition and function during a polyaromatic hydrocarbon phytoremediation field trial. Appl. Environ. Microbiol. 69: 483–489.

Strausz, O.P., M. Torres, E.M. Lown, I. Safarik and J. Murgich. 2006. Equipartitioning of precipitant solubles between the solution phase and precipitated asphaltene in the precipitation of asphaltene. Energy Fuels 20: 2013–2021.

Stroud, J.L., G.I. Paton and K.T. Semple. 2007. Microbe-aliphatic hydrocarbon interactions in soil: implications for biodegradation and bioremediation. J. Appl. Microbiol. 102: 1239–1253.

Sullivan, J.B. and G.R. Krieger. 2001. Environmental sciences: pollutant fate and transport in the environment. *In*: Clinical Environmental Health and Toxic Exposures. 2nd edition. Ed. LWW.

Sutton, N.B., F. Maphosa, J.Á. Morillo, W. Abu Al Soud, A.A. Langenhoff, T. Grotenhuis et al. 2013. Impact of long-term diesel contamination on soil microbial community structure. Appl. Environ. Microbiol. 79: 619–630.

Syed, K., H. Doddapaneni, V. Subramanian, Y.W. Lam and J.S. Yadav. 2010. Genome-to-function characterization of novel fungal P450 monooxygenases oxidizing polycyclic aromatic hydrocarbons (PAHs). Biochem. Biophys. Res. Commun. 399: 492–497.

Syed, K., C. Kattamuri, T.B. Thompson and J.S. Yadav. 2011. Cytochrome b5 reductase–cytochrome b5 as an active P450 redox enzyme system in Phanerochaete chrysosporium:

atypical properties and *in vivo* evidence of electron transfer capability to CYP63A2. Arch. Biochem. Biophys. 509: 26–32.

Syed, K., A. Porollo, Y.W. Lam, P.E. Grimmett and J.S. Yadava. 2013. CYP63A2, a catalytically versatile fungal P450 monooxygenase capable of oxidizing higher-molecular-weight polycyclic aromatic hydrocarbons, alkylphenols, and alkanes. Appl. Environ. Microbiol. 79: 2692–2702.

Tabak, H.H., J.M. Lazorchak, A.P. Khodadoust, J.E. Antia, R. Bagchi and M.T. Suidan. 2003. Studies on bioremediation of polycyclic aromatic hydrocarbon-contaminated sediments: bioavailability, biodegradability, and toxicity issues. Environ. Toxicol. Chem. 22: 473–483.

Tate, P.T., W.S. Shin, J.H. Pardue and W.A. Jackson. 2011. Bioremediation of an experimental oil spill in a coastal Louisiana salt marsh. Water Air Soil Pollut. 223: 1115–1123.

Taylor, E. and D. Reimer. 2008. Oil persistence on beaches in Prince William Sound—a review of SCAT surveys conducted from 1989 to 2002. Mar. Pollut. Bull. 43: 458–474.

Tervahauta, A.I., C. Fortelius, M. Tuomainen, M.L. Akerman, K. Rantalainen, T. Sipilä et al. 2009. Effect of birch (*Betula* spp.) and associated rhizoidal bacteria on the degradation of soil polyaromatic hydrocarbons, PAH-induced changes in birch proteome and bacterial community. Environ. Pollut. 157: 341–346.

Thavamani, P., S. Malik, M. Beer, M. Megharaj and R. Naidu. 2012a. Microbial activity and diversity in long term mixed contaminated soils with respect to polyaromatic hydrocarbons and heavy metals. J. Environ. Manag. 99: 10–17.

Thavamani, P., M. Megharaj, K. Venkateswarlu and R. Naidu. 2012b. Mixed contamination of polyaromatic hydrocarbons and metals at manufactured gas plant (MGP) sites: toxicity and implications to bioremediation. pp. 347–368. *In*: Wong, M.H. [ed.]. Environmental Contamination: Health Risks, Bioavailability and Bioremediation. Taylor & Francis Inc., London,

Tian, L., P. Ma and J.J. Zhong. 2002. Kinetics and key enzyme activities of phenanthrene degradation by *Pseudomonas mendocina*. Process Biochem. 37: 1431–1437.

Tomás-Gallardo, L., H. Gómez-Álvarez, E. Santero and B. Floriano. 2013. Combination of degradation pathways for naphthalene utilization in *Rhodococcus* sp. strain TFB. Microbial Biotechnol. 7: 100–113.

Tyagi, M., M.M.R. da Fonseca and C.C.C.R. de Carvalho. 2011. Bioaugmentation and biostimulation strategies to improve the effectiveness of bioremediation processes. Biodegradation 22: 231–241.

U.S. EPA. 2010. Development of a Relative Potency Factor (RPF) Approach for Polycyclic Aromatic Hydrocarbon (PAH) Mixtures (External Review Draft). U.S. Environmental Protection Agency, Washington, DC, EPA/635/R-08/012A, US.

Ukiwe, L.N., U.U. Egereonu, P.C. Njoku, C.I.A. Nwoko and J.I. Allinor. 2013. Polycyclic aromatic hydrocarbons degradation techniques? A review. Int. J. Chem. http://dx.doi.org/10.5539/ijc. v5n4p43.

Uribe-Alvarez, C., M. Ayala, L. Perezgasga, L. Naranjo, H. Urbina and R. Vazquez-Duhalt. 2011. First evidence of mineralization of petroleum asphaltenes by a strain of *Neosartorya fischeri*. Microb. Biotechnol. 4: 663–672.

Vidonish, J.E., K. Zygourakis, C.A. Masiello, G. Sabadell and Alvarez, P.L.L. 2016. Thermal treatment of hydrocarbon-impacted soils: a review of technology innovation for sustainable remediation. Engineering 2: 426–437.

Wang, Z.Y., D.M. Gao, F.M. Li, J. Zhao, Y.Z. Xin, S. Simkins et al. 2008. Petroleum hydrocarbon degradation potential of soil bacterial native to the Yellow River Delta. Pedosphere 18: 707–716.

Warr, L.N., A. Friese, F. Schwarz, F. Schauer, R.J. Portier, L.M. Basirico et al. 2013. Bioremediating oil spills in nutrient poor ocean water using fertilized clay mineral flakes: some experimental constraints. Biotechnol. Res. Int. 704806.

Wattiau, P. 2002. Microbial aspects in bioremediation of soils polluted by polyaromatic hydrocarbons. *In*: Agathos, S.N. and W. Reineke [eds.]. Biotechnology for the Environment: Strategy and Fundamentals. Springer Science Business Media, B.V.

Wick, L.Y., L. Shi and H. Harms. 2007. Electro-bioremediation of hydrophobic organic soil contaminants: a review of fundamental interactions. Electrochim. Acta 52: 3441–3448.

Widdel, F., K. Knittel and A. Galushko. 2010. Anaerobic hydrocarbon-degrading microorganisms: an overview. pp. 1997–2021. *In*: Timmis, K.N. [ed.]. Handbook of Hydrocarbon and Lipid Microbiology. Berlin, Springer.

Wolfe, M.F., G.J.B. Schwartz, S. Singaram and E.E. Mielbrecht. 1999. Influence of dispersants on the bioavailability and trophic transfer of phenanthrene to algae and rotifers. Aquat. Toxicol. 48: 13–24.

Xu, W., C.P. Wang, H.B. Liu, Z.Y. Zhang and H.W. Sun. 2010. A laboratory feasibility study on a new electrokinetic nutrient injection pattern and bioremediation of phenanthrene in a clayey soil. J. Hazard. Mater. 184: 798–804.

Yanto, D.H.Y. and S. Tachibana. 2013. Biodegradation of petroleum hydrocarbons by a newly isolated *Pestalotiopsis* sp. NG007. Int. Biodeterior. Biodegrad. 85: 438–450.

Yanto, D.H.Y. and S. Tachibana. 2014. Potential of fungal co-culturing for accelerated biodegradation of petroleum hydrocarbons in soil. J. Hazard. Mater. 278: 454–463.

Yrjälä, K., A.K. Keskinen, M.L. Akerman, C. Fortelius and T.P. Sipilä. 2010. The rhizosphere and PAH amendment mediate impacts on functional and structural bacterial diversity in sandy peat soil. Environ. Pollut. 158: 1680–1688.

Yu, K.S.H., A.H.Y. Wong, K.W.Y. Yau, Y.S. Wong and N.F.Y. Tam. 2005. Natural attenuation, biostimulation and bioaugmentation on biodegradation of polycyclic aromatic hydrocarbons (PAHs) in mangrove sediments. Mar. Pollut. Bull. 51: 1071–1077.

Zhang, S., Q. Wang and S. Xie. 2012. Stable isotope probing identifies anthracene degraders under methanogenic conditions. Biodegradation 23: 221–230.

Zhang, X.X., S.P. Cheng, C.J. Zhu and S.L. Sun. 2006. Microbial PAH-degradation in soil: degradation pathways and contributing factors. Pedosphere 16: 555–565.

Zhu, X., A.D. Venosa, M.T. Suidan and K. Lee. 2001. Guidelines for the bioremediation of marine shorelines and freshwater wetlands. U.S. Environmental Protection Agency, Cincinnati.

Zhuang, X., J. Chen, H. Shim and Z. Bai. 2007. New advances in plant growth-promoting rhizobacteria for bioremediation. Environ. Int. 33: 406–413.

11

Identification of a Novel Gene Through the Metagenomic Approach to Degrade the Targeted Pollutant

Suriya Jayaraman,[1,*] *Suganya Thangaiyan,*[1]
Kannan Mani,[1] *Kayalvizhi Nagarajan*[2] and
Krishnan Muthukalingan[1,3,*]

INTRODUCTION

Contamination of water, soils, and sediments by xenobiotics, which affects the human and environmental health, is a major concern. Bioremediation is an eco-friendly green technology used for the cleaning up of environmental pollutants to safe levels. Culture dependent studies shown that soil microorganisms are a large reservoir of innumerable essential enzymes (Kakirde et al. 2010). Recent estimates proven that one gram of soil may contain millions of different kinds of microbes (Wommack et al. 2008). However, 99% of these microbes cannot be cultured using conventional methods (Curtis and Sloan 2005). To overcome this bottleneck in cultivation

[1] Department of Environmental Biotechnology, School of Environmental Sciences, Bharathidasan University, Tiruchirappalli, Tamil Nadu 620024, India.
[2] Department of Zoology, School of Life Sciences, Periyar University, Salem, Tamil Nadu, India.
[3] Professor, Department of Biochemistry, Central University of Rajasthan, N.H. 8, Bandarsindri, Kishangarh-305 817, Ajmer (Dt), Rajasthan, India.
* Corresponding authors: Jsuriyamb08@gmail.com; profmkrish@gmail.com

approaches, many molecular methods have been developed. A new approach involving the total isolation of the microbial genome directly from natural environments and the cloning of this genome directly into cultivable bacterium is developed (Handelsman 2004). This technique is known as the "metagenomic approach" which excludes the culturing of microbial species and enrichment of beneficial strains under optimized conditions (Zengler et al. 2002). The metagenomic library has been constructed for small as well as large inserts, plasmids, fosmids, cosmids (utilized for small inserts), and bacterial artificial chromosomes (BACs) used for large inserts. By applying this approach, innumerable industrially important genes have been identified. Nature has its own ways to resolve imbalances in its environment. It uses microorganisms as one of the best tools to eliminate toxic pollutants. Bioremediation is termed as the process of using microorganisms to eliminate pollutants. Metagenomic is one of the promising techniques for bioremediation of pollutants. Extraction of DNA and screening of metagenomes from the polluted sites are two important steps in a metagenomic study. This chapter highlights the discovery of a novel gene through the metagenomic approach applied for the bioremediation of different contaminated environments.

History

Woese and Fox (1977) identified the 16S rRNA gene as a molecular marker to assess microbial diversity. This finding along with the Sanger automated sequencing increased the study and classification of microbial communities (Sanger et al. 1977). Staley and Konopka (1985) were the first to report the 'great plate count anomaly'. After some decades, advances in molecular techniques such as denaturing gradient gel electrophoresis and *Temperature Gradient Gel Electrophoresis* (DGGE and TGGE), polymerase chain reaction (PCR), rRNA genes cloning and sequencing, restriction-fragment length polymorphism, and fluorescent *in situ* hybridization (FISH), and terminal restriction-fragment length polymorphism (T-RFLP) have been applied to study the microbial diversity (Escobar-Zepeda et al. 2015).

In the early nineties, Schmidt et al. (1991) used 16S rRNA for the phylogenetic analysis of microbial communities. The term metagenomics was coined by Handelsman et al. for analysing a collection of similar but not identical items a decade later (Handelsman et al. 1998) and was instantly attracted by the researchers. It has unveiled new horizons in the development of biotechnology by exploiting uncultivated microorganisms. Later, the term was also used to refer the functional analysis of mixed environmental DNA for a specific activity. Tyson and Venter group explored complex microbial community using whole genome shotgun sequencing method in 2004 investigated the simple microbial community growing on the surface of an acid mine drainage where as Venter et al. (2004) reported the more complex community of the Sargasso Sea. These two studies paved the way for future metagenomic projects. At the same time, higher throughput sequencing techniques have been emerging and drawing the attention of scientific community due to their low cost, robustness, and sample size (Edwards et al. 2006). These new sequencing approaches paved the way for the findings of a new subfield of metagenomics, termed meta transcriptomics and metaproteomics. These findings initiated the Human Microbiome and Gut Microbiome Projects. The ultimate goal of these projects were to understand the human microbiome, which could help to advance the human health care system.

Mining of Metagenomic DNA from Polluted Samples

The isolation of metagenomic DNA from environmental samples is broadly classified into direct and indirect extraction methods. In the direct DNA isolation procedure, DNA is separated from the matrix after lysis of the cells (Ogram et al. 1987) whereas in the indirect approach, DNA is extracted after the separation of the cells from soil followed by cell lysis (Holben et al. 1988). The construction and screening of genomic libraries from the environment for novel genes is illustrated in Fig. 1.

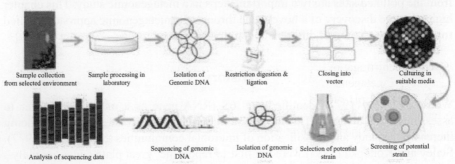

Fig. 1: A graphical picture on construction and screening of the metagenomic library form the environment for a novel gene.

Cell lysis

Cell lysis is an important step in the extraction of metagenomic DNA and it involves the breaking the cell wall and membranes to release the DNA. Enzymatic or chemical lysis is relatively mild and does not entirely penetrate sediment samples. In contrast to this approach, mechanical disruption produces uniform cell disruption by dispersing the samples for proper penetration of the lysis buffer. Therefore, the mechanical distraption is more efficient than enzymatic and chemical lysis. Ultrasonication, thermal shocks, microwave heating, bead-beating, and bead-mill homogenization are commonly used methods in mechanical distraption. In the thermal shock method, the sample suspension is subjected to continuous freezing and thawing. Based on the sample type, the incubation time, temperature, and the number of freeze–thaw cycles are determined (Porteous et al. 1997). The bacterial cells are effectively released from the samples in the ultrasonication treatment and the duration of sonication and power treatment can be varied for samples (Ramsay 1984). Sodium dodecylsulphate (SDS) has been widely used as a detergent for cell lysis (Roose-Amsaleg et al. 2001).

For lysis of Gram-positive cells and spores, microwave heating is mostly used due its high efficiency (Orsini and Romano-Spica 2001). For the complete lysis of *Streptomyces* spores, Picard et al. (1992) used ultrasonication and microwave heating followed by thermal shocks. The major drawback of mechanical distruption is the shearing of DNA. To estimate the efficiency of DNA recovery, many researchers used viable counts of bacterial cells or direct microscopic techniques before and after DNA extraction method. Tsai and Olson (1991) and Zhou et al. (1996) reported 90% cell lysis efficiencies. Comparison of different lysis methods showed that the efficiency is varied depending upon the methods we used and sample properties. Soil clay content is negatively correlated with the lysis efficiency (Zhou et al. 1996). Clay content in sample binds DNA and interferes with the extraction of DNA by precipitate DNA in the pellet.

DNA extraction and purification

Before the precipitation of metagenomic DNA, mostly deproteinisation in organic solvents, that is, chloroform–isoamyl alcohol, phenol–chloroform, and phenol have been used (Tsai and Olson 1991). To precipitate the protein from DNA extraction suspension, saturated salt solutions such as sodium acetate (Holben et al. 1988), sodium chloride (Selenska and Klingmüller 1991), potassium acetate (Smalla et al. 1993), and ammonium acetate (Xia et al. 1995) are applied. This step is followed by low speed centrifugation and then the nucleic acids are obtained from the supernatant. Recovery of metagenomic DNA highly depends on the pH of the extraction buffer. To optimize the pH of the extraction buffer, Frostegard et al. (1999) used twenty different soil samples and they found that the maximum recovery was obtained at pH 9.0–10.0. On the other hand, huge amounts of humic acid were released at pH 10.0 than at pH 9.0 and therefore, pH 9.0 has been recommended as the optimum pH for the DNA extraction buffer.

Usually, DNA is precipitated with either ethanol or isopropanol. Polyethyleneglycol (PEG) also utilized for the precipitation of DNA. Alcoholic precipitation increases the humic acid co-precipitaiton whereas PEG greatly lowers the co-precipitation of humic substances (La Montagne et al. 2002). Precipitation of DNA using 5% PEG increases the recovery of DNA significantly without affecting PCR and hence 5% PEG has been proposed for the precipitation of DNA (Arbeli and Fuentes 2007). Generally, DNA concentration and purity was determined using A_{260} value and $A_{260/280}$ ratio, respectively (Sambrook et al. 1989). However, quantification of the DNA is mostly interfered with by the humic acid content. Many researches revealed that the A_{260} value gives the amount of humic substances (Jackson et al. 1997). To estimate the humic acid substance, A_{230} values are usually used. Humic acid absorbs at 230 nm, while DNA absorb at 260 nm. Therefore, the absorbance ratio at 260/230 nm (DNA/humic acid) and 260/280 nm (DNA/protein) are mostly used to determine the purity of the soil metagenomic DNA (Roose-Amsaleg et al. 2001). Polyvinyl-polypyrrolidone (PVPP) or Hexadecyl trimethyl ammonium bromide (CTAB) addition to extraction-buffer before lysis repressed the co-extraction of humic substances and increased the purity of DNA. The usse of skim milk (Takada-Hoshino and Matsumoto 2004) or sodium ascorbate (Holben et al. 1988) also decreased the co-precipitation of humic acids. Densitometric analysis of ethidium bromide (EtBr) stained agarose gel is another method for the estimation of DNA concentration. Pico Green fluorescent dye helps in providing the effective quantification of DNA (Sandaa et al. 1998).

Commercial kits are also used to extract and purify the DNA. It is very easy to use and also yield contaminants free DNA. But the outcome of the different researches for the evaluation of different commercial kits is inconsistent. Sagova-Mareckova et al. (2008) recommended that either sample pretreatment with $CaCO_3$ or purification of DNA with $CaCl_2$ was more potent than commercial kits.

Metagenomic Library Construction

After metagenomic DNA extraction and purification, the library construction followed. It constitutes the following steps such as the generation of suitable sized DNA fragments by restriction digestion using restriction enzymes, ligation of restricted DNA into a suitable vector, and transformation of the ligated product into a suitable host. Many researchers used metagenomic approach to recover their novel gene of interest, but the vector and host have been varied depending upon their gene of interest. Metagenomic libraries are broadly classified into two groups based on their average insert size: small insert libraries and large insert libraries. Plasmids are usually used to construct small insert libraries (less than 15 kb) whereas fosmid, cosmid (both up to 40 kb), and Bacterial Artifical Chromosome (BAC) (more than 40 kb) vectors are utilized for constructing large insert libraries (Daniel 2005).

The restriction digestion of metagenomic DNA to get DNA fragments is a significant problem for constructing metagenomic libraries because the force full extraction methods applied to obtain high recovery of DNA from natural samples result in massive DNA shearing. Highly sheared DNA produces a significant loss of total metagenomic DNA which excludes the library construction using sticky ends. T–A or Blunt-end ligation is mostly used to construct a metagenomic library of sheared DNA fragments (Wilkinson et al. 2002). To construct the metagenomic libraries the commonly used vectors are Plasmid, Cosmid, and BAC. The major advantage of plasmid vectors includes: (1) Screening of weakly expressed genes owing to the production of high copy number of positive clones; (2) It is well suited to express foreign genes from vector promoter; (3) It also allows for the cloning of sheared DNA and contaminated DNA with sample matrix; and (4) It is very simple technique. The cons of the plasmid vectors are (1) the insert size should be small, (2) screening process is tedious; we have to screen large number of colonies to get a gene of interest, and (3) it is unsuited to clone large gene clusters which are involved in the functional activities or metabolic pathways (Daniel 2005). Cosmid, Fosmid, and BAC vectors are efficiently used for cloning. If we clone large fragments of DNA, it allows entire functional operons to be cloned for retrieving whole metabolic pathways of the gene of interest. This method is successfully applied for the isolation of various multigenic pathways (Brady et al. 2001; Courtois et al. 2003). The screening process is very simple; only small number of colonies has to be screened to get gene of interest. It is well suited for the partial genomic characterization of uncultured soil microbes. Fosmid vectors have been successfully applied to clone and maintain large sized inserts. Phage-vectors are used to screen the surface displayed expression products by affinity selection. This is an efficient approach to get rare gene of interest (Crameri and Suter 1993). The main drawback of large insert libraries include: (1). Technically difficulty; (2). Requires high purity DNA for library construction; and (3). It produces only low copy numbers of clones (Daniel 2005).

Innumerable successful cloning researches have been carried out in the *E. coli* host owing to its ability to grow well on chemically defined media, its rapid growth rate, transformability, genetic manipulation, and availability of numerous genetic tools (Martin 2016). The main drawback of *E. coli* as a host is the generation of a low copy number of positive clones recovered in the single round of screening. Recent studies revealed that it is not possible to obtain translational fusion products because it needs a high number of clones for screening (Gabor et al. 2004). Due to this limitation, an alternative host strains such as *Bacillus*, *Pseudomonas*, or *Streptomyces* genera have been used (Lorenz and Eck 2005; Aakvik et al. 2009). There are also few archaeal genera (*Methanococcus, Sulfolobus, Pyrococcus, Thermococcus*), which can be employed in designing the stable host-vector expression system. Yeasts such as *Saccharomyces cerevisiae* and *Pichiapastoris* are widely used for heterologous gene expression (Liu et al. 2013). The biggest disadvantage of using yeast hosts in functional screening is the poor recognition of heterologous promoters, the multiple enzymatic activities displayed by yeast, and low transformation yields (Martin 2016).

The choice of a vector system depends on the quality of the isolated soil DNA, size of the DNA, required copy number of the vector, the host and the screening strategy that will be used, all of which depend on the aim of the study (Daniel 2005). The success of metagenomic library construction mainly depends on the following factors: sample composition; collection and storage of the sample; DNA extraction approach; the vector and host used for cloning, and the screening method. Metagenomic libraries are unsuitable for the recovery of the eukaryotic gene because of the presence of introns. To overcome this bottleneck, cDNA libraries are developed (Starkey et al. 1998). However, cDNA libraries cannot be used to screen the non expressed genes.

Screening of Metagenomes for Pollutant Degradation

Screening of metagenomic libraries involves the two main approaches such as function-based and sequence-based screening. The functional activity of gene of interest, type of library constructed, resources available for characterization of library, and time determine the type of the screening strategy we have to choose. Both strategies have pros and cons.

Sequence based screening

In sequence-based screening metagenomic DNA has been directly sequenced, either with or without cloning before sequencing and then the sequences are subjected to bioinformatic analyses (Sleator et al. 2008). The booming of next-generation sequencing (NGS) technologies has paved the way for the introduction of numerous strategies for sequencing with different capabilities and cost. Despite NGS technologies being a good approach for metagenomic DNA sequencing, the short read length is a main disadvantage to assemble metagenomic sequences effectively. On the other hand, $10 is the actual cost per megabase for 454 sequencing followed by $5 for Solexa and only $2 for SOLiD (Rothberg and Leamon 2008). Due to the advancements in the NGS technologies, it reduces the cost and also increases the read length. Analysis of sequence data is time consuming and requires more resources. To overcome this limitation, a bioinformatics strategy has been developed for analyzing metagenomic data rapidly.

Tyson et al. (2004) studied the five dominant members of an acid mine drainage (AMD) by metagenomics. For retrieving novel oxygenase sequences, Jadeja et al. (2014) sequenced the metagenome of the activated biomass from a common effluent treatment plant. The SIP method coupled with shotgun sequencing was used for identification of the naphthalene degraders (Wang et al. 2012a). For the degradation of polyaromatic pollutants, 1,200 bacterial laccase-like genes were identified by Ausec et al. (2011). Zaprasis et al. (2009) utilized quantitative PCR assays to know the abundance and diversity of 2, 4-D/α-ketoglutarate and Fe^{2+} dependent dioxygenases genes responsible for the degradation of phenoxyalkanoic acid herbicides. Different environmental samples such as soils, the sea, and microcosms contaminated with nitrates, have been analysed for denitrifying enzyme such as ammonia monooxygenases (Nogales et al. 2002; Treusch et al. 2005; Bartossek et al. 2010).

Lu et al. (2012) used Geo Chip 4.0 to retrieve the functional gene involved in the hydrocarbon (HC) break down in a deep sea water sample contaminated by an oil spill, in comparison with a non-contaminated sample. The sea metagenomic library was sequenced to discover the bacterial laccases using PCR-screening and also the activity in the breakdown of azo-dyes, syringaldazine, 2, 6-dimethoxyphenol, 2,2'-Azino-bis(3-ethylbenzothiazoline-6-sulfonic acid) (ABTS), and catechol was evaluated experimentally (Fang et al. 2010, 2012). The SIP strategy was applied to find the aromatic dioxygenases gene responsible for the degradation of polychlorinated biphenyl (PCB) (Sul et al. 2009). A gene targeted metagenomic approach was used by Iwai et al. (2009) by combining PCR screening with pyro sequencing of amplicons to produce new primers designed from other conserved regions which increase the structural diversity of the targeted protein families. This approach has enabled the discovery of thousands of full length dioxygenase encoding genes, assigned to dozens of novel clusters containing still unknown sequences, from the metagenomes of soils contaminated by polychlorinated biphenyl (Iwai et al. 2009) and 3-chlorobenzoate (Morimoto and Fujii 2009).

Boubakri et al. (2006) used metagenomic shuffling approach to construct a library of more than one thousand clones and screened on a lindane containing solid medium. They retrieved 23 new genes from this library encoding lindane degrading enzymes. Vergne-Vaxelaire et al. (2013) used thirty four sequences of nitrilases as reference set which were confirmed experimentally to choose 290 nitrilase sequences from genomes of their strain collection as well as metagenome of a wastewater treatment plant.

Function based screening

Activity-based screening has successfully been used by many scientists to efficiently screen novel enzymes and also to obtain biochemical data regarding substrate specificity. The following three methods have been used for the functional screening of a metagenomic library: (1) Colored substrates are used to detect the coloration or discoloration; (2) Insoluble substrates are used to detect the reduction in the opacity of the reaction medium (Shah et al. 2008); and (3) Foreign genes are used in an auxotrophic host for heterologous complementation to allow the host growth on selective culture media (Xing et al. 2012). Positive clones have been selected by their resistance towards an antibiotic or heavy metal by excluding the clones that are unable to grow in the presence of these selective compounds (Parsley et al. 2010). However, the screening of the clones with pollutants as substrates can be challenging due to their toxic effect either to host or operator. To overcome this bottleneck, non-toxic structural analogues of pollutants are used for screening and the toxic pollutants are only utilized for secondary screening assays. GC and/or MS and HPLC are used for an analysis of reaction substrates and products.

Uchiyama et al. (2005) developed an alternative strategy, Substrate-Induced Gene Expression screening (SIGEX) for the screening of novel gene. It is based on the principle that the expression of the catabolic gene is usually accelerated by catabolic enzyme's metabolite or a specific substrate and is generally controlled by regulatory genes situated near to these genes. In SIGEX, metagenomic DNA fragments are combined with a green fluorescent protein (*gfp*) encoding reporter gene on an operon-trap vector and accelerated by a specific substrate. For the high throughput GFP clone selection, SIGEX is fused with fluorescent-activated cell sorting (FACS). Moreover, this approach excludes the incorporation of self-ligated plasmids containing clones. SIGEX is a potential approach for novel catabolic substrate-induced genes identification. However, this strategy has some limitations. Uchiyama and Miyazaki (2010) worked on SIGEX protocol to expand its abilities and developed a Product-Induced Gene Expression (PIGEX) approach. In this strategy, a transcriptional activator is placed upstream of a *gfp* gene insert which is sensitive to the desired reaction product. Appropriate substrate is added to the clone plate to select the clone of interest, if the clone has the activity of interest; transcription of the chosen transcriptional regulator is activated by the product of this reaction which in turn causes the clone to fluoresce, allowing easy detection of positive clones.

Pooja et al. (2015) identified active clones from a cow dung-derived metagenomic library based on their fluoresced in response to maltose substrate and utilized those positive clones for the production of a periplasmic α-amylase using PIGEX. *Ralstonia metallidurans* was used to express two novel compounds via functional screening from a soil metagenomic library by Craig et al. (2009). Biver et al. (2013a) evaluated the use of an *E. coli-Bacillus subtilis* shuttle vector to functionally screen for an estsoil-derived metagenomic library for antimicrobial activity and identified the novel antimicrobial agent which is active against *Bacillus cereus*. The DNA fragment revealed such activity was active only in the *B. subtilis* and didn't show any activity in *E. coli*. A series of steps such as transcription of the entire gene, translation of its mRNA, correct protein folding, and active protein secretion from the host must occur effectively to obtain an active clone. Suitable and efficient screening approaches must be applied for detecting the gene of interest from the metagenomic library. However, the probability of identifying a positive clone based on its desired activity is low; high-throughput screening (HTS) approaches may improve the chances of getting an active clone by allowing higher numbers of clones to be screened simultaneously.

Application of Metagenomic Approach

Native microbes generally expresses the gene of interest that has been involved in the biosynthetic or biodegrative pathways. The culture reliant method is not suitable for the identification of these types of genes. Researchers found an alternative approach, that is, a metagenomic approach to evaluate the functional and phylogentic diversity of microorganisms.

Xenobiotic degradation

Xenobiotics are foreign chemical or excessive substance found in an organism. The anthropogenic activities become as a critical factor for xenobiotic presence in our environment that threads ecosystem. Biotic and/or abiotic reactions have been involved in the degradation of xenobiotics. Metagenomics are applied to detect novel catabolic pathways for degradation of the xenobiotic compounds. In this chapter, we have highlighted the crucial role of some of the enzymes involved in the biodegradation of pollutants.

Oxidoreductases and Oxygenases

For breakdown of aromatic compounds, oxygenases are mostly required. The pollutants are degraded by bacteria using typical aerobic degradation pathway by two steps: (1) phenol hydroxylases are used for ring hydroxylation of adjacent carbon atoms and (2) catechol 1,2- or 2,3-dioxygenases are utilized by bacteria for ring cleavage of the resulting catecholic intermediates (Silva et al. 2013).

Ono et al. (2007) and Nagayama et al. (2015) used *Pseudomonas putida* to express naphthalene-degrading dioxygenases, phenol, and catechol degradating enzymes. Generally oxygenases enzymes are obtained from highly polluted environments like oil, PCBs, and PAH polluted soils (Suenaga et al. 2007; Sharma et al. 2012; Silva et al. 2013), or activated sludge collected from wastewater treatment facilities in the coke, petroleum, and pharmaceutical industries (Ono et al. 2007; Brennerova et al. 2009; Lu et al. 2011). For degradation of aromatic compounds, Silva et al. (2013) and Nagayama et al. (2015) used artificial enrichment approaches for bacteria. In addition, clone plates are sprayed with substrate solutions, like catechol (Brennerova et al. 2009), 3, 5-dichlorocatechol (Lu et al. 2011), phenol (Sharma et al. 2012), or 2, 3-dihydroxybiphenyl (Wang et al. 2015). Suenaga et al. (2007) used the same approach as used in the primary screening to evaluate the activities of all the positive clones by monitoring the appearance of ring cleavage products. After incubation of positive clones with substrates, GC-MS was successfully used for substrate quantification in a discrimination assay (Nagayama et al. 2015). After evaluating the target enzyme activity, the sequence of the enzyme was obtained by direct sequencing (Kimura et al. 2010). PCR based screening using a set of specific primers for eight groups of extradiol dioxygenases was used by Brennerova et al. (2009). Lu et al. (2011) obtained a novel enzyme involved in the biodegradation of fifteen substrates which retrieved from the herbicide 2, 4-dichlorophenoxyacetic acid. Wang et al. (2015) produced transgenic alfalfa plants with 2, 3-dihydroxybiphenyl-1, 2-dioxygenase acting on catechol, 2, 3-dihydroxybiphenyl, 4-methylcatechol, 3-methylcatechol, 4-chlorocatechol and were used to remove PCBs from their culture medium.

Laccases

Laccases have received much attention from scientists owing to their ability to oxidize both phenolic as well as non-phenolic compounds and also polycyclic aromatic hydrocarbons like pesticide alkenes, industrial dyes, and recalcitrant biopolymers (Ausec et al. 2011).

Ferrer et al. (2005a) and Ye et al. (2010) screened a phagemid *E. coli* library constructed from bovine rumen and mangrove metagenome for laccase enzyme. Two other laccases were characterized using sequence homology from *Bacteroides thetaiotaomicron, E. coli* (Beloqui et al. 2006). Strachan et al. (2014) obtained a pseudolaccase from coal bed metagenomes and it was screened for lignin catabolic activities.

Alkane degrading enzymes

De Vasconcellos et al. (2010) constructed a metagenomic library of oil samples collected from a petroleum reservoir. They screened the libraries on agar plates supplemented with hexadecane. Among 72 clones, only five clones revealed hexadecane degradation potential in GC-MS analysis. Dellagnezze et al. (2014) and Sierra-García et al. (2014) were further characterized when these five positive clones in GC-MS and GC were combined with flame ionisation detection to evaluate their biodegradation ability on various aromatic compounds, n-alkanes, and isoprenoids. Among the tested five clones, two clones displayed the efficient degradation of both linear and branched n-alkanes.

Hydrolases–Esterases

It is a hydrolase enzyme which catalyses the cleavage of ester bonds to produce acid and alcohol. Esterases are the key enzymes in the detoxification of herbicides and pesticides. Fan et al. (2012) collected top soil samples from 'the turban basin' in China where the surface temperature is more than 82°C and constructed the metagenomic library to retrieve the thermo tolerant esterase enzyme. To obtain halo and psychro tolerant esterases, Tchigvintsev et al. (2014) constructed a metagenomic library from the Mediterranean Sea contaminated with oil and they obtained five cold and salt tolerant carboxylesterases clones.

Li et al. (2008) and Fan et al. (2012) expressed thermotolerant esterases in *E. coli*, which are efficiently involved in the degradation of malathion and other pyrethroids pesticides. In contrast, Kambiranda et al. (2009) selected *Pichia pastoris* to express esterases after primary screening in *E. coli* owing to the antibacterial activities of the transformation product of chlorpyrifos. Kang et al. (2011) obtained a polyspecific enzyme which uses both polyurethane and diethylene glycol adipate as sole substrate. Mayumi et al. (2008) identified three new esterases after primary screening and its activity was further characterized based on other aliphatic polyesters by measuring changes in the emulsion turbidity in a liquid medium. Sulaiman et al. (2012) purified new cutinase enzyme and found their degradation potential on polyethylene terephthalate (PET) and the poly (ε-caprolactone) (PCL) by measuring the weight loss of PET and PCL films. Jiao et al. (2013) constructed a metagenomic library from biofilms of dibutyl phthalate (DBP) from a wastewater treatment and found that the positive clone is useful in the bioremediation of phthalates in cold environment.

Nitrilases

Nitrilases hydrolysed the nitriles, components of plastics, polymers, and herbicides, into carboxylic acids and ammonia (Pace and Brenner 2001). Robertson et al. (2004) constructed the metagenomic library from terrestrial and aquatic samples collected from around the world. Forest soils, oil-contaminated soil, and wastewater treatment from a refinery samples were collected by Bayer et al. (2011) to construct a metagenomic library. In both studies, *E. coli* was used as cloning host and the screening was done in a liquid medium supplemented with nitriles as sole nitrogen source. Enantio selective and region selective specificity of around 140 nitrilases were characterized after sequencing and sub cloning. The conversion of nitriles produced synthons which is used in the production of fine chemicals and pharmaceuticals. This is the main goal of the metagenomic approach; however, this strategy is useful for the discovery of enzymes involved in the catabolism of xenobiotic derived nitriles.

Tannase

Gallic acid and glucose are produced by the hydrolysis of the ester bonds in tannins by tannase enzyme. It is also utilized to remediate tannin-polluting industrial effluents and agricultural wastes. Gallic acid is an important substrate to produce propyl gallate and antibacterial drugs. Yao et al. (2011) constructed the metagenomic library from cotton field soil samples with 92,000 clones. Among 92,000 clones, only one Tan410 positive clone with tannase activity is efficiently utilized for biotechnological applications. Yao et al. (2014) immobilized these industrially important enzymes in various supports such as mesoporous silica SBA-15, chitosan, calcium alginate, and amberlite IRC 50. Among various supports used, entrapment in calcium alginate beads gives better results. The optimum pH and temperature of the immobilized Tan410 was 7.0 and 45°C whereas a free enzyme requires pH 6.4 and 30°C for its optimal activity. Immobilized enzyme retained 90% of its activity after storage for 30 days.

Industrial enzymes production

The metagenomes of uncultured microbial communities are rich sources for novel biocatalysts and they have been potentially utilized for many industrial processes. Cellulase was the first enzyme obtained via metagenomic strategy from a bioreactor "zoolibrary" (Healy et al. 1995). Palackal et al. (2007) constructed the rumen metagenomic library for retrieving multifunctional glycosyl hydrolase. Many commercial enzymes such as agarases (Voget et al. 2003), pectinolyticlyases (Solbak et al. 2005), and lipolytic enzymes such as lipases and esterases (Lee et al. 2004; Ferrer et al. 2005a) were obtained by applying metagenomics. An alkaline polluted soil sample was sieved to discover a novel β-glucosidase (Jiang et al. 2009). Unique cellulase and xylanase enzymes were discovered from bacterial flora of the hindgut paunch of *Nasutitermes* sp. and from moths using metagenomic strategy (Brennan et al. 2004; Warnecke et al. 2007). Nacke et al. (2012) and Yeh et al. (2013) isolated and characterized novel endoglucanases, cellulases and hemi-cellulases from grassland soil and rice straw compost metagenomic libraries, respectively. Alvarez et al. (2013) constructed the metagenomic library from sugarcane soil metagenome to retrieve novel cellulase. A novel bacterial lipase enzyme was isolated and characterized by Lee et al. (2006) from tidal flat sediments. Gong et al. (2013) mined cow rumen derived metagenomic libraries to get cellulase free GH10 family xylanase. Niehaus et al. (2011) and Neveu et al. (2011) isolated and characterized serine proteases from metagenomic libraries. Sharma et al. (2010) constructed a soil metagenomic library derived from Northwestern Himalayas to discover a novel amylase. Rabausch et al. (2013) identified flavonoid-modifying enzymes from a metagenomic library based on unique thin-layer chromatography (TLC) based screening method. Cheng et al. (2012a) constructed the metagenomic library from Chinese Holstein cow rumen and obtained feruloyl esterase enzyme. Cheng et al. (2012b) retrieved xylanase from the same metagenomic library which is mutually worked with the feruloyl esterase to release ferulic acid and xylooligosaccharides from wheat straw. Walter et al. (2005) mined the metagenomic library of large bowel microbiota of mouse for beta glucanase and they obtained three clones with beta-glucanase activity. Intake of glucans results in many health benefits in humans (Abumweis et al. 2010) but humans and monogastric animals are unable to break down glucans so their hydrolysis depends on bacterial fermentation. To improve the digestion of barley-based feed diets by poultry livestock, ß-glucanases are utilize by the feed industry (Von-Wettstein et al. 2000).

Extremozyme

Extremozymes have been potentially utilized for industrial applications due to their stability and activity in extreme environments. Bhat et al. (2013) obtained cold-active, acidic endocellulase from Ladakh soil, while Peng et al. (2014) obtained milk fat flavor producing alkaline stable lipase from a marine sediments derived metagenomic library. The metagenomic approach was successfully utilized by Hardeman and Sjoling (2007) and Selvin et al. (2012) to isolate low temperature active and halotolerant lipases from marine sediment and marine sponge metagenomic libraries, respectively. Various metagenomic libraries constructed by Ngo et al. (2013), Fu et al. (2013), Chow et al. (2012), and Glogauer et al. (2011) revealed thermotolerant, alkali stable, organic solvent, and halotolerant lipase enzymes for many industrial applications. Verma et al. (2013) isolated a noevel xylanase which is thermostable and alkali tolerant from soil metagenome. Oxidant stable alkaline serine protease was discovered by Biver et al. (2013b) from forest-soil metagenomic library. Pushpam et al. (2011) retrieved alkaline serine protease from the metagenome of goat skin surface. Singh et al. (2012) acquired a thermostable and thermoactive pectinase from soil derived metagenomic library. A metagenome-derived alkaline pectate lyase was accessed by Wang et al. (2013b). Vidya et al. (2011) identified a calcium-dependant and thermostable amylase from a soil metagenome. Vester et al. (2014) constructed the metagenomic libraries to obtain cold-active ß-galactosidase and suggested it can be potentially utilized by the dairy industry.

Activity based screening of non-extreme environment for cellulases displayed eight cellulase activity exhibited clones; among these eight clones only one active clone was purified and characterised (Voget et al. 2006). Despite the face, the clone was isolated from the non-extreme environment, it was stable and active at pH 9.0, 40°C, and 3M NaCl. Cellulose, pullullan, lipid as well as ethylbenzene, toluene, and o-xylene active clones were obtained by He et al. (2013) from deep-sea hydrothermal vent chimney metagenomes. Esterases EstA3 and EstCE1 were retrieved from drinking water and soil metagenome, respectively. Both esterases exhibited remarkable activity over a broad range of pH, temperature, and also a stable in the presence of organic solvents. These characteristic features make them highly suitable for biotechnological applications (Elend et al. 2006). A thermostable lipase was isolated and characterized by Faoro et al. (2012) from a metagenomic library of Brazilian Atlantic forest soil. Kang et al. (2011) isolated, purified, and characterized thermostable esterase from metagenomic library constructed from compost.

Bioactive compounds and antibiotics production

Many secondary metabolites such as antibiotics have been discovered from uncultured microorganisms. Macneil et al. (2001) and Lim et al. (2005) obtained many antimicrobial activity expressing metagenomic clones, in particular indirubin, expressing active clones from a soil and forest soil derived metagenomic library, respectively which, was utilized for human chronic myelocyticleukemia treatment (Marko et al. 2001). Scanlon et al. (2014) constructed the metagenomic library from native staphylococcal strains and co-cultured these clones with the bacterial pathogen *Staphylococcus aureus* for detecting antibiotic activity using a fluorescent viability assay. They retrieved six active clones revealed a lysostaph in activity from *Staphylococcussi mulans* against *S. aureus*. Iqbal et al. (2014) used *Ralstonia metallidurans* host for constructing a metagenomic library from Arizona soil and they identified six active clones exhibited activity against *Bacillus subtilis*. *Streptomyces lividans* was used as the host for the expression of antibiotic terragine with anti-Mycobacterium activity from metagenomic clones (Wang et al. 2000). In another study carried out by King et al. (2009) revealed the use of *S. lividans* and *S. albus* hosts for the production of erdacin, a novel polyketide. Craig et al. (2010) compared six different proteobacteria as hosts for the soil-metagenomic library. Each host expressing the activity of interest was functionally screened for pigment production as well as for antimicrobial activity. Triarylcation antibiotics turbomycin A and B (Gillespie et al. 2002), long-chain N-acyltyrosine antibiotics (Brady et al. 2004), and antifungal agents (Chung et al. 2008) were discovered from the metagenomic clones.

Lakhdari et al. (2010) constructed the metagenomic library from human fecal microbiota of Crohn's Disease (CD) patients and it obtained a positive clone with NFkB modulatory activity. Activity based metagenomics were also used to detect novel resistance mechanisms in bacteria which allow one to predict the possible routes of resistance to antibiotic therapies that could emerge. Donato et al. (2010) constructed the metagenomic library from an apple orchard soil treated with streptomycin for mining various novel resistance genes. They successfully retrieved genes encoding a multi-drug efflux pump, aminoglycoside acetyltransferases, β-lactamases, and a unique bifunctional protein has a fusion of a β-lactamase and a sigma factor. Tao et al. (2012) used functional metagenomics to identify a chloramphenicol and florfenicol resistant clones from an alluvial soil-derived metagenomic library and they identified active clones which could potentially hydrolyse chloramphenicol and florfenicol. Berman and Riley (2013) identified the positive clones which conferred resistance to trimethoprim-sulfamethoxazole, ciprofloxacin, ampicillin, aztreonam, and trimethoprim by mining spinach-derived metagenomics. Devirgiliis et al. (2014) also obtained kanamycin and ampicillin resistance clones from a Mozzarella di Bufala Campana Italian cheese derived metagenomic library.

Fouhy et al. (2014) constructed the metagenomic library from fecal samples of six-month old infants who had not been already exposed to antibiotics to determine the resistance to aminoglycoside and β-lactam antibiotics and they obtained one hundred ampicillin resistant clones and gentamicin resistant clones. Clemente et al. (2015) obtained samples from 34 Yanomami individuals of South America to construct the metagenomic library for mining antibiotics resistance.

Biofuel production

Lignocellulosic biomasses are mainly used for the production of bioethanol. Henceforth, many researchers focused on the discovery of lignocelluloses degrading enzymes for bioethanol production. Chang et al. (2011), Cheng et al. (2012a), and Jeong et al. (2012) constructed the metagenomic libraries from various environmental samples to retrieve novel xylanase to produce bioethanol using agricultural wastes as substrate. Mori et al. (2014) retrieved four unique endoglucanases and one xylanase from the soil derived metagenomic expression libraries. Nacke et al. (2012) constructed grassland soil derived three metagenomic libraries to obtain halo tolerant endoglucanase. Berlemont et al. (2011) used activity based screening for the identification of thermostable endoglucanase.

Sae-Lee and Boonmee (2014) identified a novel enzyme glycosyl hydrolase displayed both endocellulase and endoxylanase activities in a farm compost metagenomic library. Allgaier et al. (2010) incubated switch grass compost at high temperatures to discover glycoside hydrolases by sequence-based analysis. Activity based screening approaches was used for a novel endoglucanase discovery from a vermicompost metagenomic library by Yasir et al. (2013). To obtain endoglucanase and xylanase enzymes, Reddy et al. (2013) constructed a metagenomic library after high-solid incubation at 35 and 55°C. Wongwilaiwalin et al. (2013) collected samples from cow rumen fluid, activated sludge, and sugarcane bagasse-decomposed soil from a soda pulp mill to discover various endoglucanases. Verma et al. (2013) constructed the metagenomic library from hot water spring samples to obtain thermostable and alkali resistance endoxylanase. In addition, sugarcane bagasse compost and rice straw-enriched samples were utilized by Chawannapak et al. (2012) and Mo et al. (2010) to retrieve thermostable endoxylanase. Gong et al. (2013) retrieved an alkali resistant endoxylanase from a bovine rumen metagenomic library. Zhou et al. (2011) identified the first bifunctional β-xylosidase and glucosidase from the rumen metagenome. Rashamuse et al. (2014) constructed the library from the hind gut of the termite *Trinervitermes trinervoides* to obtain feruloyl esterase enzyme.

Challenges in Metagenomic Approach

The major drawback of this approach is the frequency of clones of interest can be very low. This indicates mining thousands of clones which is cumbersome and time-consuming. To overcome this bottleneck, numerous scientists worked on this issue and found many high-throughput instruments for colony picking, inoculation in microtitre plates, and screening numerous clones simultaneously. Despite this, an enrichment approach is needed for selecting a clone of interest before library construction (Entcheva et al. 2001). It is unable to obtain DNA having all the genes responsible for the degradation process. Henceforth, Schloss and Handelsman (2003) discovered a new approach for studying various clones at the same time on liquid media in which products and substrates can disperse freely. Finally, it is very crucial to potentially express the gene of interest. Hence, another bottle neck lies in choosing a suitable host for expression. In addition, a good host should have the following two important features: (1) It is relatively insensitive to toxic xenobiotics and (2) It is not able to degrade the xenobiotics in the absence of the vector.

Conclusion and Future Goals

The rise of metagenomic based technologies open access to the unseen world of microorganisms and their undiscovered catabolic genes responsible for the xenobiotic degradation. It has also remarkably increased the identification of pathways that encode numerous gene products. The development of meta-omic approaches has allowed for unlocking the potential of uncultivable microbial biodiversity. Moreover, numerous industrially and environmentally important enzymes especially produced by bacteria have been discovered using functional based screening (exclusively those originating from highly polluted and extreme environments). Diverse enzymes have been discovered to represent new tools for environmental biotechnology. In addition, some studies assessed the importance of uncharacterized proteins which are dominant in polluted environments. To completely taste the outcome of these strategies a further speeding up of discovery and development of novel enzymes and drugs is needed. To attain this target, new and improved approaches that allow the sensitive and rapid detection of the gene of interest have to be invented.

Acknowledgement

The authors are grateful to University Grants Commission—Dr. D.S. Kothari Post Doctoral Fellowship for their financial support.

References

Aakvik, T., K.F. Degnes, R. Dahlsrud, F. Schmidt, R. Dam, L. Yu et al. 2009. A plasmid RK2-based broad-hostrange cloning vector useful for transfer of metagenomic libraries to a variety of bacterial species. FEMS Microbiol. Lett. 296: 149–158.

Abumweis, S.S., S. Jew and N.P. Ames. 2010. Beta-glucan from barley and its lipid-lowering capacity: a meta-analysis of randomized, controlled trials. Eur. J. Clin. Nutr. 64: 1472–1480.

Allgaier, M., A. Reddy, J.I. Park, N. Ivanova, P. D'haeseleer, S. Lowry, R. Sapra, T.C. Hazen, B.A. Simmons, J.S. VanderGheynst and P. Hugenholtz. 2010. Targeted discovery of glycoside hydrolases from a switchgrass-adapted compost community. PLoS One 5: 1–9.

Alvarez, T.M., J.H. Paiva, D.M. Ruiz, J.P.L.F Cairo, I.O. Pereira, D.A.A.A. Paixao et al. 2013. Structure and function of a novel cellulase 5 from sugarcane soil metagenome. PLoS One 8(12): 1–9.

Arbeli, Z. and C.L. Fuentes. 2007. Improved purification and PCR amplification of DNA from environmental samples. FEMS Microbiol. Lett. 272: 269–275.

Ausec, L., M. Zakrzewski, A. Goesmann, A. Schlüter and I. Mandic-Mulec. 2011. Bioinformatic analysis reveals high diversity of bacterial genes for laccase-like enzymes. PLoS One 6(10): 1–9.

Bartossek, R., G.W. Nicol, A. Lanzen, H.-P. Klenk and C. Schleper. 2010. Homologues of nitrite reductases in ammonia-oxidizing archaea: diversity and genomic context. Environ. Microbiol. 12(4): 1075–1088.

Bayer, S., C. Birkemeyer and M. Ballschmiter. 2011. A nitrilase from a metagenomic library acts regioselectively on aliphatic dinitriles. Appl. Microbiol. Biotechnol. 89(1): 91–98.

Beloqui, A., M. Pita, J. Polaina, A. Martinez-Arias, O.V. Golyshina, M. Zumarraga et al. 2006. Novel polyphenol oxidasemined from a metagenome expression library of bovine rumen: biochemical properties, structural analysis, and phylogenetic relationships. J. Biol. Chem. 281(32): 22933–22942.

Berlemont, R., D. Pipers, M. Delsaute, F. Angiono, G. Feller, M. Galleni et al. 2011. Exploring the Antarctic soil metagenome as a source of novel cold-adapted enzymes and genetic mobile elements. Rev. Argent Microb. 43: 94–103.

Berman, H.F. and L.W. Riley. 2013. Identification of novel antimicrobial resistance genes from microbiota on retail spinach. BMC Microbiol. 13: 272.

Bhat, A., S. Riyaz-Ul-Hassan, N. Ahmad, N. Srivastava and S. Johri. 2013. Isolation of cold-active, acidic endocellulase from Ladakh soil by functional metagenomics. Extremophiles 17(2): 229–239.

Boubakri, H., M. Beuf, P. Simonet and T.M. Vogel. 2006. Development of metagenomic DNA shuffling for the construction of a xenobiotic gene. Gene 375: 87–94.

Brady, S.F., C.J. Chao, J. Handelsman and J. Clardy. 2001. Cloning and heterologous expression of a natural product biosynthetic gene cluster from eDNA. Org. Lett. 3(13): 1981–1984.

Brady, S.F., C.J. Chao and J. Clardy. 2004. Long-chain N-acyltyrosine synthases from environmental DNA. Appl. Environ. Microbiol. 70: 6865–6870.

Brennan, Y., W.N. Callen, L. Christoffersen, P. Dupree, F. Goubet, S. Healey et al. 2004. Unusual microbial xylanases from insect guts. Appl. Environ. Microbiol. 70: 3609–3617.

Brennerova, M.V., J. Josefiova, V. Brenner, D.H. Pieper and H. Junca. 2009. Metagenomics reveals diversity and abundance of meta cleavage pathways in microbial communities from soil highly contaminated with jet fuel under air-sparging bioremediation. Environ. Microbiol. 11(9): 2216–2227.

Chang, L., M. Ding, L. Bao, Y. Chen, J. Zhou and H. Lu. 2011. Characterization of a bifunctional xylanase/endoglucanase from yak rumen microorganisms. Appl. Microbiol. Biotechnol. 90(6): 1933–1942.

Chawannapak, W., T. Laothanachareon, K. Boonyapakron, S. Wongwilaiwalin, T. Nimchua, L. Eurwilaichitr et al. 2012. Alkaliphilic endoxylanase from lignocellulolytic microbial consortium metagenome for biobleaching of eucalyptus pulp. J. Microbiol. Biotechnol. 22: 1636–1643.

Cheng, F., J. Sheng, R. Dong, Y. Men, L. Gan and L. Shen. 2012a. Novel xylanase from a holstein cattle rumen metagenomic library and its application in xylooligosaccharide and ferulic acid production from wheat straw. J. Agri. Food Chem. 60(51): 12516–12524.

Cheng, F.S., J.P. Sheng, T. Cai, J. Jin, W.Z. Liu, Y.M. Lin et al. 2012b. A protease-insensitive feruloyl esterase from China holstein cow rumen metagenomic library: expression, characterization, and utilization in ferulic acid release from wheat straw. J. Agric. Food Chem. 60: 2546–2553.

Chow, J., F. Kovacic, Y. Dall Antonia, A. Krauss, F. Fersini, C. Schmeisser et al. 2012. The metagenome derived enzymes LipS and LipT increase the diversity of known lipases. PLoS ONE 7(10): 1–16.

Chung, E.J., H.K. Lim, J.C. Kim, G.J. Choi, E.J. Park, M.H. Lee et al. 2008. Forest soil metagenome gene cluster involved in antifungal activity expression in *Escherichia coli*. Appl. Environ. Microbiol. 74: 723–730.

Clemente, J.C., E.C. Pehrsson, M.J. Blaser, K. Sandhu, Z. Gao, B. Wang et al. 2015. The microbiome of uncontacted Amerindians. Sci. Adv. 1(e1500183): 1–12.

Courtois, S., C.M. Cappellano, M. Ball, F.-X. Francou, P. Normand, G. Helynck et al. 2003. Recombinant environmental libraries provide access to microbial diversity for drug discovery from natural products. Appl. Environ. Microbiol. 69(1): 49–55.

Craig, J.W., F.Y. Chang and S.F. Brady. 2009. Natural products from environmental DNA hosted in Ralstonia metallidurans. ACS Chem. Biol. 4: 23–28.

Craig, J.W., F.Y. Chang, J.H. Kim, S.C. Obiajulu and S.F. Brady. 2010. Expanding small molecule functional metagenomics through parallel screening of broad-hostrange cosmid environmental DNA libraries in diverse proteobacteria. Appl. Environ. Microbiol. 76: 1633–1641.

Crameri, R. and M. Suter. 1993. Display of biologically active proteins on the surface of filamentous phages: a cDNA cloning system for selection of functional gene products linked to the genetic information responsible for their production. Gene 137(1): 69–75.

Curtis, T.P. and W.T. Sloan. 2005. Microbiology. Exploring microbial diversity—a vast below. Science 309: 1331–1333.

Daniel, R. 2005. The metagenomics of soil. Nature Rev. Microbiol. 3(6): 470–478.

Dellagnezze, B.M., G.V. de Sousa, L.L. Martins, D.F. Domingos, E.E.G. Limache, S.P. de Vasconcellos et al. 2014. Bioremediation potential of microorganisms derived from petroleum reservoirs. Mar. Pollut. Bull. 89(1–2): 191–200.

de Vasconcellos, S.P., C.F.F. Angolini, I.N.S. García, B. Martins Dellagnezze, C.C. da Silva, A.J. Marsaioli et al. 2010. Reprint of: screening for hydrocarbon biodegraders in a metagenomic clone library derived from Brazilian petroleum reservoirs. Org. Geochem. 41(9): 1067–1073.

Devirgiliis, C., P. Zinno, M. Stirpe, S. Barile and G. Perozzi. 2014. Functional screening of antibiotic resistance genes from a representative metagenomic library of food fermenting microbiota. Biomed Res. Int. 2014: 290967.

Donato, J.J., L.A. Moe, B.J. Converse, K.D. Smart, F.C. Berklein, P.S. McManus et al. 2010. Metagenomic analysis of apple orchard soil reveals antibiotic resistance genes encoding predicted bifunctional proteins. Appl. Environ. Microbiol. 76: 4396–4401.

Edwards, R.A., B. Rodriguez-Brito, L. Wegley, M. Haynes, M. Breitbart, D.M. Peterson et al. 2006. Using pyrosequencing to shed light on deep mine microbial ecology. BMC Genomics 7: 57.

Elend, C., C. Schmeisser, P. Leggewie, J.D. Babiak, H.L. Carballeira, J.J. Steele et al. 2006. Isolation and biochemical characterization of two novel metagenome-derived esterases. Appl. Environ. Microbiol. 72: 3637–3645.

Entcheva, P., W. Liebl, A. Johann, T. Hartsch and W.R. Streit. 2001. Direct cloning from enrichment cultures, a reliable strategy for isolation of complete operons and genes from microbial consortia. Appl. Environ. Microbiol. 67: 89–99.

Escobar-Zepeda, A., A.V.P. de León and A. Sanchez-Flores. 2015. The road to metagenomics: from microbiology to DNA sequencing technologies and bioinformatics. Front. Genet. 6(348): 1–15.

Fan, X., X. Liu, R. Huang and Y. Liu. 2012. Identification and characterization of a novel thermostable pyrethroid-hydrolyzing enzyme isolated through metagenomic approach. Microb. Cell Factories 11(1): 1–11.

Fang, Z., T. Li, Q. Wang, X. Zhang, H. Peng, W. Fang et al. 2010. A bacterial laccase from marine microbial metagenome exhibiting chloride tolerance and dye decolorization ability. Appl. Microbiol. Biotechnol. 89(4): 1103–1110.

Fang, Z.-M., T.-L. Li, F. Chang, P. Zhou, W. Fang, Y.-Z. Hong et al. 2012. A new marine bacterial laccase with chloride-enhancing, alkaline-dependent activity and dye decolorization ability. Bioresour. Technol. 111: 36–41.

Faoro, H., A. Glogauer, G.H. Couto, E.M. de Souza, L.U. Rigo, L.M. Cruzs et al. 2012. Characterization of a new Acidobacteria-derived moderately thermostable lipase from a Brazilian Atlantic Forest soil metagenome. FEMS Microbiol. Ecol. 81: 386–394.

Ferrer, M., O.V. Golyshina, T.N. Chernikova, A.N. Khachane, D. Reyes-Duarte, V.A. Santos et al. 2005. Novel hydrolase diversity retrieved from a metagenome library of bovine rumen microflora. Environ. Microbiol. 7: 1996–2010.

Fouhy, F., L.A. Ogilvie, B.V. Jones, R.P. Ross, A.C. Ryan, E.M. Dempsey et al. 2014. Identification of aminoglycoside and beta-lactam resistance genes from within an infant gut functional metagenomic library. PLoS ONE 9(9): 1–10.

Frostegard, A., S. Courtois, V. Ramisse, S. Clerc, D. Bernillon, F. Le Gall et al. 1999. Quantification of bias related to the extraction of DNA directly from soils. Appl. Environ. Microbiol. 65: 5409–5420.

Fu, J., H.-K.S. Leiros, D. de Pascale, K.A. Johnson, H.-M. Blencke and B. Landfald. 2013. Functional and structural studies of a novel cold-adapted esterase from an Arctic intertidal metagenomic library. Appl. Microbiol. Biotechnol. 97(9): 3965–3978.

Gabor, E.M., W.B. Alkema and D.B. Janssen. 2004. Quantifying the accessibility of the metagenome by random expression cloning techniques. Environ. Microbiol. 6(9): 879–886.

Gillespie, D.E., S.F. Brady, A.D. Bettermann, N.P. Cianciotto, M.R. Liles, M.R. Rondon et al. 2002. Isolation of antibiotics turbomycin A and B from a metagenomic library of soil microbial DNA. Appl. Environ. Microbiol. 68: 4301–4306.

Glogauer, A., V.P. Martini, H. Faoro, G.H. Couto, M. Muller-Santos, R.A. Monteiro et al. 2011. Identification and characterization of a new true lipase isolated through metagenomic approach. Microb. Cell Fact. 10(54): 1–15.

Gong, X., R.J. Gruniniger, R.J. Forster, R.M. Teather and T.A. McAllister. 2013. Biochemical analysis of a highly specific, pH stable xylanase gene identified from a bovine rumen-derived metagenomic library. Appl. Microbiol. Biotechnol. 97(6): 2423–2431.

Handelsman, J., M.R. Rondon, S.F. Brady, J. Clardy and R.M. Goodman. 1998. Molecular biological access to the chemistry of unknown soil microbes: a new frontier for natural products. Chem. Biol. 5(10): R245–R249.

Handelsman, J. 2004. Metagenomics: application of genomics to uncultured microorganisms. Microbiol. Mol. Biol. Rev. 68: 669–685.

Hardeman, F. and S. Sjoling. 2007. Metagenomic approach for the isolation of a novel low-temperature-active lipase from uncultured bacteria of marine sediment. FEMS Microbiol. Ecol. 59(2): 524–534.

He, Y., X. Xiao and F. Wang. 2013. Metagenome reveals potential microbial degradation of hydrocarbon coupled with sulfate reduction in an oil-immersed chimney from Guaymas Basin. Front Microbiol. 4(148): 1–13.

Healy, F.G., R.M. Ray, H.C. Aldrich, A.C. Wilkie, L.O. Ingram and K.T. Shanmugam. 1995. Direct isolation of functional genes encoding cellulases from the microbial consortia in a thermophilic, anaerobic digester maintained on lignocellulose. Appl. Microbiol. Biotechnol. 43: 667–674.

Holben, W.E., J.K. Jansson, B.K. Chelm and J.M. Tiedje. 1988. DNA probe method for the detection of specific microorganisms in the soil bacterial community. Appl. Environ. Microbiol. 54: 703–711.

Iqbal, H.A., J.W. Craig and S.F. Brady. 2014. Anti bacterial enzymes from the functional screening of metagenomic libraries hosted in *Ralstonia metallidurans*. FEMS Microbiol. Lett. 354: 19–26.

Iwai, S., B. Chai, W.J. Sul, J.R. Cole, S.A. Hashsham and J.M. Tiedje. 2009. Gene-targeted metagenomics reveals extensive diversity of aromatic dioxygenase genes in the environment. ISME J. 4(2): 279–285.

Jackson, C.R., J.P. Harper, D. Willoughby, E.E. Roden and P.F. Churchill. 1997. A simple, efficient method for the separation of humic substances and DNA from environmental samples. Appl. Environ. Microbiol. 63: 4993–4995.

Jadeja, N.B., R.P. More, H.J. Purohit and A. Kapley. 2014. Metagenomic analysis of oxygenases from activated sludge. Biores. Technol. 165: 250–256.

Jeong. Y.S., H.B. Na, S.K. Kim, Y.H. Kim, E.J. Kwon, J. Kim, H.D. Yun, J.K. Lee and H. Kim. 2012. Characterization of Xyn10J, a novel family 10 xylanase from a compost metagenomic library. Appl. Biochem. Biotechnol. 166(5): 1328–1339.

Jiang, C., G. Ma, S. Li, T. Hu, Z. Che, P. Shen et al. 2009. Characterization of a novel beta-glucosidase-like activity from a soil metagenome. J. Microbiol. 47: 542–548.

Jiao, Y., X. Chen, X. Wang, X. Liao, L. Xiao, A. Miao et al. 2013. Identification and characterization of a cold-active phthalate esters hydrolase by screening a metagenomic library derived from biofilms of a wastewater treatment plant. PLoS One 8(10): 1–8.

Kakirde, K.S., L.C. Parsley and M.R. Liles. 2010. Size does matter: Application-driven approaches for soil metagenomics. Soil Biol. Biochem. 42: 1911–1923.

Kambiranda, D.M., S.M. Asraful-Islam, K.M. Cho, R.K. Math, Y.H. Lee, H. Kim et al. 2009. Expression of esterase gene in yeast for organophosphates biodegradation. Pestic. Biochem. Physiol. 94: 15–20.

Kang, C.-H., K.-H. Oh, M.-H. Lee, T.-K. Oh, B. Kim and J. Yoon. 2011. A novel family VII esterase with industrial potential from compost metagenomic library. Microb. Cell Factories 10(41): 1–8.

Kimura, N., K. Sakai and K. Nakamura. 2010. Isolation and characterization of a 4-nitrotoluene-oxidizing enzyme from activated sludge by a metagenomic approach. Microbes Environ. 25(2): 133–139.

King, R.W., J.D. Bauer and S.F Brady. 2009. An environmental DNA-derived type II polyketide biosynthetic pathway encodes the biosynthesis of the pentacyclic polyketide erdacin. Angewandte Chemie 48: 6257–6261.

La Montagne, M.G., F.C. Michel, Jr., P.A. Holden and C.A. Reddy. 2002. Evaluation of extraction and purification methods for obtaining PCR-amplifiable DNA from compost for microbial community analysis. J. Microbiol. Methods 49: 255–264.

Lakhdari, O., A. Cultrone, J. Tap, K. Gloux, F. Bernard, S.D. Ehrlich et al. 2010. Functional metagenomics: a high throughput screening method to decipher microbiota-driven NF-kappaB modulation in the human gut. PLoS ONE 5(9): 1–10.

Lee, S.W., K. Won, H.K. Lim, J.C. Kim, G.J. Choi and K.Y. Cho. 2004. Screening for novel lipolytic enzymes from uncultured soil microorganisms. Appl. Microbiol. Biotech. 65: 720–726.

Lee, M.H., C.H. Lee, T.K. Oh, J.K. Song and J.H. Yoon. 2006. Isolation and characterization of a novel lipase from a metagenomic library of tidal flat sediments: evidence for a new family of bacterial lipases. Appl. Environ. Microbiol. 72: 7406–7409.

Li, G., K. Wang and Y.H. Liu. 2008. Molecular cloning and characterization of a novel pyrethroid-hydrolyzing esterase originating from the metagenome. Microb. Cell Factories 7(38): 1–10.

Lim, H.K., E.J. Chung, J.C. Kim, G.J. Choi, K.S. Jang, Y.R. Chung et al. 2005. Characterization of a forest soil metagenome clone that confers indirubin and indigo production on *Escherichia coli*. Appl. Environ. Microbiol. 71: 7768–7777.

Liu, L., H. Yang, H.D. Shin, R.R. Chen, J. Li, G. Du et al. 2013. How to achieve high-level expression of microbial enzymes: strategies and perspectives. Bioengineered. 4(4): 212–223.

López-López, O., M.E. Cerdan and M.I. Gonzalez Siso. 2014. New extremophilic lipases and esterases from metagenomics. Curr. Protein Pept. Sci. 15(5): 445–455.

Lorenz, P. and J. Eck. 2005. Metagenomics and industrial applications. Nat. Rev. Microb. 3: 510–516.

Lu, Y., Y. Yu, R. Zhou, W. Sun, C. Dai, P. Wan et al. 2011. Cloning and characterisation of a novel 2,4-dichlorophenol hydroxylase from a metagenomic library derived from polychlorinated biphenyl-contaminated soil. Biotechnol. Lett. 33(6): 1159–1167.

Lu, Z., Y. Deng, J.D. Van Nostrand, Z. He, J. Voordeckers, A. Zhou et al. 2012. Microbial gene functions enriched in the deepwater horizon deep-sea oil plume. ISME J. 6(2): 451–460.

Macneil, I.A., C.L. Tiong, C. Minor, P.R. August, T.H. Grossman and K.A. Loiacono et al. 2001. Expression and isolation of anti microbial small molecules from soil DNA libraries. J. Mol. Microbiol. Biotechnol. 3: 301–308.

Marko, D., S. Schatzle, A. Friedel, A. Genzlinger, H. Zankl, L. Meijer et al. 2001. Inhibition of cyclin-dependent kinase1 (CDK1) by indirubin derivative sinhuman tumour cells. Br. J. Cancer. 84: 283–289.

Martin, M. 2016. Function-based Analyses of Bacterial Symbionts Associated with the Brown Alga Ascophyllum nodosum and Identification of Novel Bacterial Hydrolytic Enzyme Genes. PhD Thesis, Université de Liège, Liège, Belgique.

Mayumi, D., Y. Akutsu-Shigeno, H. Uchiyama, N. Nomura and T. Nakajima-Kambe. 2008. Identification and characterization of novel poly(DL-lactic acid) depolymerases from metagenome. Appl. Microbiol. Biotechnol. 79(5): 743–750.

Mo, X.C., C.L. Chen, H. Pang, Y. Feng and J.X. Feng. 2010. Identification and characterization of a novel xylanase derived from a rice straw degrading enrichment culture. Appl. Microbiol. Biotechnol. 87: 2137–2146.

Mori, T., I. Kamei, H. Hirai and R. Kondo. 2014. Identification of novel glycosyl hydrolases with cellulolytic activity against crystalline cellulose from metagenomic libraries constructed from bacterial enrichment cultures. SpringerPlus 3: 365–371.

Morimoto, S. and T. Fujii. 2009. A new approach to retrieve full lengths of functional genes from soil by PCR-DGGE and metagenome walking. Appl. Microbiol. Biotechnol. 83(2): 389–396.

Nacke, H., M. Engelhaupt, S. Brady, C. Fischer, J. Tautzt and R. Daniel. 2012. Identification and characterization of novel cellulolytic and hemicellulolytic genes and enzymes derived from German grassland soil metagenomes. Biotechnol. Lett. 34(4): 663–675.

Nagayama, H., T. Sugawara, R. Endo, A. Ono, H. Kato, Y. Ohtsubo et al. 2015. Isolation of oxygenase genes for indigo-forming activity from an artificially polluted soil metagenome by functional screening using Pseudomonas putida strains as hosts. Appl. Microbiol. Biotechnol. 99: 4453–4470.

Neveu, T.J., C. Regeard and M.S. Dubow. 2011. Isolation and characterization of two serine proteases from metagenomic libraries of the Gobi and Death Valley deserts. Appl. Microbiol. Biotechnol. 91(3): 635–644.

Ngo, T.D., B.H. Ryu, H. Ju, E. Jang, K. Park, K.J. Park et al. 2013. Structural and functional analyses of a bacterial homologue of hormone-sensitive lipase from a metagenomic library. Acta Cryst. D: Biol. Crystallogr. 69(9): 1726–1737.

Niehaus, F., E. Gabor, S. Wieland, P. Siegert, K.H. Maurer and J. Eck. 2011. Enzymes for the laundry industries: tapping the vast metagenomic pool of alkaline proteases. Microbial Biotechnol. 4(6): 767–776.

Nogales, B., K.N. Timmis, D.B. Nedwell and A.M. Osborn. 2002. Detection and diversity of expressed denitrification genes in estuarine sediments after reverse transcription-PCR amplification from mRNA. Appl. Environ. Microbiol. 68(10): 5017–5025.

Ogram, A., G.S. Sayler and T.J. Barkay. 1987. DNA extraction and purification from sediments. J. Microbiol. Meth. 7: 57–66.

Ono, A., R. Miyazaki, M. Sota, Y. Ohtsubo, Y. Nagata and M. Tsuda. 2007. Isolation and characterization of naphthalene-catabolic genes and plasmids from oil-contaminated soil by using two cultivation-independent approaches. Appl. Microbiol. Biotechnol. 74(2): 501–510.

Orsini, M. and V. Romano-Spica. 2001. A microwave-based method for nucleic acid isolation from environmental samples. Lett. Appl. Microbiol. 33: 17–20.

Pace, H.C. and C. Brenner. 2001. The nitrilase superfamily: classification, structure and function. Genome Biol. 2(1): 1–9.

Palackal, N., C.S. Lyon, S. Zaidi, P. Luginbuhl, P. Dupree, F. Goubet et al. 2007. A multifunctional hybrid glycosyl hydrolase discovered in an uncultured microbial consortium from ruminant gut. Appl. Microbiol. Biotechnol. 74: 113–124.

Parsley, L.C., E.J. Consuegra, K.S. Kakirde, A.M. Land, W.F. Harper, Jr. and M.R. Liles. 2010. Identification of diverse antimicrobial resistance determinants carried on bacterial, plasmid, or viral metagenomes from an activated sludge microbial assemblage. Appl. Environ. Microbiol. 76: 3753–3757.

Peng. Q., X. Wang, M. Shang, J. Huang, G. Guan and Y. Li. 2014. Isolation of a novel alkaline stable lipase from a metagenomic library and its specific application for milk fat flavor production. Micro. Cell Fact. 13(1): 2–9.

Picard, C., C. Ponsonnet, E. Paget, X. Nesme and P. Simonet. 1992. Detection and enumeration of bacteria in soil by direct DNA extraction and polymerase chain reaction. Appl. Environ. Microbiol. 58: 2717–2722.

Pooja, S., M. Pushpanathan, S. Jayashree, P. Gunasekaran and J. Rajendhran. 2015. Identification of periplasmic alpha-amlyase from cow dung metagenome by product induced gene expression profiling (Pigex). Indian J. Microbiol. 55: 57–65.

Porteous, L.A., R.J. Seidler and L.S. Watrud. 1997. An improved method for purifying DNA from soil for polymerase chain reaction amplification and molecular ecology applications. Mol. Ecol. 6: 787–791.

Pushpam, P.L., T. Rajesh and P. Gunasekaran. 2011. Identification and characterization of alkaline serine protease from goat skin surface metagenome. AMB Exp. 1(3): 1–10.

Rabausch, U., J. Juergensen, N. Ilmberger, S. Böhnke, S. Fischer, B. Schubach et al. 2013. Functional screening of metagenome and genome libraries for detection of novel flavonoid modifying enzymes. Appl. Environ. Microbiol. 79: 4551–4563.

Ramsay, A.J. 1984. Extraction of bacteria from soil: Efficiency of shaking or ultrasonication as indicated by direct counts and autoradiography. Soil Biol. Biochem. 16: 475–481.

Rashamuse, K., T. Ronneburg, W. Sanyika, K. Mathiba, E. Mmutlane and D. Brady. 2014. Metagenomic mining of feruloyl esterases from termite enteric flora. Appl. Microbiol. Biotechnol. 98 (2): 727–737.

Reddy, A.P., C.W. Simmons, P. Dhaeseleer, J. Khudyakov, H. Burd, M. Hadi et al. 2013. Discovery of microorganisms and enzymes involved in high-solids decomposition of rice straw using metagenomic analyses. PLoS One 8(10): 1–12.

Robertson, D.E., J.A. Chaplin, G. DeSantis, M. Podar, M. Madden, E. Chi et al. 2004. Exploring nitrilase sequence space for enantioselective catalysis. Appl. Environ. Microbiol. 70(4): 2429–2436.

Roose-Amsaleg, C.L., E. Garnier-Sillam and M. Harry. 2001. Extraction and purification of microbial DNA from soil and sediment samples. Appl. Soil Ecol. 18: 47–60.

Rothberg, J.M. and J.H. Leamon. 2008. The development and impact of 454 sequencing. Nat. Biotechnol. 26: 1117–1124.

Sae-Lee, R. and A. Boonmee. 2014. Newly derived GH43 gene from compost metagenome showing dual xylanase and cellulase activities. Folia Microbiol. 59: 409–417.

Sagova-Mareckova, M., L. Cermak, J. Novotna, K. Plhackova, J. Forstova and J. Kopecky. 2008. Innovative methods for soil DNA purification tested in soils with widely differing characteristics. Appl. Environ. Microbiol. 74: 2902–2907.

Sambrook, J., E.F. Fritsch and T. Maniatis. 1989. Molecular Cloning: A Laboratory Manual. 2nd ed. Cold Spring Harbor Laboratory, Cold Spring Harbor, NY.

Sandaa, R.A., O. Enger and V. Torsvik. 1999. Rapid method for fluorometric quantification of DNA in soil. Soil Biol. Biochem. 30: 265–268.

Sanger, F., S. Nicklen and A.R. Coulson. 1977. DNA sequencing with chain-terminating inhibitors. Proc. Natl. Acad. Sci. USA 74: 5463–5467.

Scanlon, T.C., S.M. Dostal and K.E. Griswold. 2014. A high-throughput screen for anti biotic drug discovery. Biotechnol. Bioeng. 111: 232–243.

Schloss, P.D. and J. Handelsman. 2003. Biotechnological prospects from metagenomics. Curr. Opin. Biotechnol. 14(3): 303–310.

Schmidt, T.M., E.F. DeLong and N.R. Pace. 1991. Analysis of a marine picoplankton community by 16S rRNA gene cloning and sequencing. J. Bacteriol. 173(14): 4371–4378.

Selenska, S. and W. Klingmüller. 1991. Direct detection of nif-gene sequences of *Enterobacter agglomerans* in soil. FEMS Microbiol. Lett. 80: 243–246.

Selvin, J., J. Kennedy, D.P.H. Lejon, G.S. Kiran and A.D.W. Dobson. 2012. Isolation identification and biochemical characterization of a novel halo-tolerant lipase from the metagenome of the marine sponge Haliclona simulans. Micro. Cell Fact. 11(72): 1–14.

Sierra-García, I.N., J. Correa Alvarez, S.P. de Vasconcellos, A. Pereira de Souza, E.V. dos Santos Neto and V.M. de Oliveira. 2014. New hydrocarbon degradation pathways in the microbial metagenome from Brazilian petroleum reservoirs. PLoS One 9(2): 1–13.

Shah, A.A., F. Hasan, A. Hameed and S. Ahmed. 2008. Biological degradation of plastics: a comprehensive review. Biotechnol. Adv. 26(3): 246–265.

Sharma, S., F.G. Khan and G.N. Qazi. 2010. Molecular cloning and characterization of amylase from soil metagenomic library derived from Northwestern Himalayas. Appl. Microbiol. Biotechnol. 86(6): 1821–1828.

Sharma, N., H. Tanksale, A. Kapley and H.J. Purohit. 2012. Mining the metagenome of activated biomass of an industrial wastewater treatment plant by a novel method. Indian J. Microbiol. 25: 538–543.

Silva, C.C., H. Hayden, T. Sawbridge, P. Mele, S.O. De Paula, L.C.F. Silva et al. 2013. Identification of genes and pathways related to phenol degradation in metagenomic libraries from petroleum refinery wastewater. PLoS One 8(4): 1–11.

Singh, R., S. Dhawan, K. Singh and J. Kaur. 2012. Cloning, expression and characterization of a metagenome derived thermoactive/thermostable pectinase. Mol. Biol. Rep. 39(8): 8353–8361.

Sleator, R.D., C. Shortall and C. Hill. 2008. Metagenomics. Lett. Appl. Microbiol. 47: 361–366.

Smalla, K., N. Cresswell, L.C. Mendoca-Hagler, A. Wolters and J.D. Van Elsas. 1993. Rapid DNA extraction protocol from soil for polymerase chain reaction-mediated amplification. J. Appl. Bacteriol. 74: 78–85.

Staley, J.T. and A. Konopka. 1985. Measurement of *in situ* activities of nonphotosynthetic microorganisms in aquatic and terrestrial habitats. Ann. Rev. Microbiol. 39: 321–346.

Strachan, C.R., R. Singh, D. VanInsberghe, K. Ievdokymenko, K. Budwill, W.W. Mohn et al. 2014. Metagenomic scaffolds enable combinatorial lignin transformation. Proc. Natl. Acad. Sci. 111(28): 10143–10148.

Starkey, M.P., Y. Umrania, C.R. Mundy and M.J. Bishop. 1998. Reference cDNA library facilities available from European sources. Mole. Biotechnol. 9(1): 35–57.

Solbak, A.I., T.H. Richardson, R.T. McCann, K.A. Kline, F. Bartnek, G. Tomlinson et al. 2005. Discovery of pectin-degrading enzymes and directed evolution of a novel pectate lyase for processing cotton fabric. J. Biol. Chem. 280: 9431–9438.

Suenaga, H., T. Ohnuki and K. Miyazaki. 2007. Functional screening of a metagenomic library for genes involved in microbial degradation of aromatic compounds. Environ. Microbiol. 9(9): 2289–2297.

Sul, W.J., J. Park, J.F. Quensen, J.L.M. Rodrigues, L. Seliger, T.V. Tsoi et al. 2009. DNA-stable isotope probing integrated with metagenomics for retrieval of biphenyl dioxygenase genes from polychlorinated biphenyl-contaminated river sediment. Appl. Environ. Microbiol. 75(17): 5501–5506.

Sulaiman, S., S. Yamato, E. Kanaya, J.-J. Kim, Y. Koga, K. Takano et al. 2012. Isolation of a novel cutinase homolog with polyethylene terephthalate-degrading activity from leaf-branch compost by using a metagenomic approach. Appl. Environ. Microbiol. 78(5): 1556–1562.

Takada-Hoshino, Y. and N. Matsumoto. 2004. An improved DNA extraction method using skim milk from soils that strongly adsorb DNA. Microbes Environ. 19: 13–19.

Tao, W., M.H. Lee, J. Wu, N.H. Kim, J.C. Kim, E. Chung et al. 2012. Inactivation of chloramphenicol and florfenicol by a novel chloramphenicol hydrolase. Appl. Environ. Microbiol. 78: 6295–6301.

Tchigvintsev, A., H. Tran, A. Popovic, F. Kovacic, G. Brown, R. Flick et al. 2014. The environment shapes microbial enzymes: five cold-active and salt-resistant carboxylesterases from marine metagenomes. Appl. Microbiol. Biotechnol. 7: 2165–2178.

Treusch, A.H., S. Leininger, A. Kletzin, S.C. Schuster, H.-P. Klenk and C. Schleper. 2005. Novel genes for nitrite reductase and Amo-related proteins indicate a role of uncultivated mesophilic Crenarchaeota in nitrogen cycling. Environ. Microbiol. 7(12): 1985–1995.

Tsai, Y. and B.H. Olson. 1991. Rapid method for direct extraction of DNA from soil and sediments. Appl. Environ. Microbiol. 57: 1070–1074.

Tyson, G.W., J. Chapman, P. Hugenholtz, E.E. Allen, R.J. Ram, P.M. Richardson et al. 2004. Community structure and metabolism through reconstruction of microbial genomes from the environment. Nature 428: 37–43.

Uchiyama, T., T. Abe, T. Ikemura and K. Watanabe. 2005. Substrate-induced gene expression screening of environmental metagenome libraries for isolation of catabolic genes. Nat. Biotechnol. 23: 88–93.

Uchiyama, T. and K. Miyazaki. 2010. Product-induced gene expression, a product-responsive reporter assay used to screen metagenomic libraries for enzyme-encoding genes. Appl. Environ. Microbiol. 76: 7029–7035.

Venter, J.C., K. Remington, J.F. Heidelberg, A.L. Halpern, D. Rusch, J.A. Eisen et al. 2004. Environmental genome shotgun sequencing of the Sargasso Sea. Science 304: 66–74.

Vergne-Vaxelaire, C., F. Bordier, A. Fossey, M. Besnard-Gonnet, A. Debard, A. Mariage et al. 2013. Nitrilase activity screening on structurally diverse substrates: providing biocatalytic tools for organic synthesis. Adv. Synth. Catal. 355(9): 1763–1779.

Verma, D., Y. Kawarabayasi, K. Miyazaki and T. Satyanarayana. 2013. Cloning, expression and characteristics of a novel alkalistable and thermostable xylanase encoding gene (Mxyl) retrieved from compost-soil metagenome. PLoS ONE 8(1): 1–8.

Vester, J.K., M.A. Glaring and P. Stougaard. 2014. Discovery of novel enzymes with industrial potential from a cold and alkaline environment by a combination of functional metagenomics and culturing. Microb. Cell Fact. 13(72): 1–14.

Vidya, J., S. Swaroop, S.K. Singh, D. Alex, R.K. Sukumaran and A. Pandey. 2011. Isolation and characterization of a novel alpha-amylase from a metagenomic library of Western Ghats of Kerala, India. Biologia 66(6): 939–944.

Voget, S., C. Leggewie, A. Uesbeck, C. Raasch, K.E. Jaeger and W.R. Streit. 2003. Prospecting for novel biocatalysts in a soil metagenome. Appl. Environ. Microbiol. 69: 6235–6242.

Voget, S., H.L. Steele and W.R. Streit. 2006. Characterization of a metagenome-derived halotolerant cellulose. J. Biotechnol. 126(1): 26–36.

von Wettstein, D., G. Mikhaylenko, J.A. Froseth, G. Kannangara. 2000. Improved barley broiler feed with transgenic malt containing heat-stable (1,3-1,4)-beta-glucanase. Proc. Natl. Acad. Sci. 97(25): 13512–13517.

Walter, J., M. Mangold and G.W. Tannock. 2005. Construction, analysis, and beta-glucanase screening of a bacterial artificial chromosome library from the large-bowel microbiota of mice. Appl. Environ. Microbiol. 71: 2347–2354.

Wang, G.Y., E. Graziani, B. Waters, W. Pan, X. Li, J. McDermott et al. 2000. Novel natural products from soil DNA libraries in a Streptomycete host. Org. Lett. 2: 2401–2404.

Wang, Y., Y. Chen, Q. Zhou, S. Huang, S. Ning, J. Xu et al. 2012a. A culture-independent approach to unravel uncultured bacteria and functional genes in a complex microbial community. PLoS One 7(10): 1–11.

Wang, H., X. Li, Y. Ma and J. Song. 2013b. Characterization and highlevel expression of a metagenome-derived alkaline pectatelyase in recombinant *Escherichia coli*. Proc. Biochem. 49: 69–76.

Wang, Y., H. Ren, H. Pan, J. Liu and L. Zhang. 2015. Enhanced tolerance and remediation to mixed contaminates of PCBs and 2,4-DCP by transgenic alfalfa plants expressing the 2,3-dihydroxybiphenyl-1,2-dioxygenase. J. Hazard. Mater. 286: 269–275.

Warnecke, F., P. Luginbuhl, N. Ivanova, M. Ghassemian, T.H. Richardson, J.T. Stege et al. 2007. Metagenomic and functional analysis of hindgut microbiota of a wood-feeding higher termite. Nature 450: 560–565.

Wilkinson, D.E., T. Jeanicke and D.A. Cowan. 2002. Efficient molecular cloning of environmental DNA from geothermal sediments. Biotechnol. Lett. 24(2): 155–161.

Woese, C.R. and G.E. Fox. 1977. Phylogenetic structure of the prokaryotic domain: the primary kingdoms. Proc. Natl. Acad. Sci. USA 74(11): 5088–5090.

Wommack, K.E., J. Bhavsar and J. Ravel. 2008. Metagenomics, read length matters. Appl. Environ. Microbiol. 74: 1453–1463.

Wongwilaiwalin, S., T. Laothanachareo, W. Mhuantong, S. Tangphatsornruang, L. Eruwilaichitr, Y. Igarashi et al. 2013. Comparative metagenomic analysis of microcosm structures and lignocellulolytic enzyme systems of symbiotic biomass-degrading consortia. Appl. Microbiol. Biotechnol. 97: 8941–8954.

Xia, X., J. Bollinger and A. Ogram. 1995. Molecular genetic analysis of the response of three soil microbial communities to the application of 2,4-D. Mol. Ecol. 4: 17–28.

Xing, M.-N., X.-Z. Zhang and H. Huang. 2012. Application of metagenomic techniques inmining enzymes from microbial communities for biofuel synthesis. Biotechnol. Adv. 30(4): 920–929.

Yao, J., X.J. Fan, Y. Lu and Y.H. Liu. 2011. Isolation and characterization of a novel tannase from a metagenomic library. J. Agric. Food Chem. 59: 3812–3818.

Yao, J., Q. Chen, G. Zhong, W. Cao, A. Yu and Y. Liu. 2014. Immobilization and characterization of tannase from a metagenomic library and its use for removal of tannins from green tea infusion. J. Microbiol. Biotechnol. 24: 80–86.

Yasir, M., H. Khan, S.S. Azam, A. Telke, S.W. Kim and Y.R. Chung. 2013. Cloning and functional characterization of endo-b-14-glucanase gene from metagenomic library of vermicompost. J. Microbiol. 51: 329–335.

Ye, M., G. Li, W.Q. Liang and Y.H. Liu. 2010. Molecular cloning and characterization of a novel metagenome-derived multicopper oxidase with alkaline laccase activity and highly soluble expression. Appl. Microbiol. Biotechnol. 87(3): 1023–1031.

Yeh, Y., S.C. Chang, H. Kuo, C. Tong, S. Yu and T.D. Ho. 2013. A metagenomic approach for the identification and cloning of an endoglucanase from rice straw compost. Gene. 519(2): 360–366.

Zaprasis, A., Y.-J. Liu, S.-J. Liu, H.L. Drake and M.A. Horn. 2009. Abundance of novel and diverse tfdA-like genes, encoding putative phenoxyalkanoic acid herbicide-degrading dioxygenases, in soil. Appl. Environ. Microbiol. 76(1): 119–128.

Zengler, K., G. Toledo, M. Rappe, J. Elkins, E.J. Mathur, J.M. Short et al. 2002. Cultivating the uncultured. Proc. Natl. Acad. Sci. USA 99: 15681–15686.

Zhou, J., M.A. Bruns and J.M. Tiedje. 1996. DNA recovery from soils of diverse composition. Appl. Environ. Microbiol. 62: 316–322.

Zhou, J., L. Bao, L. Chang, Z. Liu, C. You and H. Lu. 2011. Beta-xylosidase activity of a GH3 glucosidase/xylosidase from yak rumen metagenome promotes the enzymatic degradation of hemicellulosic xylans. Lett. Appl. Microbiol. 54: 79–87.

12

Interactions Between Nonpolar Compounds and Soil Organic Carbon Under Low Redox Potentials
Implications for Bioavailability

Jessica E. Sharpe and *Andrew Ogram**

INTRODUCTION

Hydrophobic Organic Contaminants (HOCs), including polyaromatic hydrocarbons (PAH), are often toxic and carcinogenic, and may be released into the environment from both human activities and from natural sources. Although the total amounts of PAHs in soils and sediments are unknown, a US Geological Survey study indicated that PAH concentrations tend to increase with population, energy consumption, and land use (Van Metre and Van Der See 1997). HOCs range from pesticides to by-products of fossil fuel combustion, and may persist for long periods of time in the environment (Sorensen et al. 2005). Due to their low aqueous solubilities, HOCs become sorbed by soil particles and may resist extraction and biodegradation (Mao et al. 2002); however, the impacts of an HOCs interactions with dissolved organic carbon (DOC) may be more complex with respect to bioavailability. Most

Soil and Water Sciences Department, University of Florida, Gainesville, FL USA.
Email: JessicaElaineSharpe@gmail.com
* Corresponding author: aogram@ufl.edu

research on this issue has focused on sorption under aerobic conditions, although much contaminated soil is found in anaerobic settings under reduced conditions (i.e., low redox potentials). The structure and concentrations of DOC are strongly influenced by redox potential (Pravecek et al. 2005), which may play a significant role in the solubility and potential bioavailability of HOCs. Assessing the fate of these compounds in anoxic soils and sediments is essential for accurate prediction of the exposure of humans and wildlife. This paper will address molecular models of the interaction between HOC and Soil Organic Matter (SOM), the impacts of redox potentials on the solubility and molecular structures of DOC, and the availability and biodegradation of HOCs under low redox potentials.

Interactions Between HOCs and Organic Matter

The importance of polarity

The sorption and desorption of HOCs is controlled by many factors; however, the polarity of Soil Organic Matter (SOM) and DOC typically exhibits an inverse relationship with the sorption of HOCs (Pedersen et al. 1999). Some research has suggested that the hydrophobic fraction of DOC harbors a greater number of sorption sites for HOCs (Pan et al. 2008) than the more polar fractions do; however, recent data have indicated a high sorption rate for aliphatic rich SOM. This complicates the previous notion that sorption coefficients are directly associated with the aromatic C content (Pfaender and Kim 2005). Kang and Xing (2005) found that HOC sorption is directly linked to aliphatic material in SOM, as supported by his research showing that humin fractions with the greatest aliphatic carbon contents had increased sorption ability in comparison to humic acids. Humin is the portion of humic material that is not soluble at any pH value in an aqueous solution (Stevenson 1994). Pan et al. (2008) suggested that aromatic and aliphatic carbons contribute to HOC-SOM sorption, and polarity could be utilized to understand SOM sorption mechanisms.

In recognition of the importance of SOM to sorption, the sorption of non-ionic compounds to soil is frequently described by the carbon-normalized distribution coefficient (K_{oc}) (Smernik and Kookana 2015). The K_{oc} is the ratio of the chemical mass adsorbed to the soil versus the amount dissolved at equilibrium, normalized to the mass of carbon in the soil, and can be incorporated into models for predicting the mobility of HOCs in soil. A high K_{oc} indicates a less mobile HOC, and may vary dependent on the structure of the SOM (Smernik and Kookana 2015). Binding degrees are represented by K_{oc} and are dependent on the physiochemical composition of SOM for individual HOCs. A significant amount of data have suggested that aromaticity has the greatest influence on K_{oc}, although this relationship may be complex. Smernik and Kookana (2015) found that K_{oc} increased with aromatic C content and decreased with O-alkyl C and alkyl C content, which may be due to sorption sites being obstructed by organic matter high in these functional groups.

In the case of DOC, HOC partitioning between DOC and water is typically described by a linear sorption isotherm (Caron et al. 1985); however, certain types of binding results in nonlinear sorption isotherms. Pan et al. (2007) used two PAHs (penanthene and pyrene) to demonstrate that occupation of competitive binding sites can result in nonideal binding behavior. Due to this reasoning, isotherm nonlinearity is invoked to explain HOC binding behaviors (Hur et al. 2011).

Conceptual models for HOC sorption

A critical step in the interaction between HOCs and SOM may be the diffusion of the HOC into the SOM matrix. Humic substances are known to swell during absorption of nonpolar molecules (Marschner and Kalbitz 2003), which may provide clues to the mechanism of sorption. Lyon (1995) showed that during the sorption process, some nonpolar molecules react with peat with a resulting change in volume. A noteworthy change was the volume shrinkage of one peat by 26% after reacting with propyl sulfone for eight weeks. Heating or swelling of SOM results in increased linearity of sorption, as predicted from the behavior of "glassy" (discussed below) polymers. The increase in linearity of sorption with swelling has been shown in recent studies; after addition of swelling solvents to organic soils, swelling was noted and the respective isotherms became more linear (Pravecek et al. 2006). A similar study noted isotherm linearity of phenyl ureas with a soil mixed with a known swelling solvent (dimethyl sulfoxide) (Xing and Pignatello 1997). A possible explanation for the increased linearity with swelling was introduced by Huang and Weber (1997): Sorbing molecules are able to diffuse through swollen SOM regions. The pores in these swollen regions can stretch or shrink with HOC changes in aqueous-phase concentrations. The flexible SOM areas work in pseudo-aqueous partitioning phases for hydrophobic solutes, displaying almost linear and no hysteretic sorption. Under reducing conditions, SOM domains are more rigid and follow a Langmuir-type adsorption, thus resulting in nonlinear sorption likely due to the exclusion of the molecules from access to binding functional groups. As condensed regions increase in reduced SOM, nonlinearity should also increase, and increase sorption for hydrophobic solutes. This is due to increased affinity between SOM and hydrophobic solutes and an increase in ability for hydrophobic molecules to penetrate highly condensed SOM or expand the SOM matrix.

The concepts of "glassy" and "rubbery" domains within SOM have been used to describe sorption and desorption kinetics (Pravecek et al. 2006). The glassy phase, composed of aromatic carbon, is proposed to contribute to both linear and nonlinear sorption. This glassy condensed phase is characterized by char or soot with high affinity adsorption sites, and is also considered to be more aromatic and less polar. High molecular weights, cross linking, and unsaturated bonds reduce flexibility in molecular SOM and so a glassy state will take shape (Pan et al. 2008). The so-called rubbery phase, dominated by aliphatic regions, contributes to linear sorption and is responsible for partitioning (mobility of compound). These multiple domains ultimately effect sorption rates (Pravecek et al. 2006). Rubbery domains play a dominant role in linear sorption whilst glassy domains contribute to both linear and also nonlinear sorption (Pan et al. 2008).

The dual-mode model features SOM having both partition and adsorption domains (Xing and Pignatello 1996). The glassy rigid phase includes sorption sites made of somewhat permanent nanopore structures and can be subject to co-solute

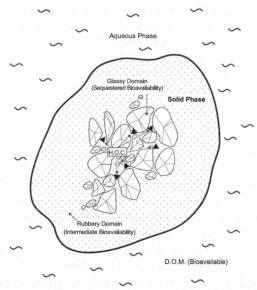

Fig. 1: Bioavailability of hydrophobic organic contaminants. Dual-mode influence determines the bioavailability of the contaminant within the soil complex. After Mao and Schmidt-Rohr (2006).

competition. White et al. (1999) showed that there is a direct correlation between bioavailability and desorption rates of HOCs in respect to aging. It has been proposed that because of the condensed structure of the glassy domain, HOCs sorbed into this fraction are poorly bioavailable in comparison to the rubbery phase (Cuypers et al. 2002). The linear components of the rubbery phase allow for HOCs to be the most bioavailable due to its prompt equilibrium with interstitial waters, while the non-linear attributes of the glassy phase impede bioavailable fractions (Burgess et al. 2003). These mechanisms work together to determine bioavailability of HOCs dependent upon the respective domains (Fig. 1).

The impacts of diagenesis on SOM molecular structure and sorption

To understand the nature of organic matter and its interaction with HOC, the system should be viewed as a complex, dynamic system of diagenesis, and not only in terms of static molecular properties. A critical driver in diagenesis is microbial activity (Schmidt et al. 2011). The initial phases of SOM formation and humification processes are highly influenced by microbial reactions (Pfaender et al. 2005). Soil factors that increase microbial metabolic activity will correspondingly increase desorption of PAHs or increase PAH aqueous concentration, and potentially escalate degradation rates (Pravecek et al. 2006). Increases in nonlinear sorption are consistent with increases in diagentic maturity and condensed regions. A study by Gunasekara et al. (2003) reported that linear isotherms were produced after aromatic carbon was removed by bleaching or hydrolysis of humic acid. Advanced diagenesis results in aromatization and condensation of fractions of the molecule that are linked to nonlinear and hysteretic sorption-desorption. Geosorbents that have undergone a substantial amount of diagenesis are reported to exhibit a higher degree of sequestration of PAHs. Furthermore, nonpolar compounds have been shown to release slowly from geosorbent material. Desorption rates are slower with high degrees of SOM condensation (a reaction where two molecules combine to create a larger one) (Käcker et al. 2002). The lower rate of desorption may be observed with higher condensation due to the molecules' chemical properties becoming more nonpolar and rigid. This phenomenon raises the issue of biostabilization or the formation of bound residues (discussed below), and if the residual HOC remaining in soil is being addressed in bioremediation efforts. Additionally, the formation of pores within SOM may directly increase sorption nonlinearity as well as sorption-desorption hysteresis (Pan et al. 2008). The kinetics of HOC desorption reflects these domain phases. The initial desorption is a fast release through the macro-mesopore network, which is then followed by a slow desorption rate due to entrapment along hydrophobic pore walls (Zhu et al. 2008).

Nonlinear sorption may result in nonideal transport through soil (as predicted by the two-domain model), which would yield erroneous retardation estimates if not accounted for. In models for prediction of the fate and transport of HOC, nonlinear sorption would more accurately depict the fact that retardation will be lower and contaminant plumes would be larger. In addition, breakthrough curves may be more asymmetrical due to nonlinear sorption (Hu and Brusseau 1998). Transport of HOCs through a soil column depends on both position and time. Contaminants move through a soil medium as they fill up available sorption sites. That movement will increase the overall adsorption through time as more sites become available at the boundary. As sites become less available (potential volume of available sites decreases in a soil medium), the rate of sorption decreases through time.

Recent research has suggested that molecular characteristics of SOM may play only a part in overall partitioning (Smernik and Kookana 2015). Environmental influences such as aeration, pH, temperature, and soil structure have all been found to play a significant part in defining the sorption of HOCs (Pravecek et al. 2006). This information is vital to biodegradation research because the aqueous phase concentration of HOCs is determined by SOM and HOC interactions. This in turn determines the mobilization and reactivity of HOCs in the soil (Kim et al. 2008).

It has been well established that sorption processes vary dependent upon the available organic matter fraction due to its reactivity with the respective HOC. Diagenetic processes may include the alteration of biopolymers to humic substances by both biotic and abiotic processes. Shale materials, that is, subsurface materials with extremely high diagenetic alteration, have been found to have both highest isotherm nonlinearity and sorption capacity (Young et al. 1995). Similarly, another study found that the glassy and rubbery domains were dependent on the origin and nature of the soil organic matter. Condensed domains in organic matter with little diagenetic alteration contain elevated amounts of aliphatic C, while condensed domains with high diagenesis harbored higher amounts of aromatic C (Cuypers et al. 2002). It can then be inferred using the dual-mode model that nonlinear sorption tendencies are related to binding of the condensed regions that were formed during diagenesis.

Formation of bound residues

Sequestration is a phenomenon where the HOC molecules diffuse into microbiologically in-accessible sites in the mineral-organic complex of the organic soil particle. Sequestration increases with time for many persistent contaminants. The fate of the sequestered contaminant is often its transformation into bound residue. This bound complex should theoretically lower bioavailability of the HOC (Ling et al. 2010). Bound residues are linked by covalent bonds between the PAH microbial metabolites and the SOM. It is hypothesized that these covalent bonds are made by enzyme-catalyzed oxidative cross-coupling reactions. The formation of bound residue was studied by Richnow et al. (2000). Utilizing [1-^{13}C]-labeled phenanthrene to follow the biotransformation of PAHs. The results showed that microbial processes promote formation of bound-residues. Another study by Käcker et al. (2002) was the first to show a PAH covalently binding to soil humic material. If PAHs and their respective metabolites are covalently bound to SOM, they are usually considered indefinitely a part of humus and no longer an environmental threat.

The Impacts of Redox Potential on HOC Sorption

Increase in solubility

Reduced environments are associated with increased amounts of DOC, and reduced DOC tends to exhibit a greater affinity than oxidized DOC for HOCs (Mao et al. 2002). Research has shown that PAH-colloid sorption increases with core depth, which is assumed to represent lower redox conditions. Higher PAH concentrations have also been observed in the aqueous phase under reduced conditions (Pravecek et al. 2005), and studies have shown that DOC promotes increased solubility of such compounds as PCBs and DDT, which typically have very low aqueous solubilities. Other compounds, such as PCP, have shown increased water-solubility in the presence of mineral colloids with DOC (D'Angelo and Reddy 2003).

Sorbed PAHs may move into the aqueous phase if they complex with DOC. Kim et al. (2008) reported that pyrene was highly sorbed to soil under oxic conditions, and postulated it could be from pyrene metabolites bound to soil constituents. Oxic soils showed a higher sorption for chlorophenols when compared to anoxic soils. It was proposed that this may be due to the enhancement of hydrophobic interactions due to the conversion of chlorophenols to neutral forms (Kim and Pfaender 2005).

The mobility and bioavailability of bound HOCs are influenced by redox conditions and, as such, are controlled by the terminal acceptor for microbial activities. Kim et al. (2008) reported that microbes utilizing sulfate as a terminal electron acceptor increased soluble SOM concentrations and thus increased PAH binding with DOC. The researchers concluded that sulfate reduction increased DOC due to the lowering of redox conditions and consequential H$^+$ utilization, allowing for SOM partitioning and leading to increased binding between PAH and DOC. They also concluded that properties of SOM are changed under microbially-mediated redox conditions. Thus, DOC liberated under anaerobic incubation was more aromatic, more condensed, and had a greater sorption capacity for PAHs.

Production of DOC under low redox potentials

DOC originates from decomposed biota and microbial byproducts (Richnow et al. 2000). Two main factors contribute to formation of DOC: microbial activity; and pH. These two influences are largely interrelated. Most studies show that disassociation of DOC is directly correlated with high pH (Kalbitz et al. 2000). With changes in pH, humic substances exhibit shifts in degree of protonation (Marschner and Kalbitz 2003). After aerobic and facultative microbes respire the available dissolved oxygen, the redox potential (E_h) begins to drop. As anaerobic microorganisms begin to reduce available soil minerals, pH rises due to consumption of H^+ ions. When pH is increased, the adsorption capacity of DOC is limited and therefore DOC solubility increases (Kalbitz et al. 2000). Several authors have found a relationship between the rise in pH due to reduced conditions and a rise in DOC dissolution (Huang et al. 2003; Kalbitz et al. 2000). It has been shown that in elevated pH conditions, more SOM is released and PAH binding with DOC was greater (Kim et al. 2008).

While most research has indicated that DOC release is directly linked to pH (Kalbitz et al. 2000), another line of research indicated that low pH can contribute to DOC formation. Lower pH values can cause the dissolution of organo-metal complexes, which may then contribute to the release of SOM from the solid matrix (Hu and Brusseau 1998). Other studies suggest that increasing pH produces an increase in DOC concentration (Kalbitz et al. 2000). Microbes control the redox potential of soils and also aid in diagenetic SOM development (Pfaender and Kim 2005). Anaerobic microbes are known to alter soil conditions such as pH which would then lead to increases in DOC (Pravecek et al. 2005). DOC solubility can usually be increased by addition of Na^+ or K^+, which expands DOC molecules (Marschner and Kalbitz 2003).

Chemical alteration of DOC under low redox potentials

DOC increases solubility of organic compounds (Marschner et al. 2003). There is research showing that supplementation of DOC increases mineralization of HOCs with respect to four to six-ring PAHs (Bengtsson and Zerhouni 2003). Kim et al. (2008b) reported that anoxic environments created by microbial activity help dissolve soil-bound OM; and DOC helps to aid in desorption of PAHs through sorption. The DOC produced under anaerobic incubation has a significant aromatic component and has a higher affinity for PAHs, resulting in an increase in concentration found in the aqueous phase.

Pfaender and Kim (2005) found that DOC with increased humic acid content exhibited higher sorption affinities for pyrene. They concluded that reduced conditions result in "humification-like alteration" of DOC and in greater sorption affinities for the PAH(s). DOC has also been reported to have higher sorption with pyrene in anaerobic conditions than when compared to aerobic conditions (D'Angelo and Reddy 2003). It has also been found that humin has relatively high aromaticity, and became more nonpolar under low redox potentials, which caused an increase in pyrene sorption rates compared to biocide treated humin (Pfaender and Kim 2005). Humin fractions are usually more aliphatic and tend to have higher K_{OC} values. Due to anaerobic incubation increasing SOM humification, which is similar to diagenesis, the higher sorption capacity for pyrene has been reported to be more nonlinear and hysteric. Similar effects were also found for DDT. DOC increases desorption rates of soil-bound DDT. The sediment-water partition coefficient showed a large decrease when humic matter was added to the aqueous phase DOC (Caron et al. 1985).

DOC under reduced conditions

The main content of SOM and DOC is humic substances, and this fraction is what controls sorption tendencies of HOCs (Pan et al. 2008). K_{OC} of fulvic acids is usually lower than humic and humin fractions. K_{OC} increases with aromaticity in humic acids. Humin is usually more aliphatic and tend to exhibit higher K_{OC} values (Smernik and Kookana 2015). K_{OC} for HOCs is proportional to the aromatic C content of humic acids (Pedersen et al. 1999). Pan et al. (2008) reported that sorption/desorption hysteresis follows the order of FA (fulvic acid) < HA (humic acid) < HM (humin). Burgess et al. (2003) found that PAH-colloid sorption tendencies increased with depth of soil, which is assumed to have a lower redox potential. They also observed higher PAH concentrations in the aqueous phase under these reduced conditions (Pravecek et al. 2005).

Chemical properties of DOC under reduced conditions related to sorption

DOC released under reduced conditions usually exhibits the following chemical properties: (1) the DOC tends to be more aromatic and polydispersed; (2) it has higher molecular weights; and (3) it is less polar and oxygenated (Pfaender and Kim 2005; Zhu et al. 2008). Pfaender et al. (2005) also reported witnessing an increase in humic acid content and a decrease in the O:C ratios under low redox potentials. Due to these molecular changes, the sorption capacity for contaminants may change. Both Pfaender et al. (2005) and Zhu et al. (2008) found that DOC under reduced conditions exhibited a high sorption capacity for pyrene compared to sorption under oxic conditions. It is postulated that anoxic degradation of SOM polar constituents aid in diagenesis of DOC, which yields increased amounts of nonpolar OM. It has also been reported that OM from reduced conditions (deep layer pore water) had a higher molecular weight and that the molecular weight is linked to the increased aromaticity of OM. Under reduced conditions, microbial activity could alter polar or "oxygenated function group-rich" DOC, creating more humified OM (Pfaender et al. 2005). The consequence of more humified material is a change in physicochemical and biochemical properties of soil to represent that of diagenesis.

Bioavailability of nonpolar compounds under low redox potentials

Due to the fact that microbial uptake mechanisms typically necessitate that solutes be dissolved in the aqueous phase (Ogram et al. 1985), DOC may be the most bioavailable part of soil. Bioavailability is a requirement for degradation and is limited if DOC is in unreachable areas of pores in soil. The dynamics that influence DOC biodegradability are: fundamental DOC features; soil and solution constraints;

and external factors. Research has shown some correlation between denitrification rates and amounts of DOC available in soils. This parallel could indicate that the accessibility of biodegradable DOC could be a necessary environmental condition for producing reduced conditions (Marschner and Kalbitz 2003). A three-phase model proposed by Mitra et al. (1999) illustrated the behavior of hydrophobic substances in a freely dissolved phase, DOC, and suspended solids. The authors stated that greater PAH pore water concentrations are a direct outcome from PAH-DOC interactions due to DOC having a high affinity for HOCs.

Many contaminants can be found in soils with varying redox potential due to microbial utilization of diverse electron acceptors (e.g., O_2, NO_3^-, Fe^{3+}). Thus, microbial redox activities have a strong impact on soil solid phase and aqueous chemical properties. Due to this, it is expected that microbial redox activities will have a large impact on the fate of soil contaminants (D'Angelo and Reddy 2003). Previous research has shown that the amount of PAH in aqueous phases increases under anaerobic conditions, and (as discussed above) that microbial activity alters pH and DOC, thus resulting in PAH release (Zhu et al. 2008). In addition to biodegradation, microbes influence PAH concentrations in solution by changing the pH and DOC alteration. In a study by Pravecek et al. (2005), the PAH was transferred from the solid into the aqueous phase at a higher rate under anaerobic incubation than under aerobic conditions (Pravecek et al. 2005). This study also confirmed that with SOM release came the release of high molecular weight PAHs (Pfaender and Kim 2005). Moeckel et al. (2014) showed a strong correlation between molecular weight of PAHs and organic carbon concentration. They found that concentrations of PAHs with five or more aromatic rings increase with greater organic carbon concentration. Pravecek et al. (2005) found that soil under anaerobic incubation with nitrate or sulfate amendment had an increase of extractable pyrene at 365 d when compared to controls. The authors postulate it resulted from oxidation–reduction potential and pH changes caused by microbial activity. They concluded that the available terminal electron acceptor will produce the most energetically favorable microbes to control this niche (Pravecek et al. 2005).

Biological degradation rates under reduced conditions relative to oxic conditions

If the increase in DOC under anoxic conditions is taken into account, it is likely that biological degradation of HOCs under reduced conditions will be higher than when compared to oxic conditions due to increased bioavailability. Some research has shown this in different contaminants. Kim et al. (2008) found that anaerobic incubation of contaminated soils can increase desorption of HOCs, increasing degradation. Some PAHs were also shown to degrade under denitrifying conditions (Hutchins et al. 1991). Fermentative degradation is also known to occur with denitrification in reduced environments. It is postulated that aromatic compounds ferment with low molecular weight organic matter due to the local microbial species acting as denitrifiers or using H_2 as the electron donor. Some PAHs, such as phenanthrene and naphthalane, have been shown to be degraded by sulfate-reducing microbes (Ambrosoli et al. 2005).

Lu et al. (2011) also found that, under anoxic conditions, denitrification is a useful solution for some areas contaminated with PAHs.

Micelle/pseudomicelle formation

The mechanisms for interactions between soluble HOC:DOC complexes and the cell surfaces of degrading microorganisms is not fully understood at this time; however, models to describe this interaction have been proposed. Humic substances possess both hydrophobic and hydrophilic properties. This amphiphilic quality allows for it to act as the main carrier in solubilization of HOCs. DOC has been shown to produce surfactant-like properties (Cho et al. 2002) which could allow for enhanced mobility of nonpolar compounds under reduced conditions. The idea of a pseudomicelle has been proposed using fluorescence-polarization. Morra et al. (1990) reported that naphthalene and 1-naphthol interact with humic acid by creating a loose cage without binding. Other studies refer to humic acid solutions as true micelles because they have similar characteristics with critical micelle concentrations (CMC) (Chiou et al. 1986). Also, once the CMC is reached, the humic acid micellar solution will solubilize PAHs (Vacca et al. 2005). This proposed "true" micelle of humic acids has been observed in concentrated alkaline aqueous solutions (Guetzloff and Rice 1994). According to this possible model, humic acid surfactants could greatly increase partitioning of HOCs into the aqueous phase after the CMC has been reached (Pan et al. 2008). The actual uptake of contaminants by the cell has little supporting evidence, but Guhu et al. (1996) proposed a model using a hemi-micelle. Their hypothesis suggested that filled micelles (with the PAH partitioned in the hydrophobic core) come into contact with cells by mixing. At this point, the cells/ enzymes are surrounded by a hemi-micellar layer containing the PAH (Fig. 2). This allows for the PAH to diffuse into the cell to be degraded. This phenanthrene-based model assumes that the system is completely mixed, equilibrium is established, and degradation kinetics are not affected by the surfactant.

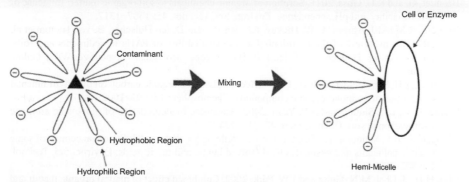

Fig. 2: Humic acid micelle formation and pseudosolubilization. Micelle formation is first formed in reduced conditions. After mixing, the contaminant is available to be transferred to the cell. After Guhu et al. (1996).

Conclusions

The fate of organic contaminants is influenced by many factors. Ecological influences as well as intrinsic physical properties determine sorption/desorption rates and thus bioavailability. Yet, DOC has been shown to be a key player in desorption mechanisms by aiding in the change of physiochemical properties under anaerobic or anoxic incubation. Under such conditions, microbial metabolism causes an increase in pH. This increase results in a drop in redox conditions, which in turn encourages a humification-like change in organic soil that causes an increase in aromaticity in SOM. This alteration increases the release of with DOC and contaminants from the solid matrix. The surfactant-like properties in DOC has led to the proposal of a micelle model for interaction between DOC, HOC, and degradating bacterial cells, in which a hemi-micelle transports the HOC to the soil microorganism for degradation. Understanding the relationship between reduced soil conditions and contaminants is critical for biodegradation and ecological health studies.

Acknowlegements

The authors thank Aaron Sotala of the University of Florida's Center for Online Learning and Technology for drawing the figures. Professors Peter Nkedi-Kizza and Christopher Wilson are thanked for critical suggestions during the writing of this manuscript.

References

Ambrosoli, R., L. Petruzzelli, J.L. Minati and F.A. Marsan. 2005. Anaerobic PAH degradation in soil by a mixed bacterial consortium under denitrifying conditions. Chemosphere 60: 1231–1236.

Bengtsson, G. and P. Zerhouni. 2003. Effects of carbon substrate enrichment and DOC concentration on biodegradation of PAHs in soil. J. Appl. Microbiol. 94: 608–617.

Bronner, G. and K.U. Goss. 2011. Sorption of organic chemicals to soil organic matter: Influence of soil variability and pH dependence. Environ. Sci. Technol. 45: 1307–1312.

Burgess, R.M., M.J. Ahrens, C.W. Hickey, P.J. den Besten, D. ten Hulscher, B. van Hattum et al. 2003. An overview of the partitioning and bioavailability of PAHs in sediments and soils. pp. 97–126. In: Douben, P.E.T. [ed.]. PAHs: An ecotoxicological Perspective. Wiley Online Library.

Caron, G., I.H. Suffet and T. Belton. 1985. Effect of dissolved organic carbon on the environmental distribution of nonpolar organic compounds. Chemosphere 14: 993–1000.

Chang, B.V., L.C. Shiung and S.Y. Yuan. 2002. Anaerobic biodegradation of polycyclic aromatic hydrocarbon in soil. Chemosphere 48: 717–724.

Chiou, C.T., R.L. Malcolm, T.I. Brinton and D.E. Kile. 1986. Water solubility enhancement of some organic pollutants and pesticides by dissolved humic and fulvic acids. Environ. Sci. Technol. 20: 502–508.

Cho, H.H., J. Choi, M.N. Goltz and J.W. Park. 2002. Combined effect of natural organic matter and surfactants on the apparent solubility of polycyclic aromatic hydrocarbons. J. Environ. Qual. 31: 275–280.

Cuypers, C., T. Grotenhuis, K.G. Nierop, E.M. Franco, A. de Jager and W. Rulkens. 2002. Amorphous and condensed organic matter domains: the effect of persulfate oxidation on the composition of soil/sediment organic matter. Chemosphere 48: 919–931.

D'Angelo, E. and K.R. Reddy. 2003. Effect of aerobic and anaerobic conditions on chlorophenol sorption in wetland soils. Soil Sci. Soc. Am. J. 67: 787–794.

Ehlers, G.A. and A.P. Loibner. 2006. Linking organic pollutant (bio)availability with geosorbent properties and biomimetic methodology: a review of geosorbent characterisation and (bio) availability prediction Environ. Pollut. 141: 494–512.

Guetzloff, T.F. and J.A. Rice. 1994. Does humic acid form a micelle? Sci. Total Environ. 152: 31–35.

Guha, S. and P.R. Jaffé. 1996. Bioavailability of hydrophobic compounds partitioned into the micellar phase of nonionic surfactants. Environ. Sci. Technol. 30: 1382–1391.

Gunasekara, A.S., M.J. Simpson and B. Xing. 2003. Identification and characterization of sorption domains in soil organic matter using structurally modified humic acids. Environ. Sci. Technol. 37: 852–858.

Hu, M.Q. and M.L. Brusseau. 1998. Coupled effects of nonlinear, rate-limited sorption and biodegradation on transport of 2, 4-dichlorophenoxyacetic acid in soil. Environ. Toxicol. Chem. 17: 1673–1680.

Huang, W. and W.J. Weber. 1997. A distributed reactivity model for sorption by soils and sediments. 10. Relationships between desorption, hysteresis, and the chemical characteristics of organic domains. Environ. Sci. Technol. 31: 2562–2569.

Huang, W.L., P.A. Peng, Z.Q. Yu and J.M. Fu. 2003. Effects of organic matter heterogeneity on sorption and desorption of organic contaminants by soils and sediments. Appl. Geochem. 18: 955–972.

Hur, J., B.M. Lee and H.S. Shin. 2011. Microbial degradation of dissolved organic matter (DOC) and its influence on phenanthrene–DOC interactions. Chemosphere 85: 1360–1367.

Hutchins, S.R., G.W. Sewell, D.A. Kovacs and G.A. Smith. 1991. Biodegradation of aromatic hydrocarbons by aquifer microorganisms under denitrifying conditions. Environ. Sci. Technol. 25: 68–76.

Käcker, T., E.T. Haupt, C. Garms, W. Francke and H. Steinhart. 2002. Structural characterisation of humic acid-bound PAH residues in soil by 13C-CPMAS-NMR-spectroscopy: evidence of covalent bonds. Chemosphere 48: 117–131.

Kalbitz, K., S. Solinger, J.H. Park, B. Michalzik and E. Matzner. 2000. Controls on the dynamics of dissolved organic matter in soils: a review. Soil Sci. 165: 277–304.

Kang, S. and B. Xing. 2005. Phenanthrene sorption to sequentially extracted soil humic acids and humins. Environ. Sci. Technol. 39: 134 140.

Kim, H.S., K.S. Lindsay and F.K. Pfaender. 2008a. Enhanced mobilization of field contaminated soil-bound PAHs to the aqueous phase under anaerobic conditions. Water Air Soil Pollut. 189: 135–147.

Kim, H.S., C.J. Roper and F.K. Pfaender. 2008b. Impacts of microbial redox conditions on the phase distribution of pyrene in soil–water systems. Environ. Pollut. 152: 106–115.

Ling, W., Y. Zeng, Y. Gao, H. Dang and X. Zhu. 2010. Availability of polycyclic aromatic hydrocarbons in aging soils. J. Soil Sedi. 10: 799–807.

Lu, X.Y., T. Zhang and H.H. Fang. 2011. Bacteria-mediated PAH degradation in soil and sediment. Appl. Microbiol. Biotechnol. 89: 1357–71.

Lu, X.Y., T. Zhang, H.H.P. Fang, K.M.Y. Leung and G. Zhang. 2011. Biodegradation of naphthalene by enriched marine denitrifying bacteria. Internat. Biodeter. Biodegrad. 65: 204–211.

Lyon, W.G. 1995. Swelling of peats in liquid methyl, tetramethylene and propyl sulfoxides and in liquid propyl sulfone. Environ. Toxicol. Chem. 14: 229–236.

Mao, J.D., L. Hundal, M. Thompson and K. Schmidt Rohr. 2002. Correlation of poly (methylene)-rich amorphous aliphatic domains in humic substances with sorption of a nonpolar organic contaminant, phenanthrene. Environ. Sci. Technol. 36: 929–936.

Mao, J.D. and K. Schmidt-Rohr. 2006. Absence of mobile carbohydrate domains in dry humic substances proven by NMR, and implications for organic-contaminant sorption models. Environ. Sci. Technol. 40: 1751–1756.

Marschner, B. and K. Kalbitz. 2003. Controls of bioavailability and biodegradability of dissolved organic matter in soils. Geoderma 113: 211–235.

Mitra, S. and R.M. Dickhut. 1999. Three-phase modeling of polycyclic aromatic hydrocarbon association with pore-water-dissolved organic carbon. Environ. Toxicol. Chem. 18: 1144–1148.

Moeckel, C., D.T. Monteith, N.R. Llewellyn, P.A. Henrys and M.G. Pereira. 2014. Relationship between the concentrations of dissolved organic matter and polycyclic aromatic hydrocarbons in a typical UK upland stream. Environ. Sci. Technol. 48: 130–138.

Morra, M.J., M.O. Corapcioglu, R.M.A. Von Wandruszka, D.B. Marshall and K. Topper. 1990. Fluorescence quenching and polarization studies of naphthalene and 1-naphthol interaction with humic acid. Soil Sci. Soc. Am. J. 54: 1283–1289.

Ogram, A.V., R.E. Jessup, L.T. Ou and P.C. Rao. 1985. Effects of sorption on biological degradation rates of (2,4-dichlorophenoxy)acetic acid in soils. Appl. Environ. Microbio. 49: 582–587.

Pan, B., P. Ning and B. Xing. 2008. Humic substances-review series. Part IV-sorption of hydrophobic organic contaminants. Environ. Sci. Pollut. Res. 15: 554–564.

Pedersen, J.A., C.J. Gabelich, C.H. Lin and I.H. Suffet. 1999. Aeration effects on the partitioning of a PCB to anoxic estuarine sediment pore water dissolved organic matter. Environ. Sci. Technol. 33: 1388–1397.

Pfaender, F.K. and H.S. Kim. 2005. Effects of microbially mediated redox conditions on PAH-soil interactions. Environ. Sci. Technol. 39: 9189–9196.

Pravecek, T.L., R.F. Christman and F.K. Pfaender. 2005. Impact of imposed anaerobic conditions and microbial activity on aqueous-phase solubility of polycyclic aromatic hydrocarbons from soil. Environ. Toxicol. Chem. 24: 286–293.

Pravecek, T.L., R.E. Christman and F.K. Pfaender. 2006. Microbial bioavailability of pyrene in three laboratory-contaminated soils under aerobic and anaerobic conditions. J. Contam. Hydrol. 86: 72–86.

Richnow, H.H., E. Annweiler, M. Koning, J.C. Lüth, R. Stegmann, C. Garms et al. 2000. Tracing the transformation of labelled [1-^{13}C] phenanthrene in a soil bioreactor. Environ. Pollut. 108: 91–101.

Schmidt, M.W.I., M.S. Torn and S. Abiven. 2011. Persistence of soil organic matter as an ecosystem property. Nature 478: 49–56.

Smernik, R.J. and R.S. Kookana. 2015. The effects of organic matter-mineral interactions and organic matter chemistry on diuron sorption across a diverse range of soils. Chemosphere 119: 99–104.

Sorensen, K.C., J.W. Stucki, R.E. Warner, E.D. Wagner and M.J. Plewa. 2005. Modulation of the genotoxicity of pesticides reacted with redoxmodified smectite clay. Environ. Mol. Mutagen. 46: 174–181.

Stevenson, F.J. 1994. Humus chemistry: genesis, composition, reactions. John Wiley & Sons.

Vacca, D.J., W.F. Bleam and W.J. Hickey. 2005. Isolation of soil bacteria adapted to degrade humic acid-sorbed phenanthrene. Appl. Environ. Microbiol. 71: 3797–3805.

Van Metre, P.C. and B.J. Mahler. 2005. Trends in hydrophobic organic contaminants in urban and reference lake sediments across the United States, 1970–2001. Environ. Sci. Technol. 39: 5567–5574.

White, J.C., M. Hunter, J.J. Pignatello and M. Alexander. 1999. Increase in bioavailability of aged phenanthrene in soils by competitive displacement with pyrene. Environment. Toxicol. Chem. 18: 1728–1732.

Xing, B. and J.J. Pignatello. 1996. Time-dependent isotherm shape of organic compounds in soil organic matter: Implications for sorption mechanism. Environment. Toxicol. Chem. 15: 1282–1288.

Xing, B. and J.J. Pignatello. 1997. Dual-mode sorption of low-polarity compounds in glassy poly (vinyl chloride) and soil organic matter. Environ. Sci. Technol. 31: 792–799.

Young, T.M. and W.J.J. Weber. 1995. A distributed reactivity model for sorption by soils and sediments. 3. Effects of diagenetic processes on sorption energetics. Environ. Sci. Technol. 29: 92–97.

Zhu, H., J.C. Roper, F.K. Pfaender and M.D. Aitken. 2008. Effects of anaerobic incubation on the desorption of polycyclic aromatic hydrocarbons from contaminated soils. Environment. Toxicol. Chem. 27: 837–844.

Index

Printed and bound by CPI Group (UK) Ltd, Croydon, CR0 4YY

24/10/2024

01778304-0003